Small-Scale Livestock Farming

A Grass-Based Approach for Health, Sustainability, and Profit

Carol Ekarius

Schoolhouse Road
Pownal, Vermont 05261

In Memory of

DOUG SPEAR

February 21, 1957, to February 2, 1999

A great friend, a craftsman and artisan, a chicken fancier, and a person who taught us a lot about sustainability. His flame will burn on, in the hearts of his family and friends.

We miss you, Doug.

The mission of Storey Communications is to serve our customers by publishing practical information that encourages personal independence in harmony with the environment.

Edited by Marie Salter and Laura Jorstad
Cover design by Meredith Maker
Cover photograph © Grant Heilman/Grant Heilman Photography, Inc.
Text design and production by Erin Lincourt
Line drawings on pp. 131–133 by Bethany Caskey; on p. 193 by Jeffrey C. Domm; on pp. 3, 4, 10, 12 (top), 13, 14, 16, 24, 29, 30, 61, 66, 73, 76 (top), 77, 85, 139, 150, 161, 162, 173, 175, 179 by Chuck Galey; on p. 193 by Carol Jessop; on pp. 8, 9, 11, 12 (bottom), 20, 21, 27, 28, 31, 35, 37, 38, 43, 44, 48, 51, 56, 57, 62, 64, 65, 70, 74, 76, 78, 82, 83, 95, 129, 163, 191, 199 by Elayne Sears; and on p. 197 by Becky Turner.
Indexed by Susan Olason/Indexes & Knowledge Maps

Printed in the United States by Vicks Lithograph and Printing Corporation
10 9 8 7 6 5 4 3 2

Library of Congress Cataloging-in-Publication Data

Ekarius, Carol.
Small-scale livestock farming : a grass-based approach for health, sustainability, and profit / Carol Ekarius.
p. cm.
ISBN 1-58017-162-1
1. Livestock. 2. Livestock—Marketing. 3. Farm management. I. Title
SF61.E465 1999
636—dc21 99-29586
CIP

Contents

Preface

While we have land to labor, then, let us never wish to see our citizens occupied at a work-bench, or twirling a distaff. Carpenters, masons, smiths are wanting in husbandry: but, for the general operations of manufacture, let our workshops remain in Europe. . . . The loss by the transportation of commodities across the Atlantic will be made up in happiness and permanence of government. The mobs of great cities add just so much to the support of pure government, as sores do the strength of the human body.

Cultivators of the earth are the most valuable citizens. They are the most vigorous, the most independent, the most virtuous, and they are tied to their country, and wedded to its liberty and interests by the most lasting bonds.

—Thomas Jefferson

SOMETIME AROUND 1983, my husband, Ken Woodard, and I were living in a ski-resort town in the Colorado Rockies and were mulling over the idea of moving to the country. We were at least a couple of generations removed from the farm and had no experience with farming or rural living, but we sure were intent on the idea.

Magazines like the *Small Farmer's Journal* and *The New Farm* were part of our regular mail; we devoured books on gardening, farming, livestock raising, and general rural living. We were preparing for our first attempt at country living.

How We Got Here

In 1984, we moved to 40 acres (16.2 ha) out of town. We bought horses. We bought ducks, geese, and chickens. We bought an orphaned calf from a local rancher. We almost killed that calf with kindness; overfeeding brought on scours, but luckily Ken's grandfather was still alive to render a fast diagnosis and recommend a treatment regimen for us. Little Fat Girl grew into a nice-size heifer.

We moved to an 80-acre (32.4-ha) commercial farming operation in Minnesota in 1989, and husbandry became our vocation as well as our avocation. The farm consisted of 53 acres (21.5 ha) of tillable ground; the remainder was in permanent pasture with a stream running through it and a pond at one end. It was in a traditional dairy area and was set up with a dairy barn, a three-sided equipment/hay shed, and some old, falling-down sheds. The soil was moderately fertile.

For as many years as the neighbors could remember (and their collective memories went back a very long time), the tillable ground had been in monocropped corn, primarily harvested as silage to feed the dairy herd. The permanent pasture became little more than a weed-choked exercise area in summer. The creek bottom was severely pugged by years of cattle hooves, and the stream had become more intermittent over the years, holding water for shorter periods of time regardless of precipitation. Fish had become nonexistent in the stream. The hillsides had large washout areas and little remaining topsoil.

During the decade prior to our purchase, four families had come and gone, trying to make a go of milking cows using "conventional" techniques. Twenty-six cows spent most of their time stanchioned in the barn. The farmer brought their feed to them, primarily in the form of silage, supplemented by purchased hay and grain. Hauling manure out was a daily job. Fans ran continuously in summer to cool the barn, and the water lines to the barn routinely froze in winter. These families fell into the too big/too small conundrum. The economics of a confinement dairy could not support them, yet they worked too many hours per day to be able to seek off-farm work. They left, economically, spiritually, and physically broken. As each family passed on, the land was left a little more degraded.

When we gave up our jobs (good, secure jobs with benefits) to become full-time farmers, everyone thought we were crazy. Our families and friends wanted to know how we thought we could possibly make it when farmers who'd been farming their whole lives were failing. They may still be dismayed with our choice, but they now acknowledge our success.

A ranch in Colorado was our first dream, but as we pursued it, we concluded that we couldn't afford full-time ranching. Colorado ranches were for rock stars and Rockefellers, so we made our move to Minnesota. During the nine years we farmed in Minnesota, we learned a lot about farming, marketing, and getting by on less. We came to believe we could get by on a small piece of land in the arid West. So in the fall of 1997, we sold the Minnesota farm, traveled through New Mexico and Arizona during the winter — taking our first joint vacation in over a decade — and returned to Colorado in the spring of 1998. We still aren't rock stars or Rockefellers, so we own just a small piece of land and lease the additional acreage we need. Leasing allows us to not tie up too much of our capital on the land.

Who This Book Is For

Demographics show that many people out there are like us; they want to escape urban/suburban life for the country. This book is for those people who are still in the dreaming phase, and for those who have recently taken the plunge. It's aimed at people who have an acre of land on the edge of town, or people with 1,000 acres (406.5 ha, which isn't a big ranch out West). Though much of the book is directed at dreamers and new farmers, there are sections that I hope may be of benefit to farmers of long standing.

Of Farms and Heroes

There are heroes in this book. Oh, not the kind of people who jump onto speeding trains to disarm a nuclear bomb aimed at a big city; I leave those kind of heroes to the Big Screen. But there are heroes nonetheless. These heroes are regular people who not only keep farming or ranching despite adversity, but do it well. They are people who show that *culture* can still be a part of *agri-culture;* that taking care of the land for future generations is compatible with making money; and that there is still a place for the little guy.

I have to say, the most fun I had while preparing this manuscript was the time I spent interviewing the farmers and ranchers whom you will read about throughout this book. These people aren't whiners; they are realists who understand well the hardships of their chosen profession, but they don't talk about "the government's got to do this, or do that." They all consider themselves to be the master of their own destinies. They have can-do, will-do attitudes — and they are doing what it takes!

Some of the families I highlight here farm part-time. In the lexicon of commercial agriculture, these people are disdainfully called "hobby farmers." Well, let me tell you, there are farmers out there who have followed the "get big or get out" adage for all it's worth, yet they still don't have enough money left to support their families. Most often, their wives take off-farm work, not because they have chosen to follow a career but because they have to in order to buy food, clothing, and other necessities. I have met many, many of these families, and despite high gross sales, often exceeding the $100,000 mark that the U.S. Department of Agriculture uses to designate a large farm operation, they struggle to buy shoes for their kids. I contend that these are the real hobby farmers, and their hobby takes a terrible toll in both material and spiritual terms. I met hundreds of farm wives in the commercial agriculture sector who said, "I pray every day that none of my children become farmers, or marry farmers!" Every time I heard this, it broke my heart.

Those in mainstream agriculture must stop disparaging part-time farmers who make a true profit (a financial, social, and environmental profit) just because they have made a choice that includes off-farm work. There is room enough for both full-time and part-time farmers.

For each of the heroes you read about here, there are thousands more. Unsung heroes who may inspire you, as they have inspired me. (To each of you, many thanks.)

Acknowledgments

Some other people also deserve my thanks. My husband, Ken Woodard, tops the list — his support, humor, intelligence, and hard work have made everything happen, and have made it all worthwhile; Elizabeth McHale for supporting this book when it was little more than a concept; and the staff at Storey Books for then making it happen.

For graciously agreeing to review this text, and provide constructive criticisms, Barbara Green, attorney and friend; Dr. Ann Wells, DVM; Jim Gerrish, research professor at the Forage Systems Research Center at the University of Missouri; Byron Shelton, rancher and certified holistic management educator.

Author's Note: Throughout this text U.S. units of measure are used, with metric equivalents given in parentheses. Where "value" discrepancies occur, they are due to rounding off errors; however, materially the work is correct. I apologize to any who may take offense at my leniency in this regard.

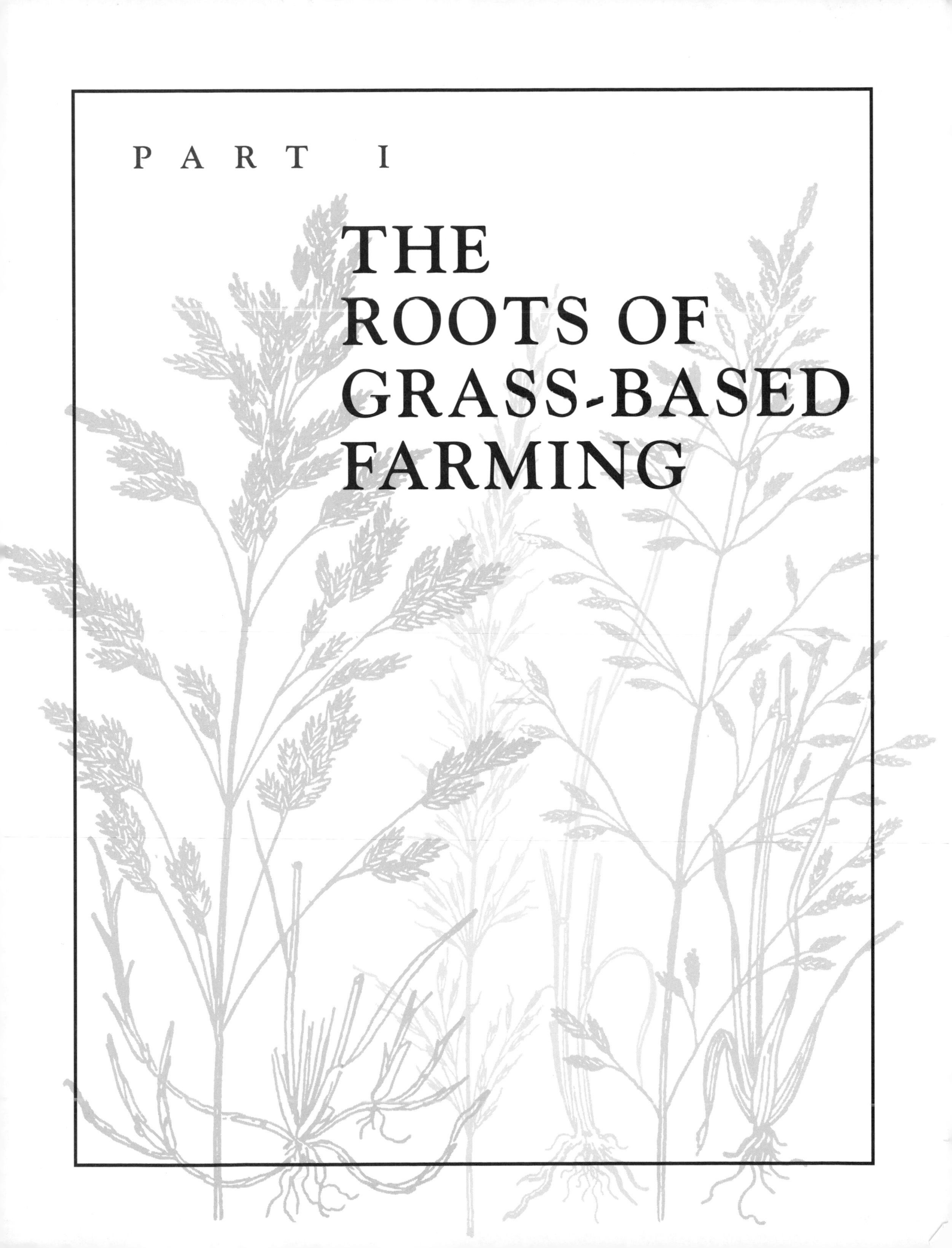

PART I

THE ROOTS OF GRASS-BASED FARMING

CHAPTER 1

Introduction

One aggressive family farm in the plains grew rapidly by investing in irrigation in the 1960s and 1970s. It grew so fast that it soon had over eight thousand irrigated acres, a cattle-feeding operation, and a farm-supply business. To acquire more working capital, the family farmer incorporated and merged his operation with an alfalfa processor whose stock was sold over one of the national stock exchanges. In a few years, the family farmer had lost so much money in the cattle feeding and futures market transactions that he had to accept a minority position in the company. Subsequently, the company was acquired by a conglomerate that held it a short time and then sold it to Texas oil interests. At the beginning of this high-speed transformation, no one would have quarreled with the claim that the farm was a family farm. At the conclusion, no one would have suggested it was anything but an industrial agribusiness. At many points along the path, however, you could have ignited a spirited debate by suggesting it was either.

— Marty Strange, *Family Farming*

THE COMMERCIAL FAMILY FARM is disappearing from the United States. Our farming system is being split into two camps: megafarms, which are corporate in nature if not in deed, and small-scale farms, often thought of as hobby farms. The middle is being squeezed out.

These changes had their beginnings around the time of World War II. The changes brought on by the war are well illustrated by the story of my husband Ken's grandfather, Clarence Woodard. During the Great Depression, Clarence ran a small farm on the outskirts of LaJunta, Colorado. He managed to support his family and a hired man with the milk from a small herd of dairy cows.

During Clarence's era, farmers had a more direct link with the consumer. In fact, 40 percent of the consumer's food dollar went directly to the farmer (Figure 1.1). In Clarence's case, the percentage was even higher; twice each day he hand-milked his fifteen to twenty cows, cooled and bottled the milk, and delivered it door to door in LaJunta. Clarence's customers were neighbors, friends, and relatives, and if things were going well, he'd take a few minutes to visit along his route.

From the beginning of the century until 1940, farm numbers hovered right around 6.4 million; these numbers began a quick descent, however, with the

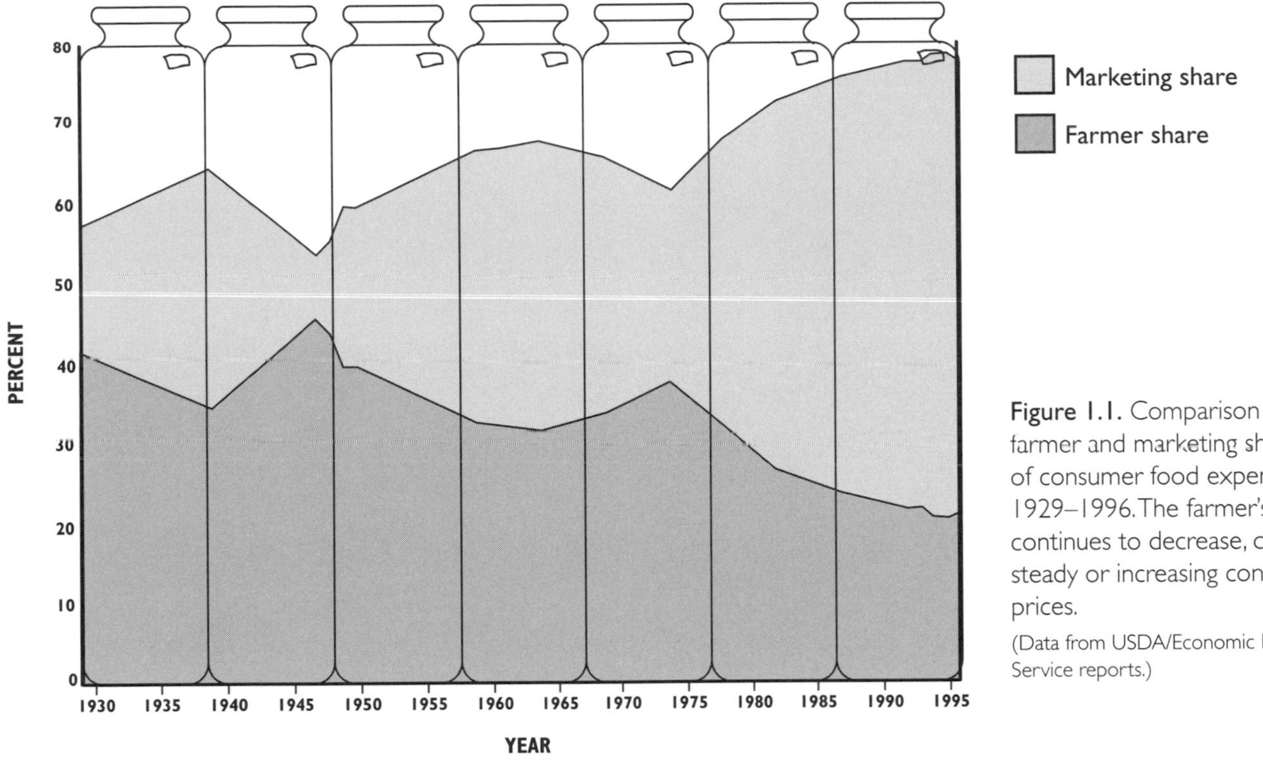

Figure 1.1. Comparison of farmer and marketing shares of consumer food expenditures, 1929–1996. The farmer's share continues to decrease, despite steady or increasing consumer prices.
(Data from USDA/Economic Research Service reports.)

coming of the war (Figure 1.2). This phenomenon was driven by many factors, but in Clarence's case, it was the abundance of good-paying construction jobs on military installations around Pueblo and Colorado Springs that led him to quit farming.

As more and more farmers left for jobs in town or were sent to war, those who remained had to produce more. The land in production stayed fairly constant, around a billion acres, but fewer farmers were working that land. Farming more land left less time for direct-marketing of crops, so the remaining farmers began counting on bulk sales of raw commodities.

After the war, industry, which had geared up for large-scale production of weapons, military transport, and other war-related goods, suddenly turned to agriculture as an open market. Chemical inputs to feed plants and fight the farmer's enemies — weeds and insects — became readily available. Initially, they produced miracles. Tractor and implement production cranked into full swing. The message to American producers was clear: Grow all you can grow, America will feed a hungry world, and you, her farmers, will reap the benefit. Some, of course, did benefit. But many were left by the wayside. Farm numbers continued to decline, as they still do today. And the farmers who are left, despite getting bigger, are continuing to struggle for their existence.

As farm sizes increased, farmers began specializing. The idea of egg money or a few pigs to pay the mortgage disappeared. Monocropping and farming fenceline to fenceline were substituted for diversity. Animal agriculture, like crop farming, moved into an industrial model, with living creatures being treated as little more than production units.

These changes have resulted not only in reduced farm numbers but also in the loss of soil productivity, reductions in wildlife, and increases in water and air pollution. They have also caused a fundamental breakdown in many rural communities: Schools consolidate, hospitals close, and small businesses disappear. That's the bad news.

The good news is that a new class of small-scale farmers are showing that things don't have to continue in this vein. They are reintroducing diversity to their operations. They are raising their animals in a more natural system that allows each critter to express its unique personality and character while eliminating or minimizing the use of synthetic hormones, antibiotics,

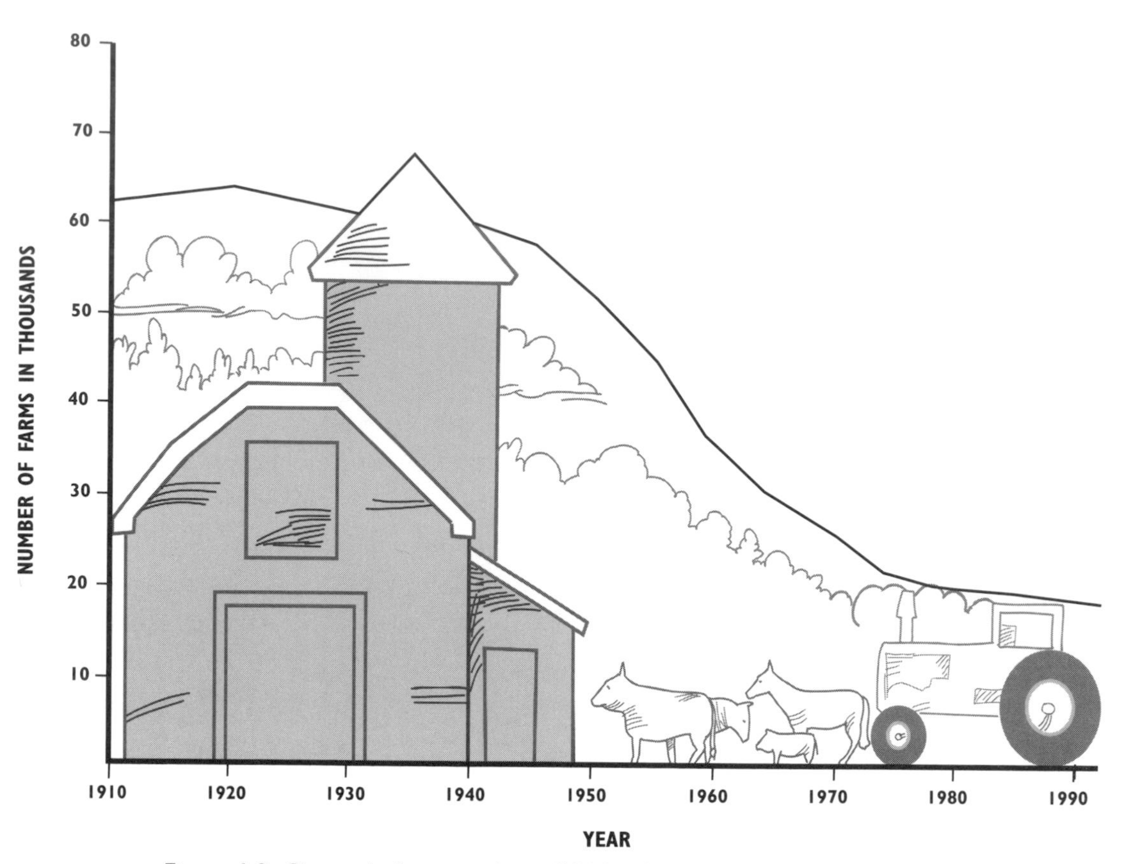

Figure 1.2. Change in farm numbers (1910–1992). Although land under production has held relatively steady during this period, farm numbers have continued to drop. (Data from USDA/Economic Research Service reports.)

and other technological quick fixes. They're learning that different methods of operating can reduce their costs while improving the environment. And they are learning to recapture some of the lost share of the consumer's dollar.

Though some of these small-scale farmers come from the traditional family-farm sector, many are new to agriculture. They bring with them a willingness to try new things and a commitment to farm in harmony with nature.

Small scale is hard to define. The U.S. Department of Agriculture (USDA) and the Census Bureau of the Department of Commerce define it as an operation with more than $1,000 of gross receipts per year from farming and farm-related industries; they define *large scale* as any farm with more than $100,000 of gross receipts. Then there are those who would define small scale based on some number of acres — 100, 500, or 5,000 (41, 203, 2,024 ha, respectively). I don't know where the magic cutoff should be placed. Ten thousand acres of prime corn ground in Iowa probably is large, but 10,000 acres (4,050 ha) of rangeland in the West isn't necessarily big.

Of the farmers and ranchers I highlight throughout this book, several would meet the USDA's $100,000 definition. Several control land in excess of 5,000 acres (2,024 ha). At the other end of the spectrum are the folks who use less than 30 acres (12.2 ha), or those whose farming operations gross far less than $10,000 per year. (And sometimes you wouldn't be able to guess which was which.)

I take a slightly different view: *Small scale* may be any operation, whether 1 acre (0.4 ha) on the edge of town that only supplies the family with a regular portion of its own food, or thousands of acres in a rural area that is owned and operated as a commercial operation that provides a family's main livelihood. In my mind, the key criteria that separate the small-scale farm from the large one are based on the labor used in the operation, and on attitude. Small-scale farms and ranches are those where the family is involved equally in management and operation. Although they may have some hired help (or interns), they don't sit in an air-conditioned office in clean chinos and a sports shirt directing the efforts of dispensable employees. If they do have employees, those employees are part of their team, included in planning and decision making, and the family's labor is applied to day-to-day operations.

Success is also hard to define. The simplistic view of success currently permeating American thought is based almost exclusively on money, or perhaps more specifically on consumptive power. In my eyes, to be deemed successful, an agricultural endeavor must be not only financially profitable but also ecologically and socially profitable. Making money while destroying the land (which is the basis of all wealth) isn't successful, and making money while destroying the local community isn't successful either.

The farmers I have highlighted here have certain things in common. Their philosophies of profit are a little different from those of the people who think profit is simply a number reported on the bottom line of an accounting form. Although they are all profit motivated, they aim for a profit that is both financial and spiritual. The profit equation for these folks balances money with quality of life, with humane treatment of their animals, with support for their communities, and with fostering a healthy environment.

There are lots of reasons for small-scale farmers to include livestock in their operations:

- You can become an active participant in the food chain, converting solar energy to grass, and grass to high-quality protein.
- The meat that comes from your own animals is usually far superior to anything you can buy at the grocery store, and at less cost. And you know how your animals were raised (no growth hormones or routine antibiotics, for instance).
- Livestock provides an opportunity for your family to learn about life, both its joys and its heartaches. Raising animals teaches children to be responsible for other creatures.
- Livestock can help you make profits from your land. When used for livestock, marginal land can provide a profit that it's less likely to provide with crops.
- Livestock can provide valuable nutrients for your soil. Manure is a fine fertilizer, and livestock not only provides it but can also incorporate it.

When properly managed (and this is the key), livestock is beneficial to the environment. Converting croplands, particularly highly erodible croplands, to permanent pastures improves water infiltration into the soil and water uptake by plants, thereby reducing runoff and erosion. Wetlands that are grazed via managed grazing show improvements in bank structure, increased variety of plants and animals, and more consistent water flows.

Having livestock around is fun! Watching a hen scratch for bugs, seeing a group of calves running across a field with their tails held high, or being nuzzled by a new lamb provides great entertainment and good feelings that you just don't get watching a plant grow. Let's face it, the refrain from "Old McDonald's Farm" doesn't run, "With a corn, corn here and a corn, corn there."

Still, despite all the good things to be said about livestock for the small-scale farmer, there are some drawbacks. Livestock ties you down; your animals' lives depend on you, so you can't just take off for two weeks and forget about them. There is some heartbreak when raising livestock; critters get sick and sometimes they die. But most of the time, the benefits far outweigh the drawbacks.

The main qualification for being a successful small-scale livestock farmer is having a deep love and appreciation of animals. If you don't really like animals, look at other aspects of small-scale farming, like market gardening. Do not go into livestock raising just because you think it will make your operation profitable: If you don't honestly care for your animals, almost to the point of obsessiveness, they won't make your operation profitable. But if you think you have the heart for it, read on!

FARMER PROFILE

Todd Lein and Annie Klawiter

Neither Todd nor Annie grew up on farms. Todd was raised in a small town in rural Minnesota, but he did have some experience with livestock: His dad raised occasional feeder pigs or calves in the backyard. The experience was a good one and left Todd with a desire to work with livestock. Annie grew up in Sioux Falls, a small midwestern city.

In 1994, Todd and Annie began farming on 37 acres of land they purchased, and 30 more they rented, near Northfield, Minnesota. Over the ensuing years, their operation has solidified into a profitable small-scale farm.

"Starting out was scary, but we started small and grew the operation as our experience grew," Todd explained. "We avoid debt, and though we have some outside income, we expect the farm to pay its own way!"

Todd and Annie's operation is centered on pastured poultry, but they also background replacement dairy heifers for a local dairy farmer, sell hay, and work in a cooperative agreement with three market gardeners who operate a Consumer Supported Agriculture (CSA) subscription garden on their farm. In each case, they have looked for unique approaches that minimize their expenses and the need for high dollar capital, while returning some profit.

"Backgrounding" of livestock refers to the practice of taking care of someone else's animals for a specific period of time. The practice is most often associated with the dairy industry, as many dairy farmers are happy to pay someone to take care of their young heifers for them. For Todd and Annie, the backgrounding operation works well, because they don't have to purchase the heifers. The dairy farmer they work with brings the heifers over in spring and picks them up in late summer or fall, depending on how long the grass lasts. He pays Todd and Annie rent, based on how many head they feed per month. Todd and Annie feed the heifers on rotated pastures and keep water and trace minerals available.

Excess hay is sold to another area farmer, who pays for the hay in the field. "The neighbor purchases the hay, but he has to cut it, bale it, and get it out of the field. We could possibly make a little more off the crop if we sold bales, but when you figure the expenses of owning and operating the equipment, and the labor involved in putting up hay, it wouldn't be worth it for us." By selling the hay as grass in the field, they don't have to worry about the weather; if the hay gets rained on, it's the neighbor's problem.

The concept of a CSA garden was interesting to Todd and Annie because they saw it as a way to bring local customers to their farm, but they were more interested in animal agriculture than gardening. By teaming up with three avid gardeners who didn't have enough land of their own to do market gardening, Todd and Annie were able to meet their own goals while helping the other three growers meet theirs. "The CSA is kind of a joint venture. The gardeners pay us rent, and it ultimately provides us with a customer base for our poultry."

Fifty area families purchase membership in the CSA at the beginning of each year. Their membership entitles them to shares of all the produce from the gardens, and they come out regularly to pick up their shares. The CSA also hosts some social events at the farm each year. "The best thing about 'hosting' the CSA is that it has created a marketing program that brings people out to our farm. When it's time to sell our chickens, we have a built-in customer base."

Todd works as an organizer for a nonprofit organization and Annie teaches at a charter school, but the farm provides succor and sustenance to both their bodies and their spirits. "The benefits of running our farm come from the things we can't go to town and buy. It fulfills our need to not only live in the country but also work the land, and grow in our understanding of what it means to be connected to a piece of land."

CHAPTER 2

Livestock & the Environment

When I arrived, I found a dozen teenagers, some in dreadlocks and nose rings, others wearing skateboard shorts and backwards hats. All were obviously as bored with sitting through a program on environmentally benign ranching as they would have been listening to a medley of Lawrence Welk's greatest hits. . . .

The first few images [slides] were what everyone expected: landscapes stripped bare of everything that wasn't too tough or too prickly to eat. Everyone sat there, arms crossed. They'd seen it before. Then came a photograph that created enough of a stir that even those who were sleeping woke up. It showed a riparian area with grasses and rushes and saplings of cottonwood and willow bordering a clear stream that was almost lost among the greenery. The vegetation was so lush it looked unreal, but it was real all right.

"Where's that?" several of the students exclaimed in honest surprise.

"Phil Knight's Date Creek Ranch, not very far south of here," I answered. "There were 500 cows in this very place two months before I took this picture. It's grazed five months out of every year."

— Dan Dagget, writing about a presentation he made to a high school class, from *Beyond the Rangeland Conflict*

FOR YEARS, MANY FOLKS within the environmental movement have blamed a wide variety of environmental woes on domestic animals and the farmers and ranchers who raise them. At best, this oversimplifies matters: It's a method of shrugging off the blame from our society's shoulders, where it truly must rest. In fact, it's not the animals that have caused the problems but how we have raised them; and how we have raised them is largely driven by our economy, our history and culture, our government's cheap-food policies, and our oligarchic (large corporate) food system. Farmers and ranchers can raise animals in a way that is actually beneficial to the environment, *and many do!*

In the winter of 1997–1998, Ken and I traveled around the country on vacation. We had more than 15 years of farming under our belts. During our travels, we studied the agricultural landscape with eyes that had learned to appreciate the nuances of the land, and generally what we saw on land with no livestock was as scary as what we saw where there was livestock. Human management of our natural resources was the problem — not domestic animals.

The issue of livestock and its negative impact on the environment, particularly its impact on public lands, is hardly a new one. In the early years of the twentieth century, as the national forest system was being developed, noted naturalist John Muir fought to keep livestock out of the new forests. He called cattle "Hooved Locusts." The main method available to farmers and ranchers who don't want hooved locusts is managed grazing. (You may also hear any of the following terms used for the concept, more or less synonymously: *intensive grazing, rotational grazing, planned grazing*, and *management-intensive grazing*.)

Managing for the Whole

Managed grazing, as I apply the term, is really just part of a broader approach to agriculture. This approach calls for looking at your farm as a whole, and attempting to manage for that whole. The whole you manage for is based on your family's quality-of-life goals, production goals, and goals about how you want your piece of land to "look" in the future — in other words, your environmental goals. Each farm family must define the goals for its own operation and land. When the family works toward these well-defined goals, its members are able to become more than just stewards of the land; they are able to become healers.

The idea of managing for the whole is often referred to as "whole-farm planning," of which there are many different models. Ken and I have been most influenced by the holistic management model developed by Allan Savory of the Center for Holistic Management in Albuquerque, New Mexico. We think it best meets the needs of livestock farmers. Chapter 3 reviews the holistic management model.

In some areas of the country, groups of farmers and ranchers are getting together with bureaucrats, environmentalists, and other interested citizens to manage whole regions with this approach. These groups often form around a given watershed or other geological feature where group efforts make sense.

Working with Nature

When managing for the whole, a farmer or rancher begins to look at how he or she can work with natural processes, instead of trying to control them. Nature is usually fairly effective in establishing balance, if given the chance. Patience is truly a virtue in these endeavors.

The pigeons that inhabited the haymow in our Minnesota barn are a good example of natural balancing. When we purchased the farm, there was a small flock of pigeons living there. Shortly after we moved in, one of our neighbors came over and told Ken he should shoot those pigeons: "Pigeons are dirty, they carry every disease known to mankind, and your animals will all get sick and die if you let them live in your barn," Joe said. This was common wisdom in our area. We chose to ignore this particular wisdom; for one thing, we enjoyed listening to the gentle coos of our resident flock.

After about four years, our flock had grown from six or seven birds to more than fifty. We began talking about shooting some of them to thin the population; after all, when a population becomes too large, disease can become a problem. Despite our talk, we didn't get around to shooting any; it was spring calving and lambing time, and we were busy milking cows. Then, around the beginning of July, we noticed that the pigeon flock seemed to be shrinking: First it was down to thirty birds, then twenty, then fifteen. It was a real puzzle; we didn't see signs of dead birds, so we didn't think that they were dying of a disease. Finally, on a very cloudy and overcast afternoon, Ken noticed a great horned owl sitting in the middle of the barnyard. It looked as though he was clutching something in his talons. When he started to fly away, we saw he had a pigeon (Figure 2.1).

Figure 2.1. When natural processes, such as the food chain, are working well, nature keeps populations in balance. Predators like the great horned owl are critical to balance in nature. An owl in our Minnesota farm kept pigeon numbers at a healthy level.

The pigeon flock had grown to the point of being very easy prey. The owl stayed in our woodlot for a couple of months that summer, picking away at the pigeons. By fall, there were about eight pigeons left to begin rebuilding the resident flock. Balance was restored.

Ecosystem Processes

To begin managing for your family's environmental goal, an understanding of nature's regularly occurring processes is helpful. Our natural world is a complex system, but it can be viewed in terms of community dynamics (living organisms), the energy cycle, the water cycle, and the mineral cycle. These four processes are the foundation upon which the ecosystem functions. They are all dynamic and interdependent, and in the healthiest environment each one is operating at its maximum efficiency.

Community dynamics can be thought of as the way living communities of organisms move toward complexity in a healthy environment. (As I discuss these processes, think of all living organisms: plants, animals, insects, bacteria, fungi, and viruses.) Depending on the type of environment you live in, the dynamics might lead to a hardwood forest or a grassland, but given favorable circumstances, they will always lead to a climax state that is both complex and stable. (Depending on your goals, you'll probably maintain your farm at some level below a climax state.)

An example is what takes place in an abandoned farm field. First annual grasses and "weeds" invade the field. During this early period, the most troublesome weeds in the area (thistles, mustard, bindweed) take over the fields. Some birds will begin coming to feed on insects, which also proliferate at first. Small animals start moving in. In the ensuing years, perennial grasses and scrub plants move in. More birds and animals, including ground-nesting birds, come to the area. Trees begin to grow, and soon shade out the grasses. Larger animals begin inhabiting the area. From farm field to forest takes only 20 years in some areas of the United States.

Living organisms didn't develop in a vacuum. Plants require animals, animals require plants; both require insects, bacteria, and viruses. The healthier the community, the more diverse the organisms that live in it, and the more stable those organisms are.

Energy from the sun is the key to all life on earth, and its transfer from one living organism to the next characterizes the *energy cycle*. This cycle is also known as the food chain (Figure 2.2).

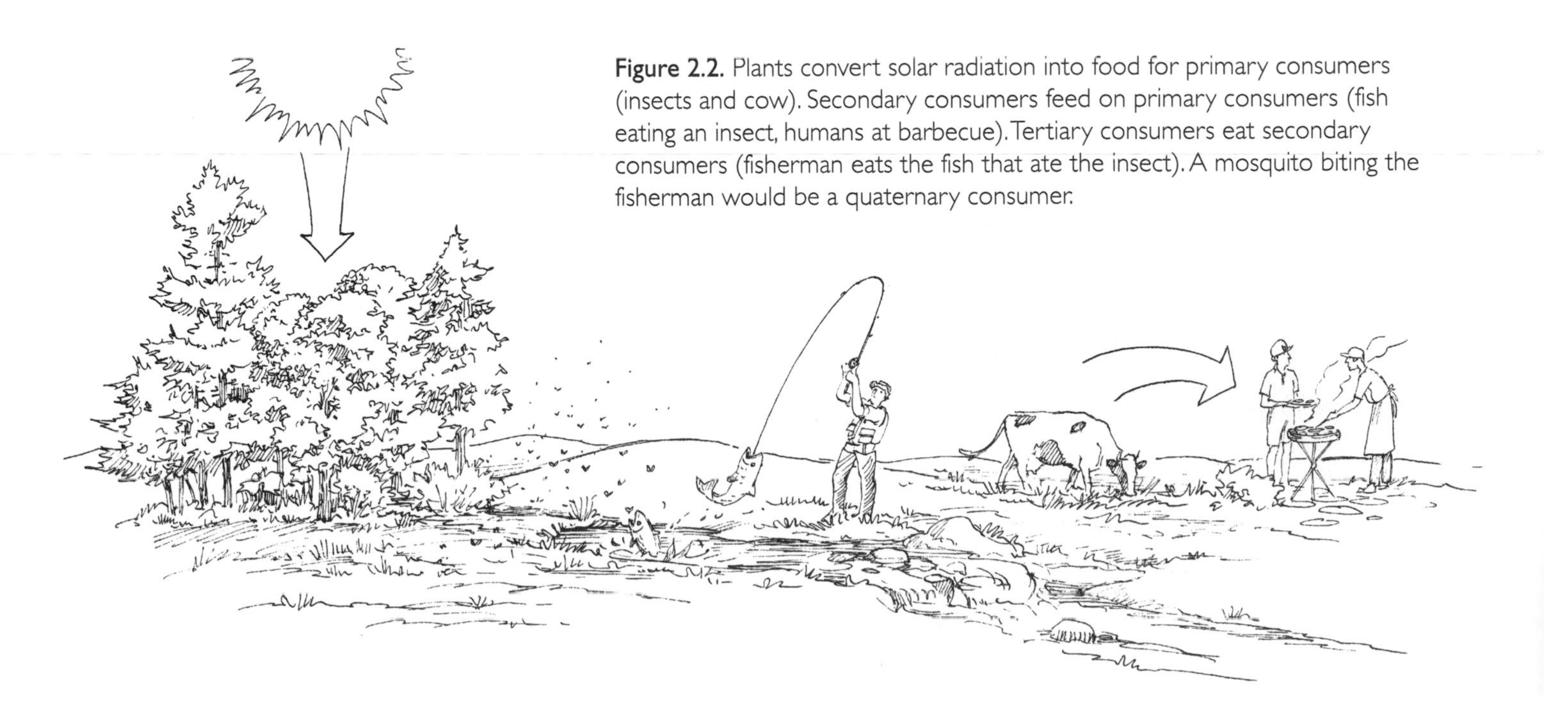

Figure 2.2. Plants convert solar radiation into food for primary consumers (insects and cow). Secondary consumers feed on primary consumers (fish eating an insect, humans at barbecue). Tertiary consumers eat secondary consumers (fisherman eats the fish that ate the insect). A mosquito biting the fisherman would be a quaternary consumer.

Plants are the primary consumer of solar energy. Through photosynthesis, they convert incoming solar radiation into organic (carbon-based) molecules, feeding themselves and providing food for the secondary consumers. Microorganisms, insects, fish, reptiles, birds, and mammals that eat vegetable matter make up these secondary consumers. The tertiary consumers eat secondary consumers. The levels continue to rise, like a pyramid, each layer feeding on the layer or layers below it (Figure 2.3). The top levels contain scavengers and organisms of decay. Many creatures feed through more than one layer. For example, humans can feed at layers two, three, and four. Canines, both wild and domestic, can feed at layers two, three, four, and five.

Plants capture just a fraction of the sun's energy, and with each move up the pyramid, less energy is available. This reduction in energy means less total biomass can survive at these higher levels. In other words, predators can't outnumber prey. At the same time, the lower levels are very unstable if there are insufficient predators in the upper levels, as evidenced by outbreaks of disease or starvation in secondary consumer populations when their numbers get too high.

The *water cycle* is the movement of water between the atmosphere and the earth (Figure 2.4). Some water runs off the land to enter streams, rivers, lakes, and oceans, but in a healthy ecosystem the soil matrix is capable of absorbing large quantities of moisture. Of the water that is absorbed by the soil, some evaporates back out of the soil, some enters the groundwater, and some is used by the living organisms in the soil, including plants through their roots. In an effective water cycle, water is readily available and used by plants. In an ineffective water cycle, most of the water runs off or evaporates from the soil.

The ability to absorb water and bank it for future plant use requires a healthy, living soil that contains plenty of humus, or organic matter, in the soil. Organic matter is made up of decaying and living organisms. Scientists estimate that tens of millions of living organisms live in a single tablespoon of healthy soil.

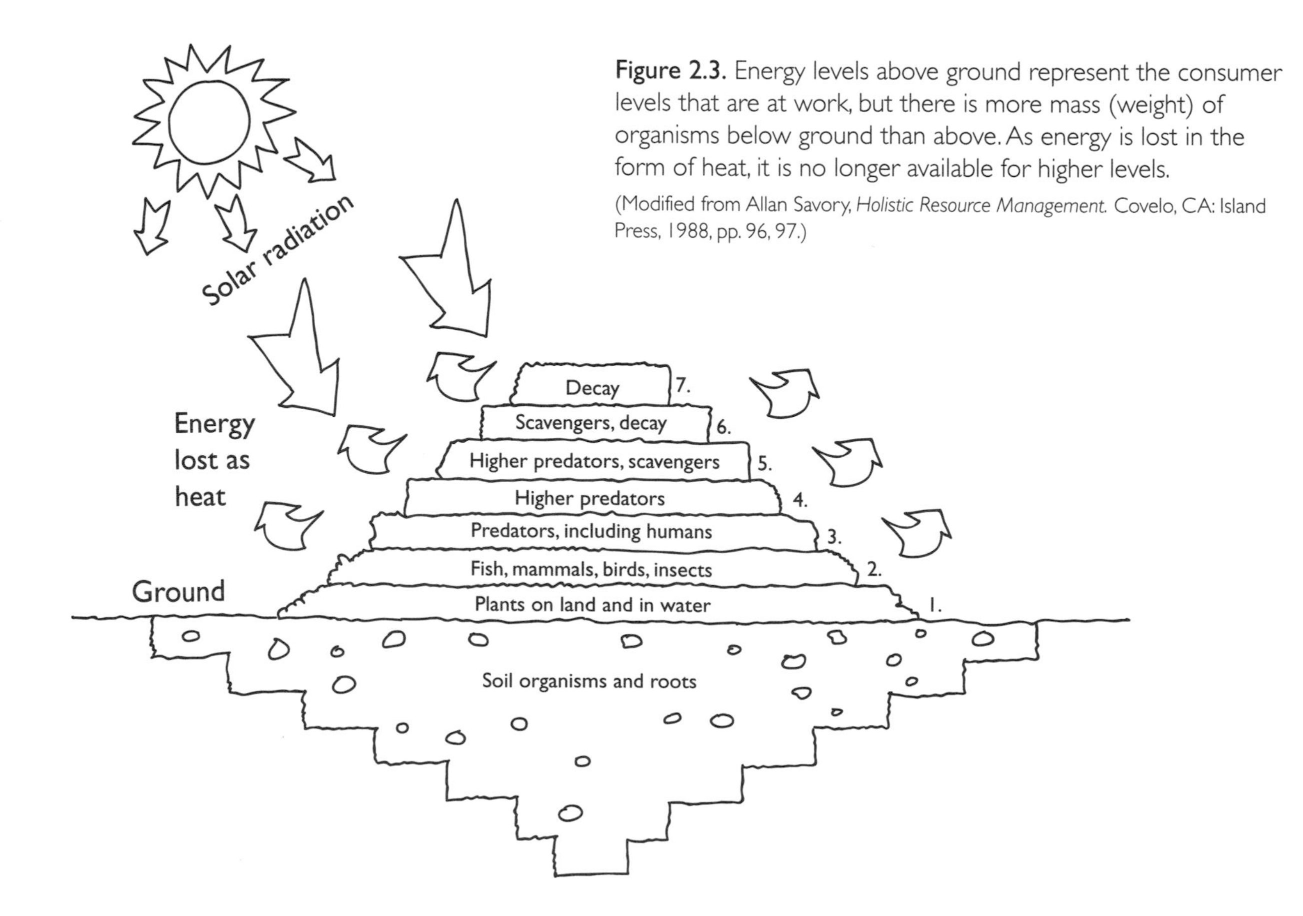

Figure 2.3. Energy levels above ground represent the consumer levels that are at work, but there is more mass (weight) of organisms below ground than above. As energy is lost in the form of heat, it is no longer available for higher levels.
(Modified from Allan Savory, *Holistic Resource Management.* Covelo, CA: Island Press, 1988, pp. 96, 97.)

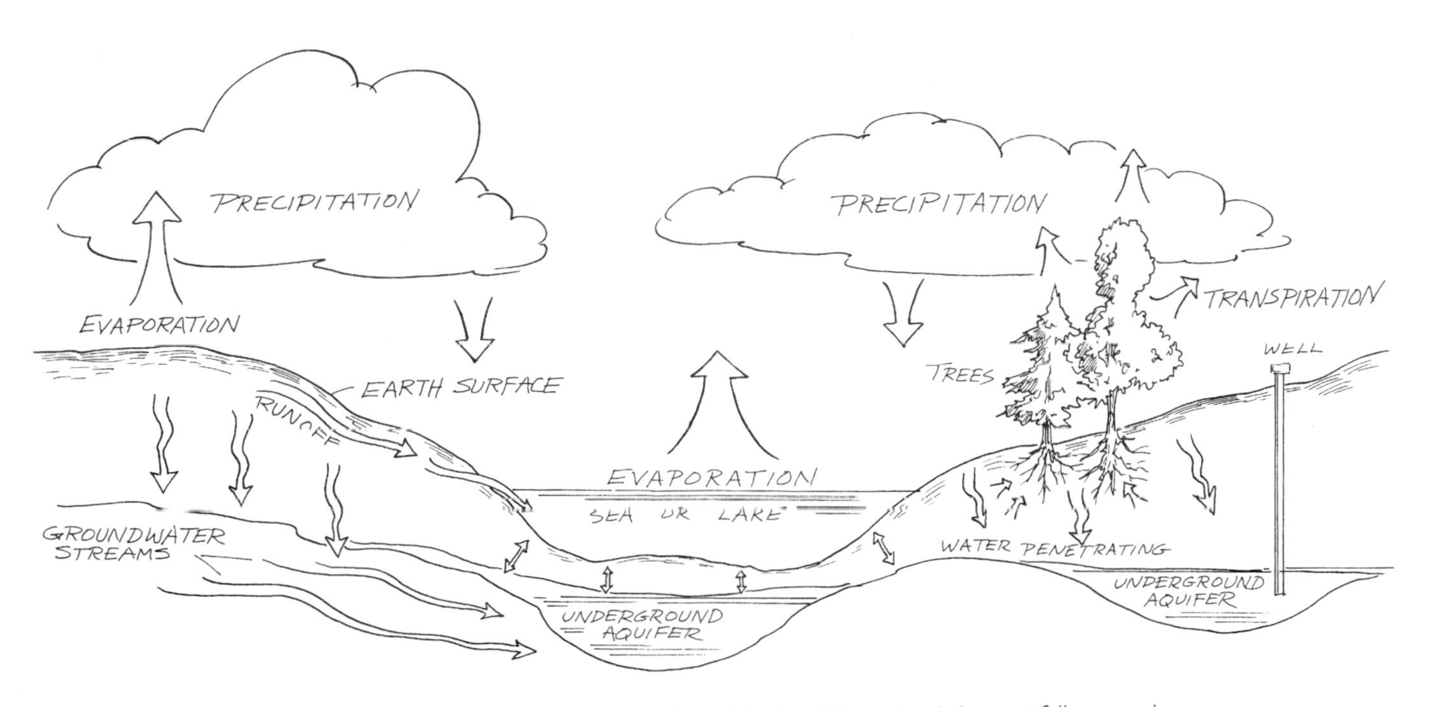

Figure 2.4. Approximately 25 percent of precipitation falls on land; the rest falls on major water bodies. Of that portion, one-third leaves the land as runoff, one-third evaporates or is transpired back into the atmosphere, and one-third enters the ground water. Ground water can move in either direction, in and out of surface bodies of water.

The *mineral cycle* is the breaking down of materials into a form that can be used by plants and animals (Figure 2.5). Through this breakdown, essential trace nutrients and minerals are made available to plants and animals. The mineral cycle includes subcycles, among them the carbon cycle, the nitrogen cycle, and the phosphorus cycle.

Carbon is the building block of all life. The implementation of the *carbon cycle* is not gentle or kind: Death is the absolute partner of life (Figure 2.6). Without death and decay, no new life is possible. One of the hardest parts of livestock farming can be coming to terms with this reality.

When all of these cycles are optimized, plants and animals thrive. Cycles that no longer operate properly result in increased water runoff, crusting of soils, erosion from both wind and water, and a variety of other symptoms. The extreme example of cycles that have gotten out of balance is seen in desertification, but even in humid environments — which tend to be more forgiving — you'll often see the same symptoms.

Figure 2.5. The mineral cycle is the process by which complex molecules are broken down into their respective parts, or elements. Manure and urine, and dead plant and animal matter are broken down by soil organisms such as earthworms and bacteria into nitrogen, phosphorus, and carbon, to name just a few elements. Then some of these elements are taken back up by plants as a source of food, beginning the process again.

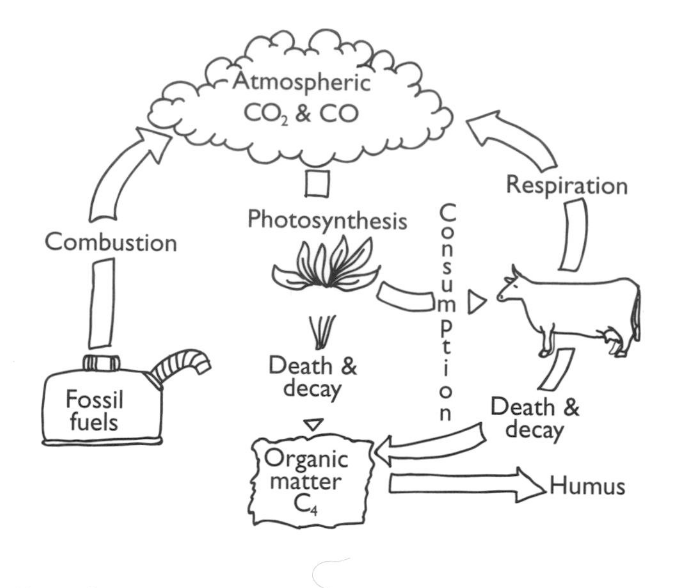

Figure 2.6. The carbon cycle is one of many subcycles within the mineral cycle. Plants remove carbon from the atmosphere with the help of solar radiation, through the process of photosynthesis, and make it available as a food source to other forms of life. Death and decay return carbon to the soil. Carbon enters the atmosphere from both respiration by living creatures and from the combustion of fossil fuels.

Brittle versus Nonbrittle Environments

All four processes (community dynamics, energy cycle, water cycle, and mineral cycle) behave somewhat differently depending on the type of environment they are operating in, brittle or nonbrittle. If you look at these two as extremes of a continuum, a true desert environment represents complete brittleness; complete nonbrittleness is seen in a rain forest (Figure 2.7). In the United States, think of Death Valley, California (with about 2.2 inches, or 5.6 cm, of precipitation per year), as the extreme brittle environment and a Louisiana bayou (with 60 to 70 inches, or 152 to 178 cm, of rain per year) as the nonbrittle extreme. Most of us live somewhere in between.

The brittleness of an individual environment is a factor not only of how much precipitation occurs but also how often it occurs, and how much humidity is in the air. In other words, Los Angeles, California, gets about the same amount of precipitation as Fort Morgan, Colorado, but the rain comes in a shorter period of the year and during the winter (the monsoon season) (Figure 2.8). In fact, during the summer months Los Angeles receives virtually no rain, whereas Fort Morgan gets most of its precipitation during the summer growing season. Although Colorado is brittle, southern California is more brittle. There are highly brittle environments in some areas of the world that receive more than 30 inches (76 cm) of rain per year, but like southern California, it all comes during a short monsoon season.

Study Table 2.1, which illustrates certain characteristics of the extremes of the brittleness scale. Think about how your land falls on the brittleness scale. Figure 2.9 might help you determine in general the brittleness of your area of the country, but remember that within any area there are more and less brittle pieces of land — like an oasis in a desert. Even on your own farm, there can be more brittle areas and less brittle areas.

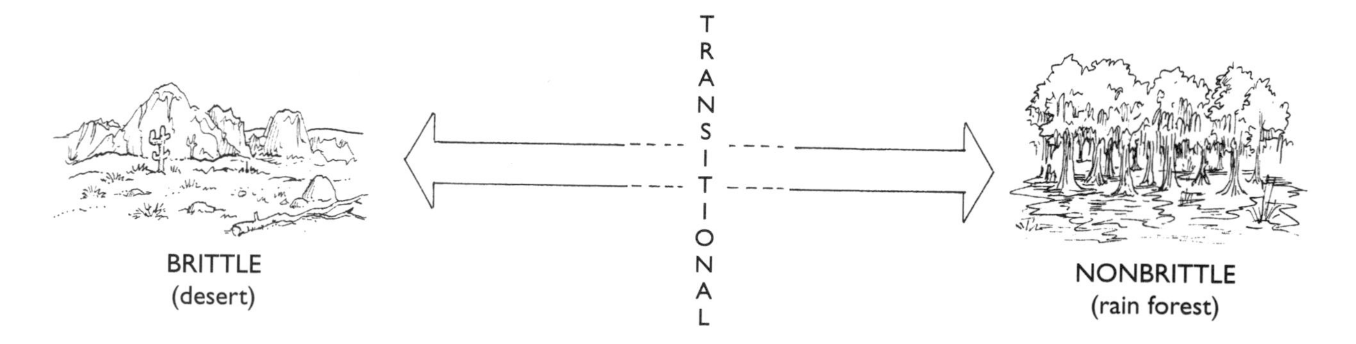

Figure 2.7. Brittleness is a measure of both the amount of precipitation that falls in a given area and the distribution of that precipitation during the year. Areas that receive little moisture, or whose moisture all comes during a very short season, are considered brittle; areas with more moisture spread out more evenly during the year are nonbrittle.

(Modified from Allan Savory, *Holistic Resource Management.* Covelo, CA: Island Press, 1988, p. 39.)

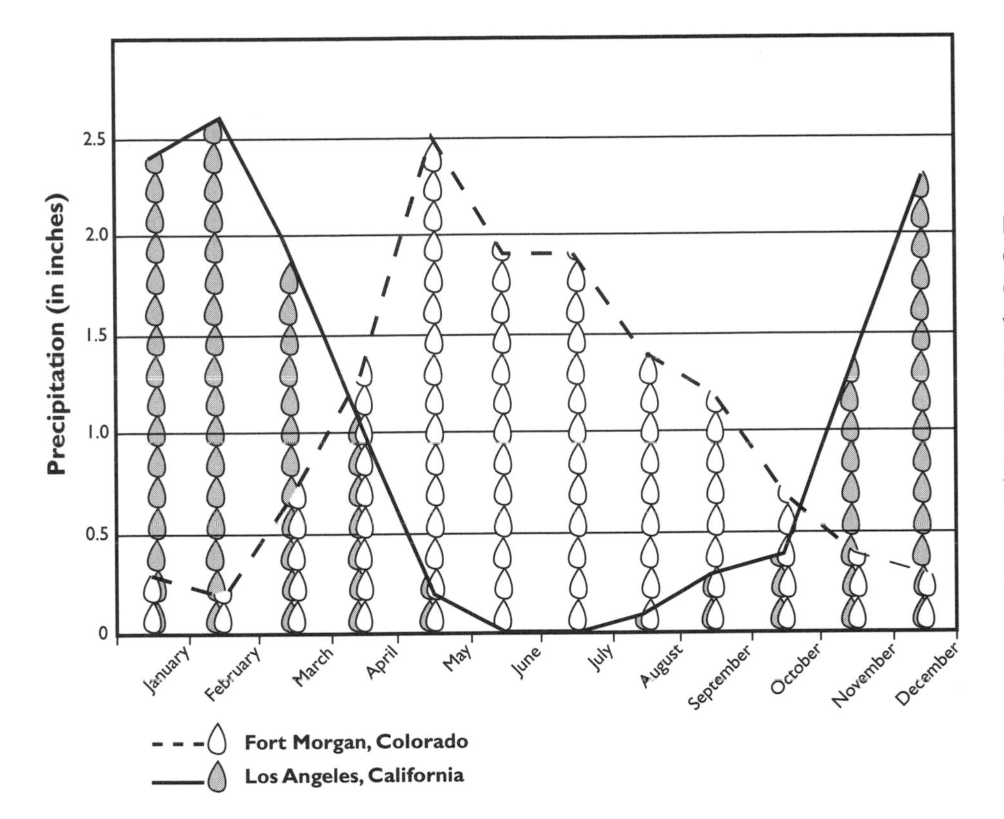

Figure 2.8. Although Los Angeles, California, and Fort Morgan, Colorado, are low precipitation areas with approximately the same annual precipitation, Los Angeles is more brittle than Fort Morgan, because its precipitation comes during a shorter period of the year. Los Angeles goes through a period in summer and early fall with virtually no precipitation; Fort Morgan's precipitation is more evenly distributed throughout the year. (Data from National Weather Service.)

Table 2.1

Characteristics of Brittle and Nonbrittle Environments

Characteristics	Brittle	Nonbrittle
Total precipitation	Generally low	Generally high
Distribution of precipitation	Poor	Good
Succession on a disturbed piece of land	Starts slowly	Starts quickly
Decay of old plant materials	Primarily physical and chemical decay (weathering and oxidation) at the top of the plant	Primarily biological decay at the base of the plant
Spacing of perennial grasses	Open spacing with distinct bunching of plants	Close with indistinct bunching; tends to form a solid sod quickly
Leaching characteristics	Tends to maintain the mineral level in soil; low leaching	Tends to lose mineral level in soil through leaching
Impact of overgrazing	Tends to expose more soil surface to forces of erosion	Tends to thicken the grass cover and tighten the plants
Effect of low animal impact	Tends to expose more soil	Tends to maintain close plant spacing
Effect of high animal impact	Tends to tighten plant spacing	Tends to tighten plant spacing
Impact of rest, or removing animals from the land	Community becomes less complex	Community becomes more complex

Conclusion

Livestock is capable of turning grasses that are not edible by humans into a high-quality protein for human consumption. By covering more soil with long-term grass crops for livestock forage, we feed ourselves and improve the environment.

Grass plants reduce erosion by breaking the size of raindrops that strike the soil and by acting like tiny dams, slowing the water movement along the surface of the soil. The blades of the grass plants also reduce wind speed at ground level, thereby reducing wind erosion. The plants, along with livestock manure and urine, add nutrients to the soil and build up organic matter. When grass is a long-term crop, the plants' roots tend to open the soil structure for air and water to move easily.

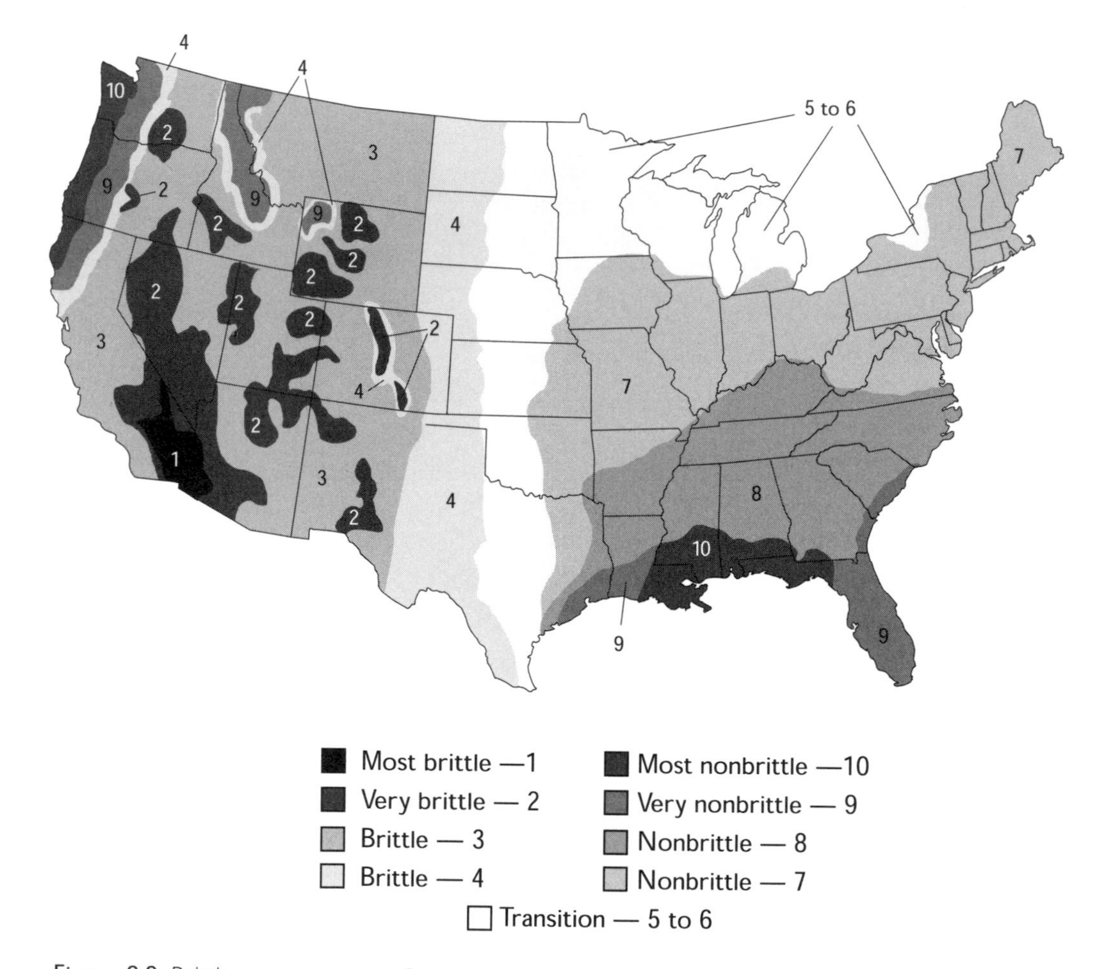

Figure 2.9. Brittleness areas vary. On a scale of 1 to 10 (1 being the most brittle and 10 being the least brittle areas), this map provides some general indication of brittleness throughout the United States, but even within these areas there are more brittle and less brittle spots. To get a sense of the variability that's possible within an area, think of an oasis in a desert.
(Data from Oregon Climate Service and USDA/Natural Resources Conservation Service.)

C H A P T E R 3

The Holistic Management Model

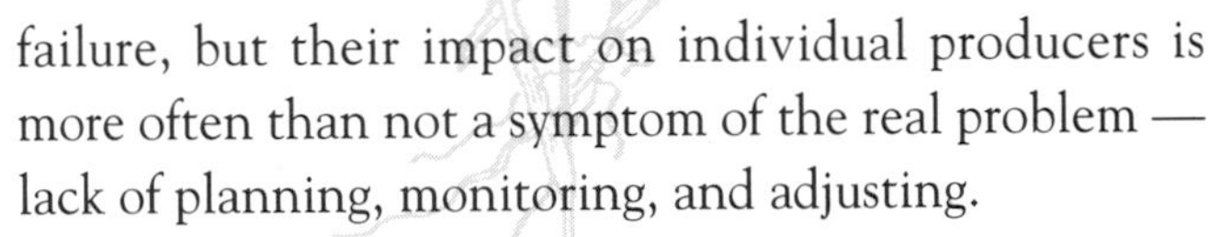

We may be very busy. We may be very efficient. But we will also be truly effective when we begin with the end in mind.

— Steven Covey, *The Seven Habits of Highly Effective People*

Whether or not a farmer becomes a successful producer depends largely upon his ability to judge true values in his land, crops and livestock. To judge correctly, he must see each object or enterprise as a whole with all its essential parts and understand their relation to each other as well as to the desired result.

— Wilbur J. Fraser, *Profitable Farming and Life Management*

NO ONE PLANS TO FAIL. Yet so many people, particularly in agriculture, fail at what they try to do. Beginning farmers and third-generation family farmers often suffer from the same types of failures, and those failures usually lead to jobs in town. Farmers have lots of things to blame their failure on; weather, prices, government regulations, and markets are all popular. Granted, all of these may be contributing factors to failure, but their impact on individual producers is more often than not a symptom of the real problem — lack of planning, monitoring, and adjusting.

In the holistic management model, planning is not simply a process to be completed once. It is a continuous process that involves multiple parts: setting realistic, broad goals; developing plans and making decisions that move you toward those goals; continually monitoring what is happening with respect to your plans and your goals; and making appropriate adjustments to your plans if things are not working out the way you anticipated. And make no mistake about it, no matter how good your planning and monitoring, things have a way of not working out the way you thought they would. Planning should be started now. Whether you have yet to buy your farm or have been farming all your life, don't put it off for another day.

Part of the reason Ken and I succeeded at all over the years was that we were very goal oriented. Still, we didn't arrive at a structured approach to planning until after we'd made many costly mistakes. As we look back, none of these mistakes would have been as bad had we been planning in a logical and methodical manner. I'll give one example: We raised pigs one year. Unfortunately, it happened to be a year when corn prices hit their all-time record high. Irregular summer weather (hotter and wetter than normal) and gray leaf spot, normally a minor disease in corn that didn't behave in a minor fashion that year, triggered the record prices. We knew corn prices were high in fall and going to get higher during the winter, but we kept feeding our pigs for finished weight. We justified the decision based on our direct-marketing, for which we were paid more than the regular commercial return for finished hogs. It didn't pay us enough more.

Because of our direct-marketing, we were hurt less than farmers selling through the typical commercial venues were, but when the numbers were all in, we just broke even. Hundreds of unpaid man-hours went into an enterprise that didn't return any profit because of prices and weather. But the prices and weather weren't the real problem; our failure to monitor the situation and make early adjustments was the problem.

By following the holistic management model, we began making changes to our lives and our farming operations that moved us toward our broad goals.

Overview of the Holistic Management Model

The holistic management model (Figure 3.1) has four major parts:

1. ***The holistic goal,*** which takes into account quality of life, forms of production, and future resource base.
2. ***The ecosystem processes,*** which I talked about in chapter 2. These include community dynamics, energy flow, water cycle, and mineral cycle.
3. ***The tools,*** including human creativity, money, and labor, along with rest, fire, grazing, animal impact, living organisms, and technology.
4. ***The guidelines,*** including techniques for testing and managing your system.

The holistic management model is based on the concept that you plan, monitor, control, and replan. The model provides logical ways of doing this. The full planning process, as developed by Allan Savory, is time consuming and complex. In part IV, Planning, I'll present a somewhat easier and more compact version. This easier version will serve well enough for farmers living in a moderate to nonbrittle environment, particularly those with smaller acreages and those who do not have a high debt ratio. If you are living in a very brittle environment, are trying to manage a very large piece of land, or are trying to make your living off your farm despite a high debt ratio, I urge you to study the full methods of planning. (See appendix E for sources of information on the full method.) Even if you choose to use the more simplified planning procedures outlined in part IV, a brief overview of the full model is helpful.

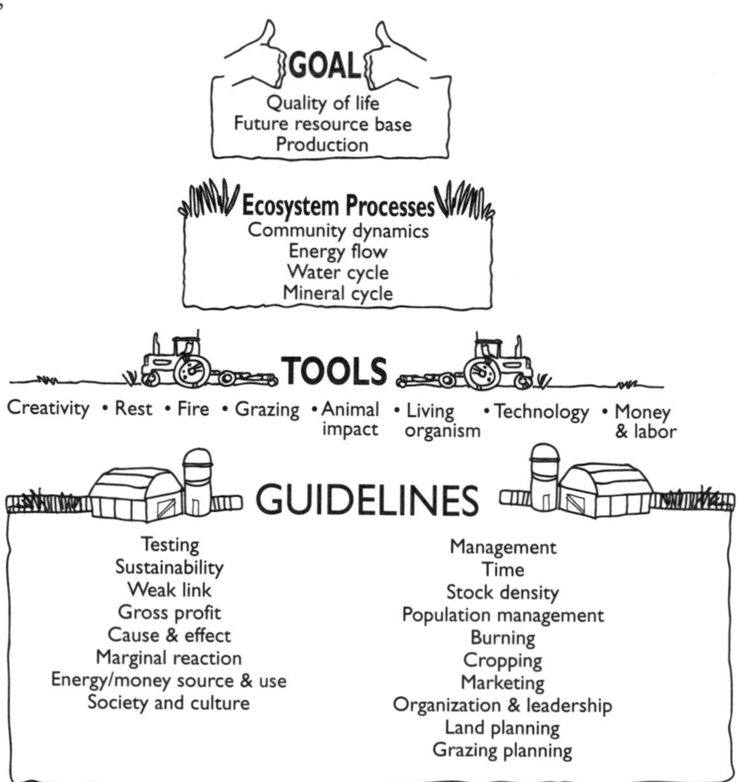

Figure 3.1. The holistic management model. Use these guidelines to decide which tools to apply to a given situation, so that you move the ecosystem processes toward better community dynamics, energy flow, water cycling, and mineral cycling, and to move yourself closer to your goal.

(Modified from Allan Savory, *Holistic Resource Management.* Covelo, CA: 1988, p. 5.)

FARMER PROFILE

Jim Weaver — North Penn Holistic Management Network

Learning is often easier in a group; Jim Weaver can attest to that fact. Jim is a member of the North Penn Holistic Management Network.

For Jim, 1992 was an important year; that was when he first heard of holistic management. "One of my neighbors had been to a conference where Allan Savory spoke. He was excited by what he heard. He brought a copy of the *Holistic Management* textbook back with him, and began passing it around."

Jim admits that his first introduction to the textbook was confusing. Parts of what he read made perfect sense, but other parts left him with more questions than answers.

"I wasn't involved with farming at the time, though I grew up on a farm. I am an aquatic biologist by training and operate a consulting business that deals with aquaculture and wetlands management. I saw connections to my own work when reading the textbook, but I needed to learn more."

A small group invited Allan Savory to Pennsylvania to give an overview of holistic management, and things started to become clearer. Next, the group scheduled an introductory training session with Ed Martsof, a certified holistic management educator.

"We finished the intro class and formed the North Penn Holistic Management Network. There are about ten of us who have met every two weeks ever since, and other people who attend sporadically." Group members act as a support network for each other; they host field days and pasture walks, and they are trying to get involved with nonfarmers to create a dialogue about agriculture, the environment, and "where our food comes from." Members help each other run ideas through the testing guidelines. Some outcomes of the group process: One dairy farmer installed an outdoor milking parlor for the summer grazing months; another installed a solar dairy barn; still others diversified into pork and pastured poultry; and Jim began experimenting with on-farm aquaculture, an endeavor that meshed well with his aquatic biology interests.

As a result of his involvement in holistic management, Jim and his family — wife Beth and three sons — also decided that they needed to get into livestock agriculture. His first approach was to raise stocker cattle, but as he says, "Stockers beat me up pretty good. I didn't study the markets enough, and prices hit bottom. That convinced me that I didn't want to do 'commodity' agriculture — I needed to get into value-added approaches."

Today, Jim and his family background dairy replacement heifers for another member of the network, and graze dry cows as well. They raise some beef and hogs for a fledgling direct-marketing meat program, and are trying to develop the aquaculture into a viable on-farm business.

"When I began getting involved with holistic management, it dawned on me that I never really thought about life goals. Developing a holistic goal has been good for us as a family. I think it's helped us communicate with our kids, and involve them in something positive."

Jim goes on to tell the story of the day his middle son, Seth (who is in charge of the family's pig project), said to him, "If I raised more pigs, maybe I wouldn't have to get a regular job when I grow up."

"It was like a light had gone on for Seth. He saw that life was about choices and decisions."

Jim also says it's been interesting to watch the network evolve. "Our perception of reality has changed. When we first began the group, we wanted to change the world, but we learned that that's a big job, and it takes patience. We are, at least, changing our own lives. And we are beginning to make connections to the nonfarming public. At our last pasture walk of the season, we had many nonfarming visitors. It was great!"

Setting Your Broad Goal

As I mentioned above, the holistic goal I'm talking about is actually a three-part affair:

1. Quality of life. What is it you want from your life? What would you like for your children and for your community? This goal is about happiness, health, wisdom, aesthetics, and culture.

Our decision to sell our farm in Minnesota and move back to the West was largely based on our quality-of-life statement. We wanted to be closer to family. We yearned for the vistas of the West, where a hundred-mile view was not out of the question. We missed stream fishing for brook trout, elk hunting, and downhill skiing.

2. Forms of production. What do you expect to produce from your land? Are you primarily interested in producing your own food, or are you looking for profits from livestock and crops? Do the profits have to provide the family's living and pay for the land, or are you primarily interested in just a little extra cash above what your endeavor costs? Are you trying to preserve something — for example, historic buildings on your site, or an endangered species? Is recreation an important product of your land, or perhaps timber? Is an environment conducive to writing, music, or art part of what you're looking for?

Don't be too specific at this point about which crops or animals you might grow, particularly if you'll need sufficient profits to support your family. When you develop your plans in part IV, you'll identify the type of animals or crops.

Unrealistic expectations have sunk many of the beginning farmers I've met over the years. If your goal is to provide your family's living from the farm, but a minimally acceptable living includes new vehicles, clothes ordered from Eddie Bauer, dinner out twice a week, and so on, then plan on keeping your day job and letting the farm be a fun hobby. Hobby farms can return more than they cost to run, so your operation can make money without being your primary occupation.

Your money expectations may be realistic, but not your time expectations. Each endeavor you add to your operation will add a great deal of time to your day, especially at first. Trying to manage five different classes of livestock at once will probably leave you managing none of them well. Taking on too large of an operation from the start may also leave you time stressed. Start slowly. Build up slowly.

3. Future resources. If you could look at this land 100 years from now, what would you want to see? A mixture of forest and pasture? A series of ponds? Are there gullies to be repaired, or bare patches of dirt? Do waterways need improvement? What kinds of wildlife would you like to be present?

Future resources can also include people that are important to you: your extended family, community, and so on. For example, if you want to see a strong, local community with thriving small businesses, numerous cultural amenities, or fine institutions, then list these in this part of your goal.

Goal setting should involve all the decision makers (the entire family, employees); it may also benefit from participation by others who might have a stake in your land or success (perhaps friends, relatives, local environmentalists, or bureaucrats from the National Resources Conservation Service (NRCS). Don't bypass this step, thinking that everyone is in agreement. Write your holistic goal down, and then put it in a conspicuous place, such as taped to the bathroom mirror or on the front of the refrigerator.

Although for most people, setting a goal involves a discussion of money and profits, try to get beyond money for its own sake. Paper money is the grease that keeps the economic cogs turning in this age, but people whose goal is money in its own right will probably never be happy. A recent survey indicated that people in cities make more money than their rural counterparts, but are less happy with their incomes; more highly educated people make more money, too, but they are also less happy with their incomes. Too many of these people — and we know some well — fall into the trap of thinking money will buy happiness. Money is a tool, not a goal.

Remember that your goal statement is broad, so the whole thing will probably take up just a page or two. The process is somewhat evolutionary: Over time, some aspects of the statement may be refined or clarified, but in general your goal statement will represent longer-term thoughts. Our first version was very formulaic, but as we have continued grappling with the questions of what we really want out of life it has become shorter and more succinct, yet also more lyrical. The most important thing is to get something written down to start with. Don't expect it to be just right, the first time around. As you live with it, it will grow and flower.

Family Goals. Each family's goals will be different. For example, Sherry O'Donnell and Virgil Benoit are both tenured university professors. They want their operation to be profitable, but it doesn't have to pay all the bills. Their goals include maintaining the farm that Virgil's grandfather homesteaded, using sheep in part to support Sherry's hand spinning, having a nice place to live in the country, and being able to provide most of their own food, either directly or through barter with other farmers they know.

On the other hand, Tom and Irene Frantzen are third-generation farmers from Iowa. Their goals definitely include a form of production that can support their family from the land without outside income. Another important goal for Tom and Irene is making the farm into a place that one or more of their children may want to run after they do. They want to leave the land in better shape than when they got it, and they want to "be able to look their animals in the eye, and know they are having a good, happy life while it lasts."

The Tools

We humans have three major tools to make things happen. The first is the mind. Through our creativity and ingenuity, we are often able to solve problems or make things happen. The second major tool is our labor. Through physical work, we can create or build something. The third tool is money. With money, we can buy something already done. I think of the three major tools as Brains, Brawn, and Pictures of Dead Presidents.

Of course, money is nothing more than a trade medium. Its availability is related to either our brains or our brawn (though sometimes it may be the result

KEN AND CAROL'S HOLISTIC GOAL

First, we want to live in places off the beaten track. We want to feast our eyes on spectacular views of rugged mountain vistas. We want a place with lots of sun, abundant wildlife, and minimal light pollution. We want to sit out at night lost in a sea of stars, listening for the serenade of song dogs (coyotes). Beauty is, in itself, worthy of effort.

Second, we want to be as independent as is feasible in this modern age. We want a significant portion of our energy to come from renewable sources, and much of our food to come from the work of our own hands, whether hunting, gathering, or growing. We want the bulk of our living to come from our own endeavors, not jobs.

Third, we want to be honorable and honest in our dealings with each other and with other people. We want to be compassionate and respectful in our treatment of other living creatures. We want our actions on the land to be beneficial to the ecosystem; we want to understand and work with ecosystem processes. We want to be contributors, not just takers, from society and the earth. And we want the world to be a better place for our having passed through.

Fourth, we want to do meaningful work, both physically and mentally. Our work needs to keep us in the outdoors, and to keep us surrounded by animals. We enjoy work, and take pride in it, and believe that work keeps us healthy and happy.

Finally, we have little interest in material possessions, but financial security with little or no debt is important to us. We want a small but comfortable home. More important to us than money is time: We want to make time for reading, writing, art, music, travel, and a loving relationship. We want to keep our minds open and expanding; our hearts full and content.

of someone else's efforts, as in the case of an inheritance). Unless your hobby is collecting pictures of dead presidents, having money for money's sake isn't usually part of your goal, yet in this day and age money is integral to everything we do.

The other tools listed in the model (rest, fire, grazing, animal impact, living organisms, and technology) are what we apply with our creativity, labor, or money. These tools are used to move us toward our holistic goal.

Rest

Rest refers to allowing an environment to be undisturbed — by human activity or grazing animals — for extended periods of time. In a crop-farming system, this would mean allowing a field to lie fallow for a season. In a livestock system, it means keeping the grazers off, or keeping their numbers so low that many plants mature without ever being bitten. *Rest* in this context refers to fairly long periods, not to the short time required for a plant to regrow.

Rest is sometimes good for the land in a nonbrittle environment, but overrest may cause deterioration, particularly in a brittle environment, where perennial plants that are allowed to reach maturity become extremely coarse and shade out the basal growth point near the base of each plant. As the plants die at the end of the growing season, the tops accumulate dead material, which is slow to decompose.

The dead leaf matter that accumulates begins to reduce growth in subsequent years, thus weakening the plant's roots. Eventually, some plants die, and more bare earth shows up between the remaining plants, reducing all four ecosystem foundation blocks — community dynamics, water, energy, and minerals.

Fire

Fire is a natural rejuvenation method for land. Early peoples used fire not only as a source of heat and light, but also as a tool to drive away wild animals and freshen grasslands. Fire is now overused in some areas, while others have adopted a complete "no-burn" policy — even for naturally occurring lightning fires. Both these extremes have negative consequences.

In the Southwest ranchers are beginning to let natural fires burn themselves out. Historically, this method controlled brush and refreshed the grass, but for the better part of this century, fires have been completely suppressed, allowing brush like mesquite to proliferate.

The problem with fire comes when it becomes a habit. For instance, where we lived in central Minnesota, spring burning was a practiced ritual and had negative impacts on all ecosystem blocks. Used repeatedly, fire favors certain woody plants, reduces the organic content of soils, and kills many soil microorganisms.

Grazing

Grazing — and when I talk about grazing you could substitute the term *browsing* — is one of the tools we are most interested in as livestock farmers (Figure 3.2). Most animals can get a significant, if not complete, portion of their diet from fresh forage (grass and leaves off shrubs and trees). (See chapter 6 for a detailed discussion of feeding livestock.) Forage plants protect our soil from erosion, convert solar energy into

Figure 3.2. Grazers prefer grass; browsers prefer trees and shrubs; intermediate species prefer forbs. Despite this, there is regular cross over between the three types of feeders.

food, balance carbon dioxide and oxygen levels in the atmosphere, and provide an aesthetically appealing environment (green is soothing and beautiful).

Effective grazing requires balance, and when that balance is reached, the four ecosystem processes move to a higher and more stable level. Balanced grazing provides for good gains by stock while also improving the land. Grass that is overgrazed, like grass that is overrested, moves the ecosystem processes to a lower and less stable level and, most importantly, reduces profitability.

Overgrazing, or overbrowsing, occurs when a plant is bitten a second time before it has had a chance to regain the store of energy it lost from the first bite. Figure 3.3 shows what happens to two plants, each bitten for the first time at approximately the same moment. Plant A is only bitten the one time and then allowed to regrow to its full energy potential. Plant B is bitten again before that full regrowth takes place. If a plant is either severely overgrazed, or overgrazed lightly, often, it will weaken and eventually die. The time it takes for a plant to regrow to its full energy potential is called the *recovery period.*

Traditionally, the definition of *overgrazing* was simply too many animals, or overstocking. But although overgrazing is often caused by too many animals, it isn't always; it may also be caused by animals being allowed to selectively feed, even when overstocking isn't a concern. Carefully timing the grazing of plants so plants have an adequate recovery period is the key to preventing overgrazing.

In the most common grazing method, the animals are let out into a big area and kept there for long periods of time. Like kids in a candy store, they first go around eating the things they like best. Then, before their feed of choice has had a chance to regrow to its full energy level, they come along and bite it again. Meanwhile, a plant they don't like quite as well becomes overly mature. The paradox is that both plants continue to lose energy — one plant because it is bitten too often, the other because it isn't bitten often enough. This method of *set stocking* results in overgrazing and overresting of plants in the same pasture, at the same time. Sometimes the overrested plants do well in the short term, and many overrested plants are noxious weeds, so weed infestations usually increase with set stocking.

In managed-grazing scenarios, the animals are moved before they have a chance to regraze the same plant, and kept out of the paddock until the plant has had sufficient time to recharge its batteries. Timing becomes critical to maintaining a balance between livestock gain and moving the ecosystem foundation blocks forward. During the highest flush time of the growing season, the recovery period for plants may be as few as 10 days; during the drier periods of the growing season, a plant may take 90 days to recover; and during the dormant period of the year, it may be 180 days or more before a plant can recover.

To control timing, large fields are broken up into smaller paddocks. Generally speaking, the more paddocks are available, the better. Paddocks can be permanent or temporary; chapter 4, Grass-Farming Basics, goes into more detail.

Stocking rates increase with managed grazing. Depending on your management level and the health of your land, doubling the stocking rate from the level that's typical in your area is not unusual, even early in

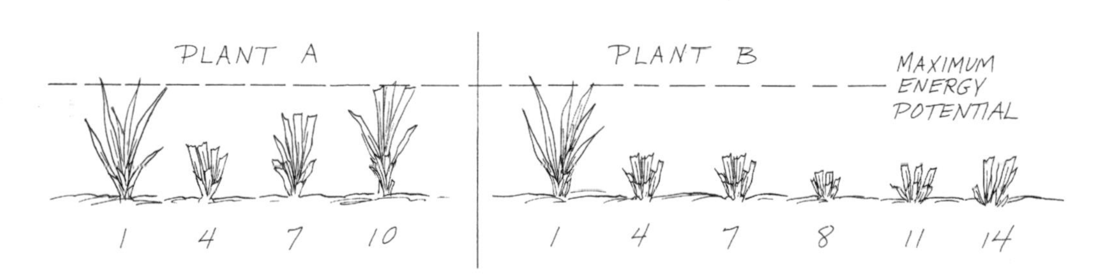

Figure 3.3. Plants A and B are both bitten the first time on day 1; plant A is not rebitten during the 10-day recovery period. On day 8, plant B is rebitten. Its recovery is stunted and would require more than 20 days to fully recover.

FARMER PROFILE

David and Deb Bosle

Goals can change over time — and that's exactly what happened to David and Deb Bosle. David and Deb grew up near Hastings, Nebraska. After they married, David farmed in partnership with his dad on their home place, and Deb owned and operated a child-care center. The farm was a typical commercial grain operation. "We grew corn and soybeans," David recalls.

The farm was a struggle. Poor prices, high taxes and operating expenses, and long hours soon tarnished the dream of continuing the family farm, at least as it was. In 1994, David attended a meeting hosted by the local County Extension Agent. "He was trying to spur some interest in some alternative enterprises to bolster farm income. At the meeting, the agent talked about a farmer named Joel Salatin, in Virginia, who was raising chickens on pasture in portable cages, and direct-marketing them." That meeting resulted in a small number of growers getting together to form a pastured poultry group.

The group jointly purchased a mobile poultry-processing unit from two brothers in Iowa. The brothers had started their own poultry operation by building a processing facility in an old "refer trailer" (a refrigerated box trailer, like those used to transport food over the road), but their business had grown to the point that they had built a permanent facility.

In 1995, the initial group of four growers began raising and marketing chickens. Each farmer was responsible for selling his or her own birds, but they helped each other at butcher time. In 1998, David and Deb marketed 2,500 birds — "less the seven that we kept out for ourselves" — and could have marketed more.

By 1997, David and Deb were going through some major changes. "Getting into raising the birds on pasture, naturally, helped us begin to see the connections that exist between health and the environment. We 'retired' from farming, except for raising the chickens, and Deb sold her child-care business. We began marketing alternative health products and environmentally safe cleaning products, direct from our home via mail and the Internet. These products meshed with what our chicken customers were interested in."

Today, David and Deb are rethinking their goals. "We can direct market the products we're selling from anywhere in the country, and we can raise pastured poultry on a small piece of land anywhere. As it is, we're only using about 10 acres for the poultry production.

"We're seriously considering selling our poultry customer base to a young couple who want to continue farming here, and then looking around for someplace else to live. Hastings was once our dream, but it's changed, and so have we."

---

the managed-grazing process. As the land becomes healthier, even higher stocking rates are possible.

Animal Impact

Animal impact is another tool in the model. The grazing tool simply looks at the relation between the animal and the plant through the eating process. The animal impact tool, on the other hand, looks at the relation between the animal's behavior and its impact on the land.

When discussing animal impact, it helps to think about how a wild herd of animals behaved before modern human practices changed the natural processes significantly. Take a herd of bison roaming the Plains a couple of hundred years ago. Herds — some estimate that individual herds once numbered hundreds of thousands of animals — came through an area and ate everything in site. Because of predators, including Native Americans, these large herds of herbivores ran in tight formations: It's harder for a predator to pick off a prey animal from a tightly bunched group. When a herd came through the animals not only ate all the available forage, but also trampled the soil surface and deposited massive quantities of manure and urine. When the herd moved on, the area looked decimated, but the animal impact stimulated

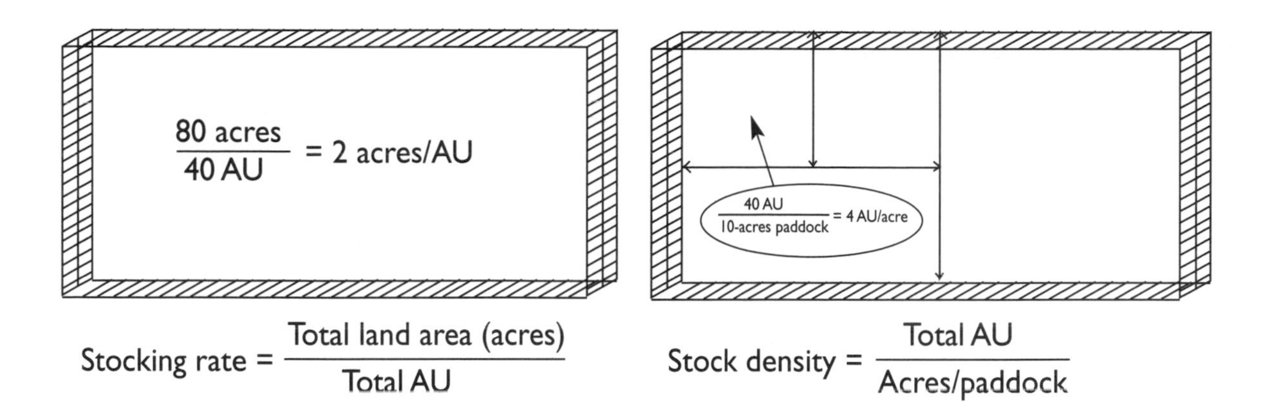

80 acres / 40 AU = 2 acres/AU
40 AU / 10-acres paddock = 4 AU/acre
Stocking rate = Total land area (acres) / Total AU
Stock density = Total AU / Acres/paddock

Figure 3.4. *Stocking rate* is a measure of the total acres available per animal; *stock density* measures the animals (as 1,000-pound or 454-kg animal units [AU]) in an individual paddock at any one point in time. A farm with 80 acres (32 ha) and 40 AU would have a stocking rate of 2 acres (0.8 ha)/AU. If those animals were concentrated on a 10-acre (4-ha) paddock, the stock density would be 4 AU per acre (10 AU per ha).

new growth. The herd would not return for a long period, allowing the plants adequate time for a full recovery. When they did return, the process began again.

Animal impact is the result of stock density and herd effect. Stock density differs subtly from stocking rate. The *stock density* is not measured in animal units (AU) per acre, but on incremental acres in a paddock for a short period of time. The *stocking rate* refers to the total acres per AU on the farm. One animal unit is approximately 1,000 pounds (453.6 kg) of live animal (Table 3.1). Figure 3.4 shows examples of stock-density and stocking-rate calculations.

Herd effect is achieved by animals moving excitedly. In the example above, the herded bison were bunched tightly, and when predators were near they moved in an excited and animated fashion. Picture a stampede, and you picture a high-level herd effect. When humans significantly reduce predator impact, both wild and domestic, animals tend to graze placidly, and that placid behavior results in low herd effect. Luckily, playful and happy animals, not just scared animals, can create high herd effect.

Living Organism

The living organism tool seeks to use biological systems to move toward goals. The great horned owl eating our pigeons is a good example of the tool of living organisms, even if we didn't realize we were employing the tool until the job was almost done.

By fostering a complex environment with lots of biodiversity, we encourage living organisms to do a good bit of work for us. At times, however, we may go out of our way to use one particular living organism. Introducing a beneficial organism to control a pest (say, ladybugs to control aphids) and enhancing the environment for a beneficial organism (placing bat houses around to reduce mosquitoes and other insects) are good examples. Planting windbreaks is another good example of using living organisms as a tool.

Technology

The last tool at our disposal is that two-edged sword, technology. Technology is nothing new; the first tool made from a piece of rock was a form of technology. Now our technology has reached almost unimaginable and at times, dangerous levels.

Those of us interested in sustainable agriculture are often accused of trying to turn back the clock and do away with technology. This isn't quite true, but we don't want technology to be the only answer humankind can come up with, either, because often enough, it isn't the best answer to a problem.

Sadly, our current reliance on and fascination with technology have evolved out of a "science is absolute

Table 3.1

Animal Units

Class of Livestock	Approximate Animal Units	Approximate Weight in lb	Approximate Weight in kg
Beef — Mature bull	1.5	1,500	681
Beef — Young bull	1.2	1,200	545
Beef — Cow with calf	1.3	1,300	590
Beef — Cow, nonlactating	1.0	1,000	454
Beef — Pregnant heifer	1.0	1,000	454
Beef — Yearling	0.7	700	318
Beef — Weaned calf	0.5	500	227
Dairy — Bull, Holstein/Brown Swiss	1.9	1,900	863
Dairy — Cow, Holstein/Brown Swiss	1.5	1,500	681
Dairy — Heifer, Holstein/Brown Swiss	1.0	1,000	454
Dairy — Bull, Guernsey/Ayrshire	1.5	1,500	681
Dairy — Cow, Guernsey/Ayrshire	1.2	1,200	545
Dairy — Heifer, Guernsey/Ayrshire	0.8	800	363
Dairy — Bull, Jersey	1.3	1,300	590
Dairy — Cow, Jersey	1.0	1,000	454
Dairy — Heifer, Jersey	0.7	700	318
Sheep — Ram	0.4	350	159
Sheep — Ewe with twins	0.3	300	136
Sheep — Ewe, nonlactating	0.2	200	91
Sheep — Weaned lamb	0.2	150	68
Goat — Buck	0.4	350	159
Goat — Doe with twins	0.3	250	114
Goat —Weaned kid	0.1	140	64
Horse — Draft	1.5	1,500	681
Horse — Saddle	1.3	1,250	568
Horse — Colt	0.5	500	227
Horse — Pony	0.5	500	227
Pig — Boar	0.4	400	182
Pig — Sow with litter	0.4	400	182
Pig — Sow, nonlactating	0.3	250	114
Pig — Feeder	0.1	100	45

Note: One animal unit equals approximately 1,000 pounds (454 kg) of live animal. Use animal units to help determine animal impact and to establish proper stocking rate.

and perfect" paradigm. Yet no matter how small a particle physicists can define and study, they still haven't explained the "nature" of nature; in other words, they haven't looked at the whole. In 1932, the physicist Max Planck eloquently summed up this idea when he said, "Science cannot solve the ultimate mystery of nature, because, in the last analysis, we ourselves are part of nature, and therefore part of the mystery to be solved." Interestingly, scientists in many fields are beginning to come to terms with this, through their work on chaos theory. These scientists are trying to put the pieces back together as a whole.

Technology is a tool that has a place on your farm and in your planning. For example, recent improvements in fencing technology make it possible for you to manage livestock so that you balance the animals' needs and the plants' needs, while emulating herd effect, with minimal labor. When you use the model, technology, like any other tool, is evaluated for its appropriateness to your goals, and used accordingly.

In a grass-based livestock system, few technological tools are really necessary. So many farmers tie themselves to payments on "heavy metal" such as tractors, plows, disks, planters, and harvesting equipment. These things cost a great deal of money to purchase, they depreciate while you own them, and they require significant amounts of time for maintenance.

When we first purchased our Minnesota farm, we went out and bought an extensive line of equipment. We could have done without most of it; we would have made more money if we'd put the funds that we tied up in equipment into some type of financial investment instead. It pays to rent equipment when you need it, or else to use the services of contractors to do a good deal of your work. This time around, our only equipment is our four-wheel-drive truck and a stock trailer.

Grass-based agriculture also allows you to avoid, or minimize, the use of chemicals on your farm. In Minnesota, the only chemical we ever applied to the land was lime. We rarely used antibiotics, wormers, or other chemicals directly on the livestock. We were able to use natural products such as homeopathic medicines and diatomaceous earth instead. (See chapter 8, Health & Reproduction.)

The Guidelines

The guidelines are the final part of the holistic model and are used for testing and managing ideas during planning and implementation. As Allan Savory says in *Holistic Resource Management* (Island Press, 1988):

> *On first exposure, the guidelines appear a bit confusing. Some titles seem absurdly abstract and others narrowly concrete. . . . Unlike the broad principles discussed earlier, they were not born out of systematic theoretical analysis, but from the day-to-day demands of connecting neat theories to the messy realities of practical life.*

The *testing guidelines* — sustainability, weak link, cause and effect, marginal reaction, gross profit analysis, energy/money source and use, and society and culture — allow you to evaluate ideas, enterprises, and tools during the planning stage. Your goal is to choose the ideas and enterprises that most effectively move you toward your broad goal.

Once you've decided on an idea, enterprise, or tool, the management guidelines help you implement and monitor the project. The *management guidelines* include: time, stock density, herd effect, population management, burning, cropping, marketing, organization and leadership, financial planning, land planning, and grazing planning. In my simplified planning process, I don't apply the management guidelines directly — though they are incorporated into the process.

I'll discuss the testing guidelines in more detail in chapter 12. For now, just remember that each guideline is intended to move you toward your holistic goal.

CHAPTER 4

Grass-Farming Basics

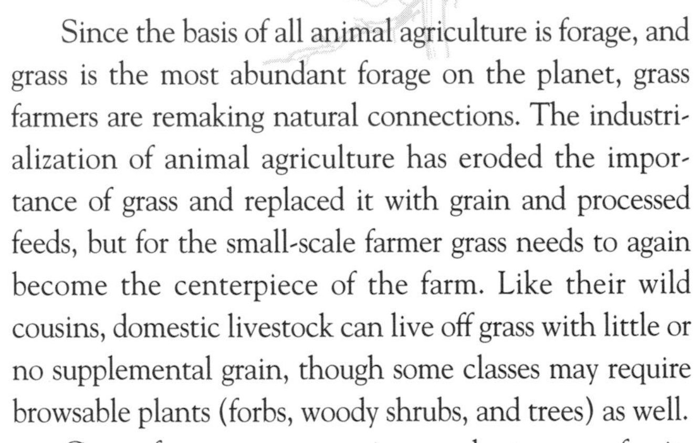

The grazed meadow, or pasture, or grassland — call it what you will — is the ecological and economic foundation of farming.

— Gene Logsdon, *The Contrary Farmer*

WE THINK OF OURSELVES as "grass farmers." We collect solar energy in the grass and then convert it to a product in the form of livestock. One of the best things about grass farming is the way it puts you in touch with your piece of land and your animals — both domestic and wild. Except for the thick of winter, you spend some time walking your land almost every day. We came to differentiate a hundred shades of green during the seasons: the first light bright greens of spring, the brilliant dark greens of summer, and the pale golden greens of fall. One spring afternoon, we lay down in the grass and watched the sun fade away, a great red ball sinking into the pink horizon. The hazy light of late afternoon revealed an infinite universe of ground-spider webs waving on the breeze at the interface of grass and sky. All the grass farmers we know speak in tones of reverence about the experiences they've enjoyed on their land and with their animals.

Since the basis of all animal agriculture is forage, and grass is the most abundant forage on the planet, grass farmers are remaking natural connections. The industrialization of animal agriculture has eroded the importance of grass and replaced it with grain and processed feeds, but for the small-scale farmer grass needs to again become the centerpiece of the farm. Like their wild cousins, domestic livestock can live off grass with little or no supplemental grain, though some classes may require browsable plants (forbs, woody shrubs, and trees) as well.

Grass, for our purposes, is any plant grown for its forage production, including both the true grasses and the legumes. The ideal forage for grazing is a mixture of true grasses and legumes, not a monoculture of any one plant; however, a monoculture may be planted for special purposes, such as establishing a new field or providing winter feed.

Many "weeds" are actually great forage plants. Crop farmers despise quack grass, but early settlers called it miracle grass because it produced good pastures, and animals savored it. Dandelions are always an early-season favorite for our cows and sheep.

Remember that all energy comes from the sun. It's the conversion of solar energy to chemical energy through photosynthesis that makes animal life possible. Grass farmers recognize this conversion, and work to maximize it.

Grass Reproduction

Grass plants are *spermatocytes*, or seed-bearing plants. Each begins life as a seed, and their ultimate purpose is to produce new seeds. True grasses (such as timothy, brome, bluestem, fescue, and even corn) grow from one initial leaf, whereas the legumes (clover, alfalfa, bird's-foot trefoil, peas, and beans) grow from two initial leaves. The plants that start from a single leaf are called *monocotyledons*, or monocots for short; those starting from two leaves are called *dicotyledons*, or dicots. The monocots tend to have a more fibrous root system, and the leaves have parallel veins running through them. The dicots have thicker taproots that look like a carrot, and the leaves have patterned veins, like a web or net (Figure 4.1).

Some plants are annuals, and must reproduce each year from seeds. Others are perennial plants, which live from year to year. Of the annual plants, some are self-seeding, so for practical purposes they behave like perennial plants.

Some plants reproduce only from seeds (sexual), but for other plants new growth may also come from existing roots (asexual). Sexual reproduction produces unique genetics in each plant, but asexual production results in genetically identical plants. For example, a grove of aspen or poplar trees is often genetically one plant; each tree simply sprouts from the roots of an original tree. When you purchase annual hybrid seeds from a seed dealer, you are getting the same type of identical genetics that occurs in asexual reproduction. Identical genetics was a key condition for the 1995 outbreak of gray leaf spot in the corn crop: One hybrid was planted extensively throughout the country that summer, and it had low tolerance to the disease.

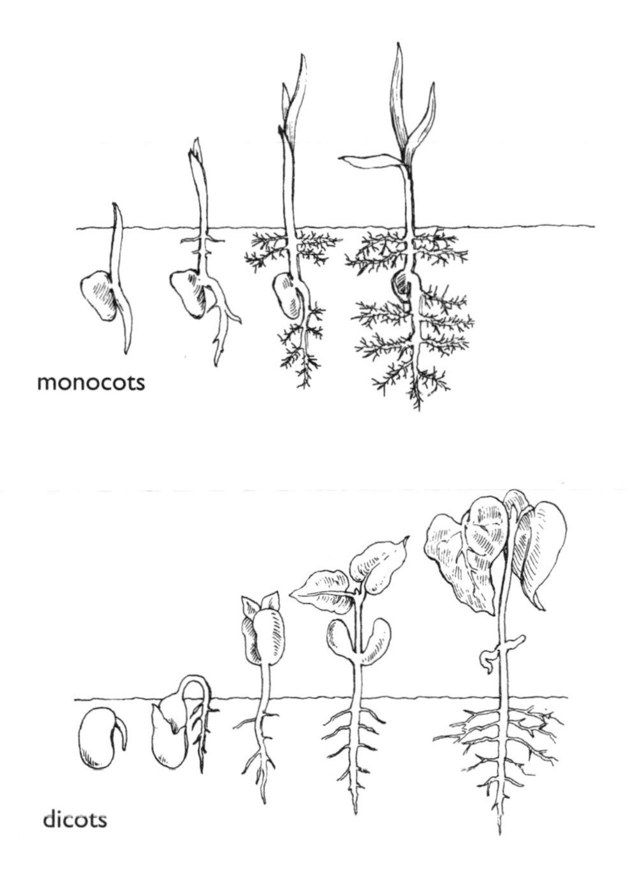

Figure 4.1. Plants that sprout with a single leaf are monocots; those that sprout with two leaves are dicots. Most grasses are monocots, and most legumes are dicots. The monocots tend to have a shallower and more fibrous root mass, whereas the dicots have a deeper, tap-rooted mass.

Feeding the Grass

Like animals, plants need to eat and drink. The primary nutrients they require to grow healthy and vigorous are nitrogen, phosphorus, potassium, calcium, and magnesium. A vast array of other trace nutrients is required as well: sulfur, copper, and iron, to name a few (Figure 4.2).

True grass plants get all of their nutrients directly from the soil, but legumes can acquire part of their nitrogen from air molecules captured in the soil. This ability to take nitrogen out of the air is called *nitrogen fixation*. The legumes are able to fix nitrogen due to a symbiotic (mutually beneficial) relationship they have with a group of bacteria known as rhizobium. These bacteria live in nodules on the roots of a legume plant, allowing the legume to capture nitrogen not regularly available to plants and convert it to a soilborne form that the plant can reuse as it is needed. The beauty of this process is that most legumes fix far more nitrogen than they need, yielding additional nitrogen in the soil for other plants to feed on. Fertilization with inorganic forms of nitrogen reduces the activity of these beneficial bacteria, thereby requiring even higher levels of fertilizer in the future.

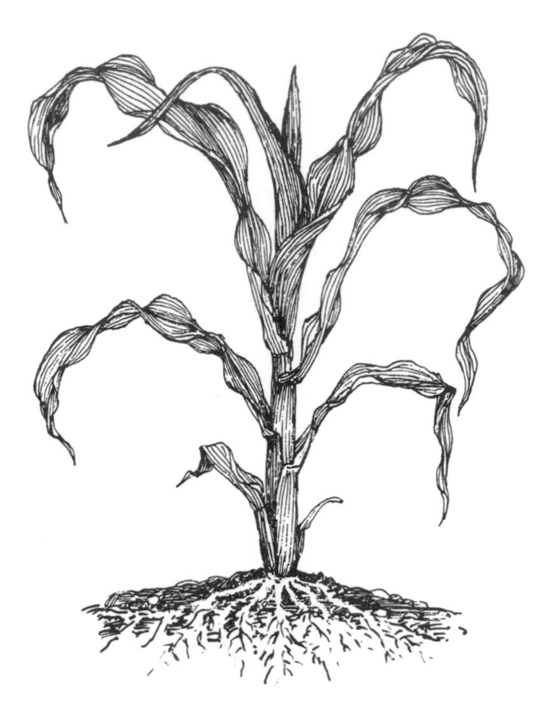

Figure 4.2. Plants acquire most of their food and water through their roots, though some can be absorbed through their leaves (foliar feeding). Plants regularly absorb carbon dioxide through their leaves and discharge oxygen and water through their leaves.

All plants require water for growth, but some require more than others do. Plants that grow in very wet or waterlogged soils are called *hydrophytic*. These plants tend toward thick, waxy surfaces. The *xerophytic* group (plants that survive in dry soils) have narrow leaves and tend to be spiny or barbed: nature's method of protecting them from overgrazing by herbivores. *Mesophytic* plants grow in well-drained and well-aerated soil with moderate moisture. These are favored as forage producers, but livestock can use some xerophytic and hydrophytic plants as well.

Most grasses like to grow in a well-buffered, or neutral pH soil; *pH* refers to a soil's relative acidity or alkalinity, and is reported on a scale of 0 to 14, with 7 being neutral. Numbers below 7 represent acidic pHs; numbers above 7, alkaline pHs. The ideal pH for most plants is between 6.5 and 8.0. A quick test of soil pH can be made using litmus paper, which can be purchased at drugstores. Place a piece of litmus paper in a ball of barely moist soil for a minute, then pull it out and compare the resulting color to the chart on the back of the package. This test is especially worthwhile if you farm in a nonbrittle environment, where excessive moisture leaches alkalinity from the soil.

With a healthy polyculture of grasses and legumes, on soil where the four ecosystem foundation blocks are operating well, you'll need little external fertilizer to feed your grasses. However, there will be times when a little extra fertilizer may help, particularly lime if your soil's pH or calcium levels are low. Part IV, Planning, will help you decide when external fertilizers are the best place to put your money.

If you do decide to fertilize, several small applications throughout the growing season are far more beneficial than one big application, especially for nitrogen. Small applications don't harm the soil's microorganisms the way a big application does, and they are less likely to run off or leach from the soil, because plants take them up quickly or they become stable in the soil.

Managing Peak Growth

Because grazable forage is the primary source of feed in a pastoral system, learning to keep it growing well is critical. The ability to manage grass for peak growth is both an art and a science. It takes a keen appreciation and awareness of both the plants' and the animals' needs, which only comes with time and practice. Still, some general observations might just help.

Growth primarily takes place at each plant's basal growth point, which is located near the soil surface (Figure 4.3). Seeds provide the nutrients required when plants first sprout, but rather quickly the roots must develop and begin supplying all the nutrients for continued growth. As the season progresses, the nutrients are used for the development of a flower and seed head. (Some perennial and biennial plants don't try to make a seed head each year. Biennial plants are a special class of plants that need two years to do what an annual plant does in one.) As the seed head develops, the root system becomes depleted of its energy supply, and the plant will be vulnerable to additional stress.

If plants are clipped — either by mechanical means or by an animal biting them — before they begin developing the seed head, the roots keep supplying energy for leaf growth instead of seed production. However, clipping too much or too often reduces the energy in the root system too much, thereby slowing leaf growth. Understanding these principles and learning to work with them will allow you to manipulate plant growth for maximum leaf (forage) growth. It also allows you to control weeds: Mechanically clip perennial weeds, such as thistles, just as the seed head begins to form, but before it opens. This is the moment when the plant's energy reserves are at their lowest point of the growing season. If you clip before the seed head begins to form, the plant just keeps making more leaves, and still goes into seed production as soon as it has enough. If you wait and clip after the seed head has opened, it's too late — root energy levels increase rapidly after seed dispersal. You may not kill the plant the first year you do this, but given a couple of years of timely clipping you will significantly reduce weed pressure.

So how much grazing or mechanical clipping do you want to do to maximize leaf production? It depends on the time of year, temperature, precipitation, soil fertility, and other mysterious forces (Figure 4.4). But don't give up hope, because there are some rules of thumb that you can use initially, and with practice you'll get to know how your grass is doing. As André Voison, an early pioneer of managed grazing, said in his book *Grass Productivity* (republished by Island Press, 1988), "In the long run, it is the eye of the grazier, supported by his experience, that is the judge." Chapter 14, Biological Planning, will go into more detail on manipulating plant growth through grazing, but for now mull over the rules of thumb.

Rejuvenating Old Pastures

When we first looked at the farm we hoped to buy in Minnesota, it was early May. The grass was just beginning to green up, but the weeds hadn't really sprung into full gear. The fences around the old permanent pasture needed some work, but we could visualize a herd of animals out grazing. By the time we went out for the closing in August, we were sure we'd made a terrible mistake: The pasture was so heavily infested

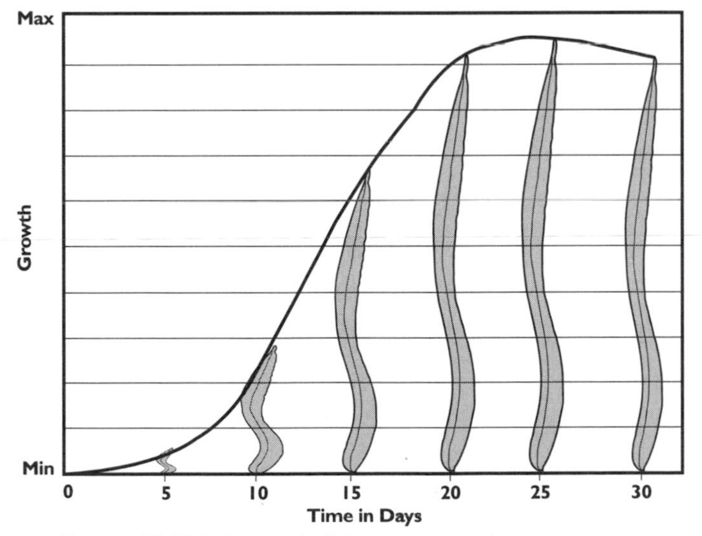

Figure 4.3. This is a typical S-type curve that represents growth of most living organisms. Note how growth begins somewhat slowly, then in the middle period it speeds up (as evidenced by the steep climb of the curve in the center), and then finally slows and tapers off. This is similar to the growth spurts children go through.

(Modified from André Voison, *Grass Productivity.* Covelo, CA: Island Press, 1988, p. 12.)

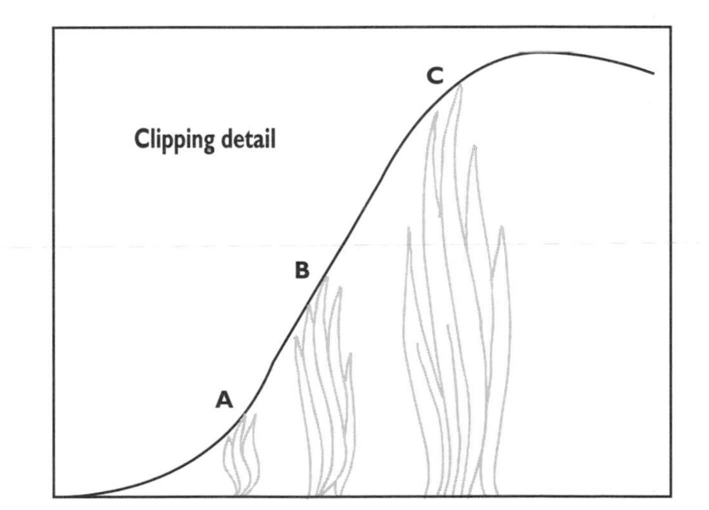

Figure 4.4. The objective of clipping is to take off growth in the high growth part of the curve. Ideally, you would clip grass when it reaches point C, and take it down to point B. Never clip below point A.

11 RULES OF THUMB FOR GRAZING AND CLIPPING

Rule 1. In a fairly well-developed pasture, graze plants that have reached at least 6 inches (15.2 cm) of growth. In a newly seeded field, wait until plants have reached 8 or 9 inches (20.3 to 22.9 cm), and avoid grazing if the field is wet.

Rule 2. Graze before plants reach a height of 12 inches (30.5 cm). Most plants begin going to seed when they're somewhere between 1 and 1½ feet (30.5 to 45.7 cm) tall. One exception to this rule is Sudan and Sudex grasses (see the discussion on page 32), which aren't grazed until they're at least 18 inches (45.7 cm) tall.

Rule 3. Leave an absolute minimum of 3 inches (7.6 cm) of green leaves on plants after grazing in a well-established pasture, and 4 inches (10.2 cm) in a newly seeded field. This is a rule that most farmers seem to break when cutting hay. They take the hay off right at ground level, resulting in lower-quality hay (instead of sitting up on a cushion of plant stems, it sits on the ground and takes longer to dry) and less hay production during the year.

Rule 4. Maximum forage production occurs when each grazing period removes 40 to 50 percent of the plant's leaf matter.

Rule 5. The more paddocks you have created, the easier they are to control growth with livestock. This improvement is remarkable for the first eight to ten subdivisions, then remains impressive up until about thirty subdivisions. This increase in productivity with higher paddock numbers is the result of increased recovery periods, better utilization of available forage, and more even distribution of manure. Using temporary subdivisions is an easy way to increase paddock numbers quickly, and with minimal expense (Figure 4.5).

Rule 6. When the grass is growing fast, move animals fast; when the grass is growing slower, slow down the movement. This strikes most people as backward, but it works.

Rule 7. If the grass is getting too far ahead of your animal's ability to graze it (overmature) during the early part of the growing season, mechanically clip some for hay, for silage, or just to be left on the field as green manure. Overmature grass loses its nutritive value as well as its palatability.

Rule 8. When deciding if it's time to move your animals, look at the most severely bitten plants in the field. If they are being eaten down below the 2-inch (5.1-cm) mark, move the herd.

Rule 9. When growth is slow, rest periods lengthen; when growth is fast, rest periods shorten.

Rule 10. Under most circumstances, your best initial investment will be fencing.

Rule 11. To best manage both grass and livestock, look at the paddock you last grazed to determine growth rate; look at the paddock you're currently in to assess stocking rate; and look at the paddock you plan to go to next to see if it has gotten enough rest.

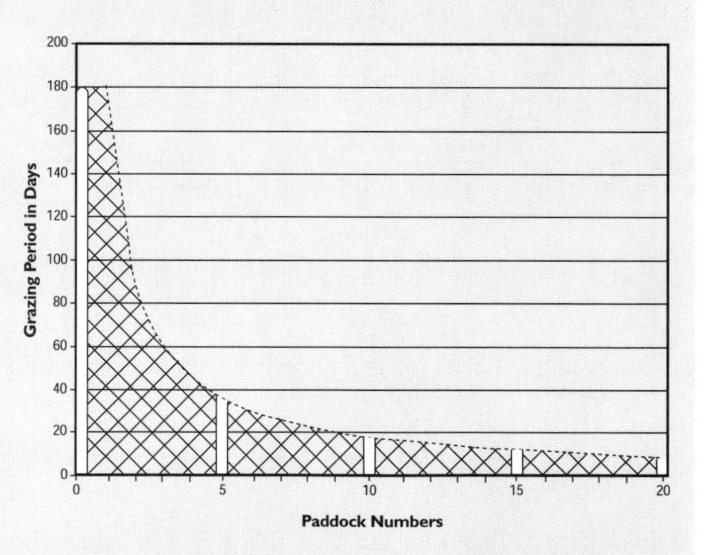

Figure 4.5. As the number of paddocks in a field or farm increases, the time spent in each paddock, or *grazing period* (in days), decreases. For example, if a field was traditionally grazed for 180 days with all the animals placed in at the beginning of the 180-day period and removed at the end, the grazing period would be 180 days. Divide that field into two paddocks, and the grazing period drops to 90 days in each paddock; divide it into three paddocks and the grazing period becomes 60 days.

with thistles that we couldn't even make out the fencelines, much less walk through the pasture.

We were stoic, and went through with the closing. The neighbors had their advice for us: Apply 2,4-D amine at twice the recommended dose. We were in for an uphill battle.

Mechanical Clipping

The first step we took was mechanical clipping. Along the fencelines, we used a walking tractor with a sickle bar attachment. Walking tractors are like rototillers, but they allow for attachment of a variety of implements, not just tillers. Away from the fences, Ken used our Farmall M (circa 1948) and a semimount sickle mower that ran off the PTO (power take-off). In the permanent pasture, we created two permanent subdivisions the first summer. We used polywire to create additional temporary subdivisions. (For more information on fencing, see page 33.)

Whenever we walked around, we carried some seeds in our pockets to throw on the bare spots left where the weeds had been thickest. During the winter, we fed overly mature hay, which had lots of seeds in it, out on the permanent pasture. Timely clipping continued each year, with fewer thistles all the time. By the time we moved, the old pasture was separated into six permanent subdivisions, and we rarely used temporary fencing in those areas.

Frost Seeding

An excellent technique for adding more varieties of plants — particularly legumes — to an existing pasture is *frost seeding*. We used the technique often over the years. Frost seeding works in any part of the country where regular freezing and thawing occurs. Go out in late winter or early spring and spread seeds in a thin layer over the surface of the soil. They can be spread on snow, but if there's too much snow a quick spring melt will wash away most of the seeds, so it's best to wait until winter's accumulated snowpack is gone. We used a hand-cranked seeder to walk the seeds onto the frozen ground (Figure 4.6).

Cultivating

Another tool that came in handy for us while rejuvenating the old pasture was an ancient cultivator (Figure 4.7). This one I purchased at a farm auction. It was out back with the rusted junk equipment that only the scrap-metal dealers were interested in. It was mine for $22, and the scrap guys thought I got took.

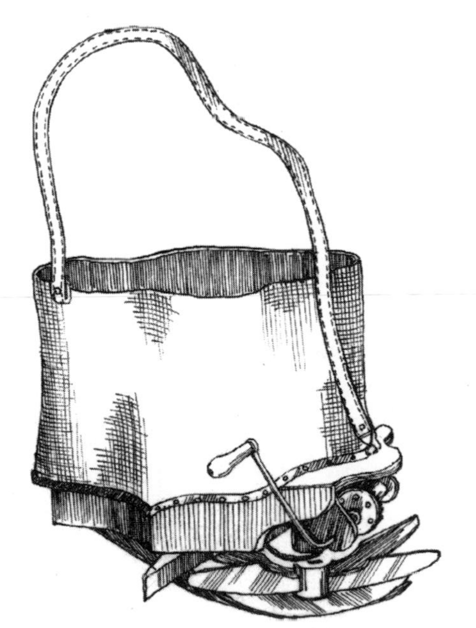

Figure 4.6. A hand seeder is a helpful and inexpensive tool for increasing the diversity of plants in existing pastures. Use it to spread seed on bare areas or to seed a new plant variety into a pasture. In cold-winter areas, spread in late winter or early spring and the frost action will help "plant" the seeds for better germination.

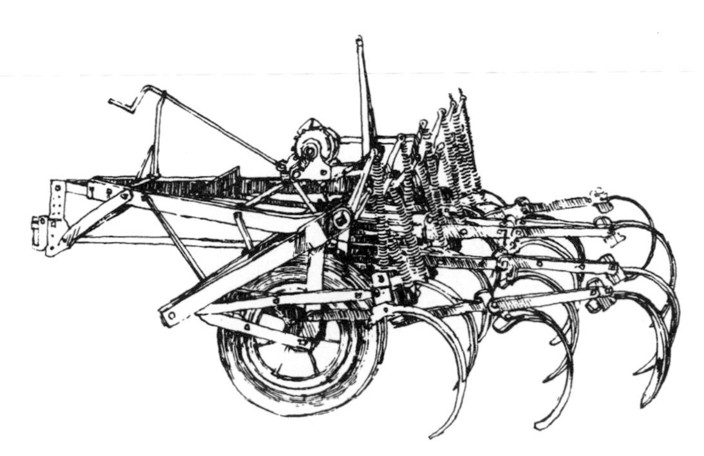

Figure 4.7. Harrows are the traditional tool for spreading manure patties out, breaking up thatched sod, and aerating the topsoil surface, but an old cultivator can be used for the same job (and can often be picked up for scrap prices at farm auctions).

Ken would lightly drag the pasture with the cultivator to help break up the soil surface and spread out matted old growth and dung. After a few years, this treatment became unnecessary, but in the first year or two the pasture really responded to it. Breaking up the surface allowed air and water to penetrate into the soil better.

Converting Farmed Fields to Pasture

One of the first requisites for creating a quality pasture from a previously farmed field is patience: It takes a few seasons to get a good pasture. Think of it like art: An original painting takes more time than a paint-by-numbers. The original work doesn't come with anything already filled in — it's just a blank palette or a bare piece of ground. But an old sward, like the paint-by-numbers, already has some space filled in. It has some kind of a green background from the start.

The quality of the pasture is measured by a composite of features: plant height, plant density, plant diversity, and plant age. At the same height, the old sward will have more tons of dry matter. Not just a little more, either! There are more plants per square foot, and there is more variety of grasses in an older sward than in a new one. These characteristics take time to achieve.

Preparing the Seedbed

Procedurally, planting a new pasture is like planting a hay field, and the field may serve both purposes. The first step depends in part on the previous crop. If the field was used to grow corn, or some other crop that was grown on ridges, you may have to plow the field first and then disk it. If it was used for small grains, or a crop that left a relatively flat surface, begin with just a light disking. Follow disking with a harrow to smooth out the seedbed. The rule to follow in preparing the seedbed is: Do the least amount of plowing, disking, and harrowing that you can get away with to yield a relatively even seedbed. After the seedbed is prepared, the crop needs to be drilled in, but be careful not to drill seeds too deeply. More new stands fail to take because of drilling too deeply than from any other factor.

If you're a new farmer, try to hire a neighbor or a custom operator to do this work. Investing in the necessary equipment will take too much money, and you won't need the equipment after your first couple of years.

Choosing What to Plant

Choosing what to plant will depend in part on past crops. When we got to Minnesota, our tilled fields had been in corn for years, and the farmers had used atrazine to control weeds in their corn. Atrazine kills most grasses, and residual amounts can stay in the soil for a year or two after the last use. With a little research, we found that Sudan and Sudex (a Sudan/sorghum cross) are closely related to corn, so they can tolerate the residual atrazine. Sudan became our choice the first year.

If you do need to use Sudan or Sudex, there are some considerations. Both these grasses can be toxic to livestock if they are grazed — or cut for hay — before they reach 18 inches (45.7 cm) in height, or right after a frost. The toxicity comes from prussic acid that's produced in young plants, or by frost. If the plants are killed by frost, wait about two weeks before grazing or cutting. This allows the prussic acid to dissipate.

If you don't need to be concerned with prior herbicide use, your choice of seeds is pretty much wide open. Plant a variety of seeds, including about 30 percent legumes. Check with local graziers, a County Extension Agent, or a seed dealer to find out what types of seeds do well in your area. In this day and age, most seed dealers carry pasture mixes, which makes things easy, but you can save some money by making up your own mixture. Appendix D, Grasses & Legumes, discusses a variety of grasses and legumes; it should help you decide what will be most appropriate for you to use.

Fencing Materials and Techniques

As the traditional saying goes, "Good fences make good neighbors," and a few years in farm country teaches you the absolute truth of this. Fencing is one of the most important investments a small-scale livestock farmer will make, and it's best made early in the process. As our friends Erik and Heather Olson learned when their pigs rooted up the neighbor's lawn, good fences are best put in place *before* the animals arrive. This small patch of disturbed lawn cost them about $600 (which would have gone a long way toward fencing), not to mention a good deal of aggravation. Erik now says, unequivocally, "Fence has to come before animals."

Exterior fences are crucial, since they protect your animals from the outside world, and the outside world from your animals. What type of fence you need around the exterior depends on the type of livestock you're keeping, how much money you want to spend, and, to a lesser extent, your proximity to neighbors and highways. It also depends on how badly your livestock wants to get out; if the feed is very good inside the fence and water is available, they don't usually have a burning desire to leave.

Wooden fences are beautiful but very expensive. Barbed-wire fences are difficult to construct and maintain, and animals that get caught in one often hurt themselves badly. High-tensile fences are moderately expensive but work well when combined with electric, especially for exotic animals such as bison, deer, or elk. For most small-scale farmers, smooth-wire electric fences are the least costly and most effective choice.

Permanent fences represent a large investment; they should be well constructed and require minimum maintenance. The investment is depreciable on your taxes (ask your accountant). Table 4.1 reviews various fencing materials. Check appendix E, Resources, for a list of suppliers of fencing materials.

Electric Wire

We fenced the perimeter of the farm with smooth electric wire, and once animals were trained to it, two strands — one about 1½ feet (45.7 cm) off the ground and the other about 3½ feet (107 cm) off the ground — kept in horses, pigs, sheep, and cows. Baby pigs and lambs could walk under the bottom wire, but they never strayed too far from Mom. We used a single low strand (6 inches, or 15.2 cm) around the garden to keep animals out of it.

FENCING GEESE

Our first summer on the farm, we had a flock of geese. Being kind of naive about geese and gardens, we thought, "Oh boy, we'll use the geese as weeders." The problem that no one mentions when they write about weeder geese is that when garden plants are very small — new transplants, for instance — the geese eat them readily. I spent one day transplanting all kinds of plants I'd raised from seeds, only to look the next morning and see virtually none left. The mystery of the disappearing transplants resolved itself as soon as Ken let the geese out of the barn: They marched right for the garden and cleaned up what few transplants they'd missed the previous day. That's when we ran the electric wire, about 6 inches (15.2 cm) off the ground.

The geese marched over. One goose grabbed the wire in its beak and couldn't let go because the shocks kept it locked on, so I turned off the fencer for a minute, and the goose dropped the wire. Said goose went squawking away. I expected to have to repeat this procedure more than once, but geese are smart, and the entire flock learned from the one goose's experience. After that, the flock would walk over and stare at the fence, but none of them ever grabbed the wire or entered the garden while it was up. When the plants were big enough for the weeders to come in, I'd drop the wire for them.

Table 4.1

Fencing Materials

Type of Fence	Materials	Posts/Spacing	Cost Comparision — Materials Only, per ¼ Mile (0.4 km)	Approximate Life in Years — Humid Climate	Upkeep
Board*	Four 1 x 6 boards	Wood posts spaced 8 ft (2.4 m)	$3,381	20	Low
Barbed wire	3-strand, 12-gauge	T-posts spaced 25 ft (7.6 m)	$411	30	High
	4-strand, 12-gauge	T-posts spaced 25 ft (7.6 m)	$451	30	High
	5-strand, 12-gauge	T-posts spaced 25 ft (7.6 m)	$491	30	High
Woven wire	39 in. (1 m) tall, 12.5-gauge	Wood posts spaced 16 ft (4.9 m)	$504	19	Medium
	47 in. (1.2 m) tall, 12.5-gauge	Wood posts spaced 16 ft (4.9 m)	$516	33	Medium
Permanent electric	2-strand galvanized, 12-gauge	T-posts spaced 60 ft (18.3 m)	$199	25	Medium
	3-strand galvanized, 12-gauge	T-posts spaced 60 ft (18.3 m)	$228	25	Medium
	4-strand galvanized, 12-gauge	T-posts spaced 60 ft (18.3 m)	$257	25	Medium
	5-strand galvanized, 12-gauge	T-posts spaced 60 ft (18.3 m)	$298	25	Medium
	2-strand aluminum, 12.5-gauge	T-posts spaced 60 ft (18.3 m)	$233	25	Medium
	3-strand aluminum, 12.5-gauge	T-posts spaced 60 ft (18.3 m)	$277	25	Medium
	4-strand aluminum, 12.5-gauge	T-posts spaced 60 ft (18.3 m)	$320	25	Medium
	5-strand aluminum, 12.5 gauge	T-posts spaced 60 ft(18.3 m)	$377	25	Medium
Temporary electric	1-strand polywire	Heavy-weight step-in posts spaced 75 ft (22.9 m)	$130	10	Medium
	2-strand polywire	Heavy-weight step-in posts spaced 75 ft (22.9 m)	$196	10	Medium
	Electroplastic Net**		$836	10	Medium

Note: This table is intended only to provide some idea of relative costs. Your actual fence costs could be a little bit less, or quite a bit more, depending on the region of the country where you live and when you purchase your materials. These comparisons include corner posts, gates, and chargers, where applicable, and are based on what it would have cost me to purchase materials in Colorado in the fall of 1998.

* Painted with oil-based paint.

** This is normally used in small areas, e.g., to fence a garden or to move a flock of sheep.

On permanent interior fences, we used just one strand of electric wire placed about 2 feet (61 cm) off the ground. Temporary subdivisions were always created using a single strand of polywire.

Training animals to an electric fence can take some patience on your part, but once trained, most animals will respect the fence. To train animals, have a small and well-fenced area. If your animals are already fairly tame, the training pen can be made up of just two strands of smooth electric wire. On the other hand, if you're training animals that are kind of wild, fence an area securely with wood, barbed wire, woven wire, or stock panels, and then use electric wire inside it. Don't force animals into the fence: Let them investigate it on their own. Placing small dabs of peanut butter or molasses on the wire, or on aluminum cans connected to the wire, will help attract the animals to it. An animal that touches the fence with its nose won't forget the experience; the damp end of a nose conducts very well.

A tactic that Mike and Keri Salber, dairy farmers in Browerville, Minnesota, use to help train animals to a new fence is to cut plastic bread bags into strips about 1 inch (2.5 cm) wide and tie them on the fence every 50 feet (15.2 m) or so. Movement of the strips helps alert the animals to the fence's location. Mike says this also helps keep the deer, which are thick around his farm, from taking down the fences.

Aluminum vs. Galvanized Wire

Deer take down the fence when they run through at night and don't see the fence until it's too late. This is less of a problem if you use aluminum wire instead of galvanized. Aluminum wire is brighter after dark, and the slightest breeze causes the wire to whistle, which seems to make the deer aware of its location so they simply jump over it. We started out using galvanized, because it's quite a bit cheaper, but aluminum is much lighter and easier to work with, conducts better, and lasts longer; the fact that animals — both domestic and wild — always seem to know where it is makes it worth the extra expense.

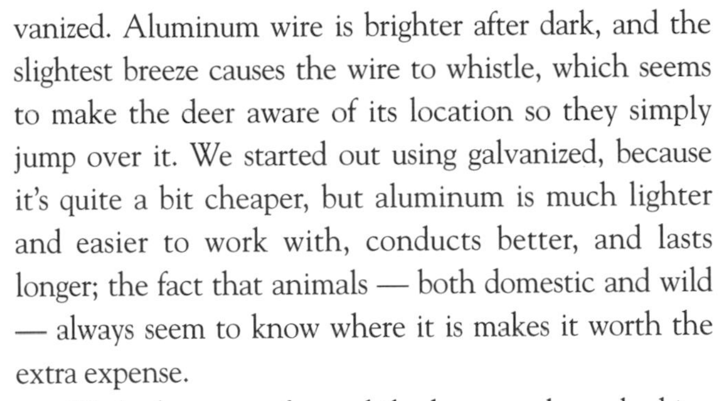

We had one ewe for a while that was always looking for a way out, and leading the rest of the flock with her if she found it. Like this ewe, some animals make a habit out of checking the fence to see if it's on. If you get one of these critters, it may be best to send it down the road or put it in the freezer, rather than having continual trouble with all your animals.

The Fence Charger

When you use electric fencing, one of the most important components you'll purchase is the fence charger. This is one device that you should plan to splurge on: Buy the best low-impedance fence charger you can afford. Look for a model that has a good warranty and replaceable components. The biggest cause of problems for high-quality chargers is lightning: A strike anywhere near your fence can take out the charger, but there are lightning arresters available from a number of companies. Our charger warranty includes repairs for lightning damage if the arrester is installed on the fence (Figure 4.8). We keep a cheaper model as a backup, but it isn't designed to handle miles of fence. It can keep a small area secured in an emergency.

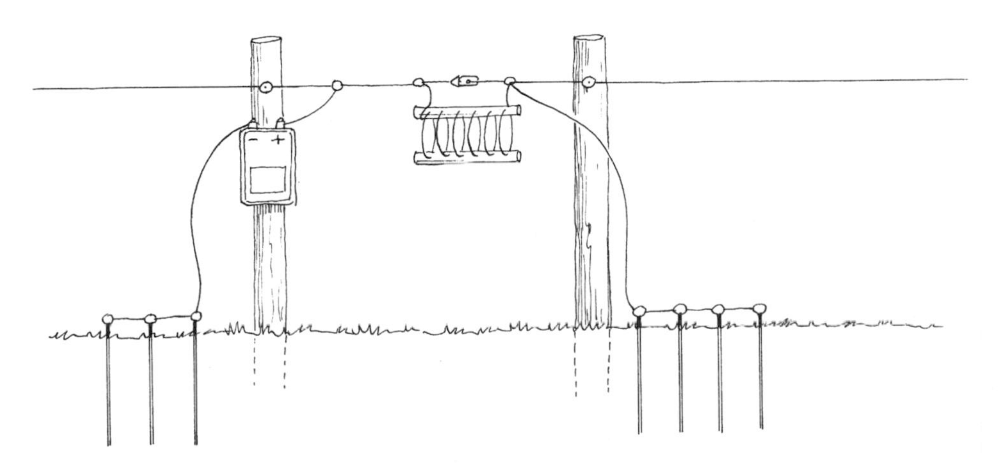

Figure 4.8. One major hazard to electrical fence chargers is lightning strikes. A lightning arrester wired into your fence system will take the surge of power that lightning sends through fence wires to earth before they reach — and fry — your charger.

(Redrawn from David Pratt, "Grounding Electric Fences," *Livestock and Range Report #914*, Fall 1991.)

Most modern chargers are designed to allow a pulsed flow of energy. Like a reservoir, an internal capacitor stores electrons until it's full; then the capacitor spills out what it has stored as a pulse of electricity. A short, intense pulse is better than a long, less intense pulse. Avoid the cheapest chargers; they run a continuous charge or have long pulse times. Either can injure an animal caught in the wires, or start grass fires.

Voltage. Ideally, voltage on your wires should run between 4,000 and 5,000. Voltage is the electrical pressure that must be applied to cause electrons to flow. Amperage is a measure of the current flow through a conductor (wire) of known resistance.

volts x amps = watts, or the power used

Voltage can be checked with a voltmeter.

Joules. When shopping for a charger, a more important measure to look for than volts is joules. A *joule* is the effective power the unit supplies, or a unit of energy delivered at a specific point in time (watts per second). As you increase the joules, you increase the "jolt" of the fence. How many joules you need depends on your total length of wire. Table 4.2 shows recommended joule ratings for different lengths of wire. (Remember that if you are running multiple strands, you must multiply the length of fence by the number of strands to get the length of wire.) A good way to comparison-shop for chargers is to calculate the cost per joule.

Table 4.2

JOULES

Minimum Recommended Joule Rating	Miles of Electrical Fence Wire on the System
1	6
2	12
3	18
4	24

Impedance. *Impedance* refers to the internal resistance in the charger. Low-impedance chargers don't leak as much energy as high-impedance models. The impedance is related to the duration of the pulse and the voltage.

Chargers come ready to plug into a 110-volt AC service, or in battery style. (*Note:* Voltage and frequency of incoming electric supply vary from country to country, but chargers are available to meet the needs of international users.) Some battery chargers are available in solar models. In our experience, battery chargers worked well for small areas, but weren't adequate for the miles of wire that ran around our farm. However, some of the newer battery-operated chargers are supposed to be quite a bit more powerful.

Grounding. Chargers *must* be well grounded. When the ground is wet, even a poorly grounded charger will deliver a good jolt, but if the ground is very dry or frozen a poorly grounded charger is like no charger at all. We learned this lesson the hard way one winter, when our pigs kept walking right through the wire to go and help themselves to our haystacks. Unfortunately, every time the pigs took the wire down, the sheep joined them. Self-serve feeding became the order of the day. Drive at least three 8-foot-long (2.4 m) ground rods, spaced 10 feet (3.1 m) apart (Figure 4.9).

Tautness

Wire for electric fence doesn't need to be run real tightly. In fact, leaving the wire a little bit slack is better. This slackness may not look quite as spiffy as people are used to, but if an animal runs into the fence the wire will act like a rubber band if it isn't too tight. An animal that runs into a very tight wire either breaks the wire or pops it off the insulators.

Posts

Fence posts act as the skeleton for your fencing system. They provide the structural integrity that keeps everything in place. For permanent fences, a little preliminary planning can save a lot of time and effort down the line. (See chapter 14.)

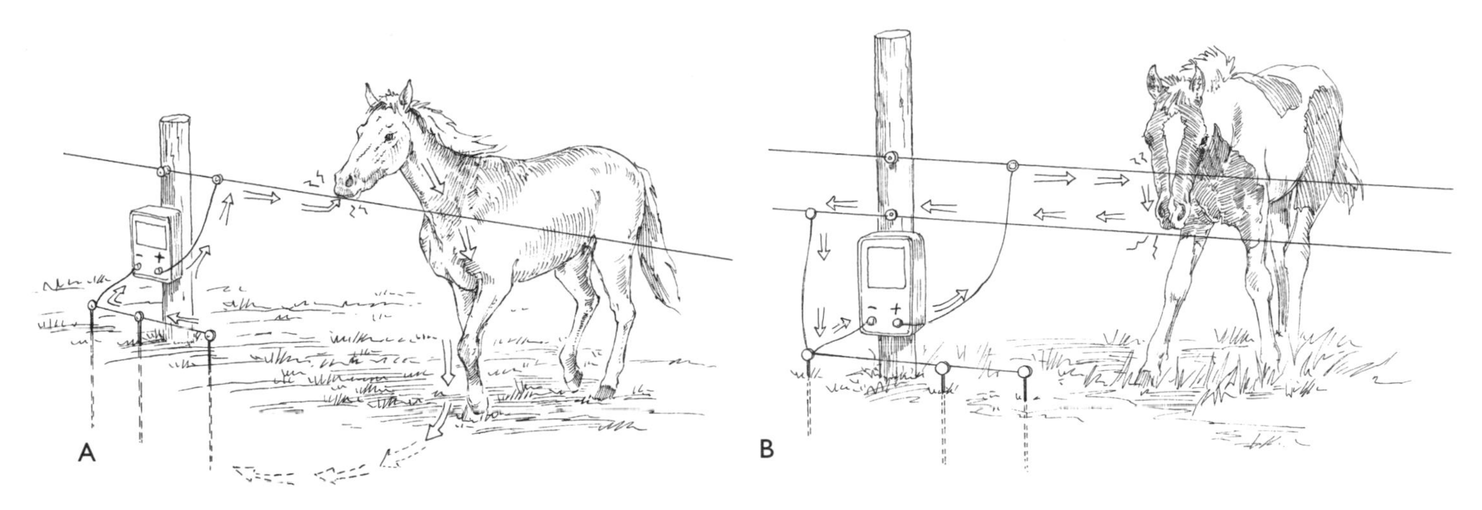

Figure 4.9. Electric fencing works by sending a pulse of electrons down the wire. Those electrons must somehow return to the negative side of the charger to complete a circuit for a fence to deliver a jolt. **(A.)** A single hot wire works well in areas where there is normally some natural moisture in the soil; the charge can simply pass through an animal's body and go to ground, returning to the negative side of the charger through the ground rods. **(B.)** A hot wire *(top)* and a ground wire *(bottom)* complete the circuit. This is often needed in areas with extremely dry soils or areas where soils freeze solid and deep.

(Redrawn from David Pratt, "Grounding Electric Fences," *Livestock and Range Report #914,* Fall 1991.)

Corner posts. Corner posts do most of the work, so they should be strong, well-buried, wooden posts. For short runs of fence — say, less than ¼ mile (0.4 km) — a single corner post will probably be sufficient. Longer runs require the use of braced corner posts (Figure 4.10).

T-posts. Between corner posts on exterior or property-line fences, we exclusively use metal T-posts with good plastic insulators. These T-posts should be well driven so that the bottom plate is completely buried by an inch or so (2.5 cm) of soil. Between corner posts use metal T-posts, placed about 60 to 75 feet (18.3 to 22.9 m) apart. Leaving a fairly good space between the posts helps the rubber-band effect.

T-posts can also serve as corner posts for small, or temporary, enclosures. If you plan to take the T-post out again in the near future, leave about ½ inch (1.3 cm) of the top plate exposed above the soil.

Fiber posts. Fiber posts are a real boon for livestock farmers. They are made from fiberglass and are designed to be either "stepped in" or easily driven. The step-in style has a little footpad attached at the bottom for you to push down with your foot. The convenience of step-in posts is worth the extra expense. Fiber posts are ideal for temporary fencing. We also use them on interior permanent fences in place of two-thirds of the T-posts: Alternate one T-post with two fiber posts between corner posts.

Shorts

A regular problem with electric fencing is that it can short out. In a short, the flow of electrons is lost to the ground, meaning that the fence has no jolt at all. Shorts can either be a *dead short* (the wire carries no jolt) or a *minor short* (the wire still carries a jolt, but it is not as strong as usual).

The most common cause of dead shorts is a charged wire touching an old woven-wire or barbed-wire fence, or touching a metal T-post. The hardest short we ever had to find was one that occurred where a plastic insulator had split, so the charge was being lost to the T-post through the insulator.

When you're trying to secure an old metal fence with electric wire, run a new fence on the inside of it, leaving about 3 feet (91 cm) between. Simply offsetting an electric wire onto an existing fence is a sure

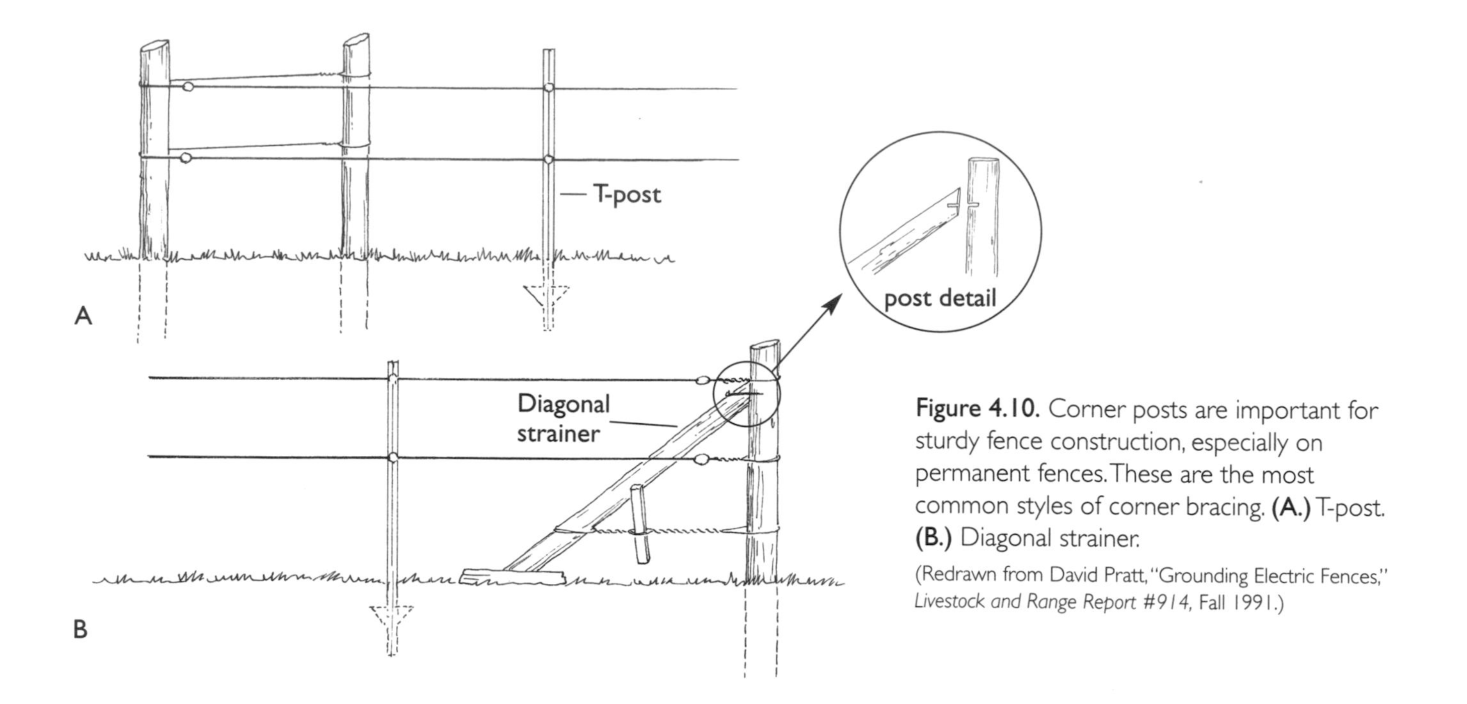

Figure 4.10. Corner posts are important for sturdy fence construction, especially on permanent fences. These are the most common styles of corner bracing. **(A.)** T-post. **(B.)** Diagonal strainer.

(Redrawn from David Pratt, "Grounding Electric Fences," *Livestock and Range Report #914*, Fall 1991.)

way to spend half your free time looking for shorts. In this case, the new electric fence can be run on fiber posts, with T-post corners.

Minor shorts are most often the result of too much vegetation growing up around the fence. Keep an eye out for vegetation that looks "burned" from touching a wire.

Finding shorts can be a real challenge. One trick is to walk along the fence with a transistor radio tuned between stations. When you near the short, the radio will begin clicking as the pulses arc across the metal to ground.

Purchase a fence tester. This doesn't have to be expensive or fancy; a tester that lights a bulb when one end is pushed into the soil and the other touches the wire will work, although a voltmeter works well also.

Finding a short requires a process of elimination, and designing your fencing system so that you can "drop" one portion of the fence at a time facilitates the process. This is accomplished by setting up branches. Drop one branch at a time to isolate which branch has the short on it. Then keep working down the branches, isolating sections of line until you find the short. Branches can be set up using insulated handles; also, some fencing manufacturers now offer a style of disconnect that acts like a switch for just this purpose.

How to Run a Straight Fence

Most beginning farmers start building a fence by placing their first corner post, walking out so many feet to place their second post, then connecting the wire to it, then to the third post. When the fence is done, the posts zig and zag along the field, and beginners can't understand why they didn't get a nice straight fence. Our first couple of fences were done like this; then we learned how to run a straight fence. First, place the corner posts at both ends of the run; second, run one wire and attach it between the corners (don't make it supertight; a little slack is okay). Now the wire will act as your guide for installing the T-posts in a nice straight line. Drive all the T-posts directly under the wire, then walk back down the line attaching the wire to the T-posts with plastic insulators. Attach any additional wires to the corners, and then connect them to the T-posts.

Multispecies Grazing

In nature, succession continually moves environments toward complexity. The industrial agriculture model has moved farmers away from complexity and toward specialization, but that move hasn't helped either the farmers or the land. Small-scale farmers can reap tremendous benefits by moving toward complexity in their own operations.

One excellent method of moving a farm toward complexity is multispecies grazing. Consider these benefits:

1. The fact that different species prefer different plants can be used to your advantage. Grazing multiple species can actually increase the carrying capacity of your land. For example, sheep prefer forbs, goats prefer browse, and cattle prefer grasses, so they complement each other on pasture. When running complementary species on diverse pastures, you can almost double the total stocking rate of your land. If you ran ten cows before, for instance, you can now run ten cows and thirty or forty sheep!
2. By diversifying the species you run, you can effectively buffer your bank account. Different species' markets run in different cycles. Typically, when the cattle market is strong and prices are good, the sheep market isn't doing as well; when the sheep market is strong, cattle prices are down. For a small farmer, adding grazed poultry can be an excellent move, and pastured poultry is easily direct-marketed for top prices.
3. Disease and parasites are often reduced where multispecies grazing is taking place. The parasites that affect one species don't usually affect another. Reducing parasites, flies, ticks, and mosquitoes can be accomplished by poultry (or fowl) following herds. In fact, after poultry roamed around our farm in Minnesota for a few years, the ticks virtually disappeared from our place. (If our dogs ran across the neighbor's fields, though, they'd come home full of ticks.) Poultry break up manure piles and eat the larvae of many of these pests.
4. Meat production per species is higher where species graze together. University studies have shown that meat production per acre (hectare) can be increased by as much as 125 percent, due to the increased carrying capacity of the land and increased individual animal performance.

There are a few different strategies you can employ for multispecies grazing: mob grazing, leader/follower grazing, and alternate grazing. In *mob grazing*, all the animals are run as one group. In *leader/follower grazing*, one herd is run through a paddock first, than a second herd runs through the same paddock immediately after. *Alternate grazing* is a variation of leader/follower, with one type of animal run during one period over a group of paddocks, then some other type run at a later period. Another alternate-grazing scheme that some stocker operators are experimenting with is running stocker cattle one year, then stocker sheep the following year. Stocker operators purchase animals in spring (after they're weaned) and graze them for the summer forage months, then they sell the animals in fall.

As we added different species to our mix in Minnesota, we initially ran them in an alternate-grazing system: one portion of the farm dedicated to horses, one portion to sheep, one portion to cattle. Then we moved into a leader/follower system. In the end, though, we ran a mob system most of the year. Managing one large herd instead of multiple herds was easy. These mixed groups are sometimes called "flerds" (a cross between a *flock* and a *herd*).

After we went to mob grazing, we learned to keep the equines out of the big herd when calves and lambs first started being born. Our young mules would try to steal newborn calves or lambs! They didn't seem to want to hurt the babies, they just wanted to play Mom, but it was always quite a traumatic experience for the first couple of calves or lambs, as well as for their mothers. Once the bulk of the babies were born, we'd reintroduce the equines to the mob; by then, the mules seemed to have lost interest.

When you run a mob, remember that the animals must be familiar with each other before they begin having their babies. While we were still running leader/follower, we went to check on the cows one afternoon. Orphaned bottle lambs will follow you anywhere, and Junior was no exception; he followed us right out to the paddock where the cows were. At that time, the cows weren't used to sheep, so they perceived Junior — all 10 pounds (4.5 kg) of him — as a big threat to their calves and chased him out of the paddock. If he hadn't been able to run under the fence, I think they would have killed him. Once the cattle are used to the sheep, their maternal instincts help protect the lambs as well as their own calves from intruders. Our dogs always watch their backs in the paddocks when the babies are little.

FARMER PROFILE

David and Leianne Wright

"We've always grazed our dairy cattle. My dad always grazed his before me," David tells me. "But about a decade ago, we began learning how to intensify our grazing."

The Wrights own a 200-acre farm in Alabama and rent an additional 450 acres from his dad. "We do the milking here on our farm, and use Dad's farm for raising our young stock."

Over the years, David has dedicated a great deal of time to developing his grazing system. He has traveled to New Zealand, Ireland, and Africa to study grazing in those countries. "I've learned things in each place that, even if they didn't apply directly to my operation, could be tweaked to work here."

The Wrights have fenced the entire perimeter of their farm with permanent electric fence. They've also built internal roads and lanes, as well as bridges over the creeks, which are fenced off. The water system is made up primarily of garden hoses now — though David feels that a permanent water system will be the next improvement he'll make to his grazing system. "We got a heck of a deal on the hoses. I went into a Wal-Mart near here at the end of the season, when they were trying to make room for the Christmas stuff. The manager really wanted to get rid of what he had left — so I bought the whole lot, by the foot."

All subdivisions are created using polywire, a plastic wire that contains interwoven strands of conducting wire. "I've designed a cart that pulls behind a four-wheeler with all my fencing supplies on it. It carries 5 miles of polywire, posts, insulators; you name it and it's on the cart. There's a reel on it that can let polywire be drawn off, or returned to the reel quickly, by one person. With this system, I can run a mile of interior fence in about an hour — though usually it takes about 15 minutes after each milking to move the fences."

Because the Wrights are trying to produce milk with as little purchased grain as they can, they need to manage their grass carefully. "We move the cows twice a day, after each milking. They only have access to as much grass as they'll eat in an eight-hour grazing period."

To improve the quality of their pastures, the Wrights use a variety of grasses and legumes. "We use lots of annual ryegrass, which works pretty well during our mild winters. The annual ryegrass is a cool-season grass that provides great grazing for us in fall, part of winter, and spring. Other cool-season grasses we use are the fescues and Matua bromegrass. For summer grazing, we use millet, Bermuda grass, and grazing alfalfas."

For the Wrights, improving the management of their forage resource has allowed them to increase total animal numbers without increasing acreage. It has resulted in healthy, happy cattle, and a healthier environment. "It's paid off."

Note: The Wrights are marketing their fencing cart and reel system (patent pending). See appendix E.

PART II

ANIMAL HUSBANDRY

CHAPTER 5

Genetics, Breeding, & Training

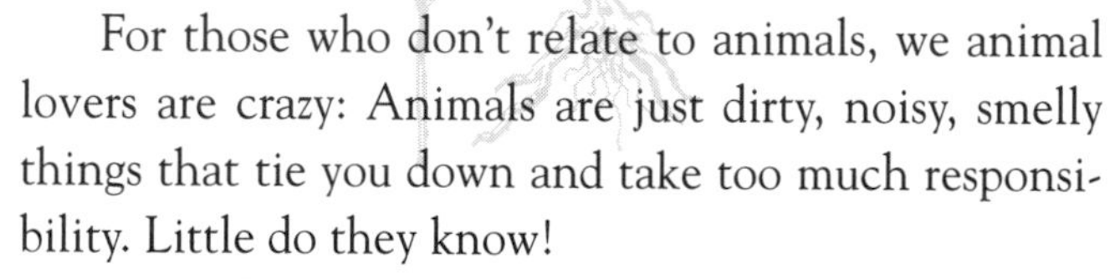

Animals are creatures of instinct. Do not think that what nature has taught them over the long, long centuries you can unteach in your puny little lifetime. Don't try. Observe instinct instead and learn from it. There is a wisdom to it far beyond the ken of humans.

Animals are creatures of habit, too. Domesticated, they will tolerate being trained to a routine somewhat foreign to their instincts, but once that routine is established, they do not look kindly upon an abrupt change. Change only confuses and alarms them. You must be patient, or suffer the consequences.

— Gene Logsdon, *Practical Skills*

Nature teaches beasts to know their friends.

— Shakespeare, *Coriolanus*

SOME OF US ARE DRIVEN BY A NEED to have animals be a part of our lives. We are soothed by their presence, fascinated by their behavior, and amused by their antics. We relate to them as fellow creatures.

For those who don't relate to animals, we animal lovers are crazy: Animals are just dirty, noisy, smelly things that tie you down and take too much responsibility. Little do they know!

Animal companionship is an elixir, a fountain of youth. Having animals around reduces stress, lowers blood pressure, and floods the body with endorphins: measurable, positive, physiological responses. We're the lucky ones!

Domestication

The human need for animals isn't new; it goes back many thousands of years. Our needs for food, clothing, traction, and companionship led early peoples to domesticate animals.

Domestication wasn't simply human dominance over another species; it was an evolutionary codependence in which both human and animal found benefits from living together. As Stephen Budiansky puts it in *The Nature of Horses*, "If it were simply a matter of human will, it would be hard to explain why we should have domestic dogs, sheep, goats, cows, pigs, horses, asses, camels, rabbits, and cats — but not deer, squirrels, foxes, antelope, or even hippos and zebras." In other words, the animals that became today's domesticated species were the wild animals that found the greatest biological advantages in associating with humans, including a secure source of feed, protection from other predators, and leadership.

Archaeological evidence indicates that we were still hunter-gatherers at the time that dogs became the first domesticated species, about 15,000 years ago. Anthropologists speculate that dogs were first domesticated to aid with hunting, but they may also have been the first animals domesticated because they are care-soliciting creatures, bringing out our nurturing side. One thing is certain: They share similar traits with all the animals that have been domesticated since. They are generalists that can live in a wide variety of ecosystems; they tend to organize in groups, and within the group they tend to follow a leader, or an alpha animal; one male will breed with multiple females; and they're curious.

As humans settled into agriculturally based societies and gave up the nomadic life of hunter-gatherers, a sudden boom took place in the domestication of both plants and animals. Initially, most progress was made with plants, but livestock domestication followed quickly as some wild herbivores discovered that living in proximity to these human settlements wasn't all bad. Between 10,000 and 12,000 years ago, humans domesticated goats, sheep, cows, and pigs. Horses came kind of late in the process, a mere 8,000 years ago.

Figure 5.1. Przewalski's horse is the wild horse from which modern horses are believed to have been bred. These horses are considered extinct in the wild, though several zoos around the world are working to save the species from complete extinction through captive breeding programs.

With the exception of pigs, the wild ancestors of most of our current livestock species have either disappeared completely or are found in very small numbers. Modern cattle descended from the auroch, a large wild animal that was hunted to extinction in the seventeenth century. Przewalski's horse, considered by many to be the wild forerunner of our modern horse, survives only in small, captive populations (Figure 5.1). Wild sheep and goats are endangered or threatened throughout the world.

Breeds and Genetic Diversity

In more than 15,000 years, humans have succeeded in domesticating less than fifty species of animals. From these species, though, thousands of distinct breeds have been developed worldwide. Early in the process, breed development was fairly informal, a luck-of-the-draw kind of affair. But over the millennia, breeding became a serious endeavor. Given time, genetics can be manipulated to produce animals with desired traits. Some breeds were acclimated to northern weather, some to southern. Some breeds thrived in arid environments, some in humid. Some breeds were prized for the color of their coats, the length of their feathers, or the size of their horns.

Unfortunately, this genetic heritage is shrinking as many minor breeds disappear. More than one-third of the livestock and poultry breeds in the United States are currently considered rare or endangered; 80 percent suffer from insufficient numbers of breeding animals to maintain a healthy diversity within the breed. Worldwide, things don't look any better; 30 percent are in imminent risk of extinction, and about six breeds are lost each month. A look at the American dairy industry provides a clue to what has happened to the livestock gene pool: 83 percent of all dairy cows in the United States are Holsteins, and because of artificial insemination, more than 80 percent of Holstein cows are bred to one of only twenty sires or their sons (Figure 5.2).

Figure 5.2. Dutch Belted cattle are an old and very rare breed of cattle. Considered a dual-purpose animal (i.e., good for both meat and milk production), they make an ideal homestead cow.

Losing minor breeds results in a loss of genetic variation, which can have tragic consequences. When most animals come from the same stock, they're more vulnerable to disease, parasites, and environmental changes. The predominant breeds in agriculture today are single-purpose breeds: the Holstein in dairy production, the White Rock in poultry, the Landrace in pork, and the Suffolk in sheep. These are industrial animals, bred for one trait only — production. Their performance when measured strictly in terms of production is outstanding, but they require high inputs of expensive feeds and veterinary services, and they tend to burn out quickly. The average dairy cow in this country lives less than three years before she is shipped to a packer; in a less intensive system, she can easily be milked for a decade or more.

Minor Breeds and Small-Scale Farms

The loss of minor breeds can have especially grave impacts on small-scale farmers. Many of the breeds that

TERMINOLOGY

▶ ***Crossbred.*** Crossbred animals are known to have more than one breed in their lineage. Many crossbred animals perform well, due to an effect known as *hybrid vigor.* Constant inbreeding can result in bad traits coming through, but when you cross two breeds you tend to get a vigorous, healthy animal.

▶ ***Exotic.*** Exotic animals are those species that are not generally raised for commercial agricultural purposes in the United States. Elk, red deer, pot-bellied pigs, and camels are a few examples of exotic species.

▶ ***Grade.*** Grade animals may or may not be purebred. If you look at a Jersey cow from a nonregistered herd, she may very possibly be a purebred Jersey, but there are no records of her breeding so she's called a grade Jersey.

▶ ***Minor breeds.*** Minor breeds are those breeds of livestock that have fallen from favor in commercial agriculture. As the breeds lose popularity in the commercial sector, their numbers decrease — sometimes to such an extent the breed becomes extinct.

▶ ***Papered.*** This is the same as registered. It just means the owner has registration papers on the animal.

▶ ***Purebred.*** Purebred animals have 100 percent of their bloodlines coming out of one breed of animal. A purebred Black Angus cow, for example, has no other breed in her lineage.

▶ ***Registered.*** Registered animals are all purebred or bred in accordance with an official breed association's standards, and their lineage is recorded with a breed association. A registered Quarter Horse comes from either two registered Quarter Horse parents or, as the American Quarter Horse Association (AQHA) rules allow, the cross of a Quarter Horse with a Thoroughbred. Its lineage is documented with the AQHA, and the owners' receive a registration certificate for the horse from the association. Most major breeds and many minor breeds have breed associations that record the registrations of animals within their breeds. Registration costs money; the amount varies from one breed association to the next. Registered animals aren't necessarily better than nonregistered animals; there is simply a detailed record of their breeding going back many generations.

EXOTIC SPECIES AND SALVAGE VALUE

Exotic species are all those different critters that are not part of the standard marketplace for livestock. Currently, exotic species being raised in the United States include elk, various types of deer, bison, llamas, alpacas, emu, ostrich, rhea (an ostrich family member that comes from South America), and pot-bellied pigs.

Exotic animals typically go through a period in the marketplace when their prices go sky high. We've watched this happen time and again. At one time, llamas were selling for $50,000 dollars; you can purchase one for $500. A few years back, emu were going for $15,000 for a breeding pair; now people give them away. One person who is familiar with the livestock markets likened what happens with exotic species to a pyramid scheme, an excellent analogy.

The first people who breed these exotics sell them for breeding stock. There aren't many around, so if the initial breeders can stimulate some demand the price flies. A spiraling effect takes place: The more the animals sell for, the more people think they can make when they buy their breeding pair. But sooner or later, the breeding market becomes saturated. Finally, the animals must be sold based on what their ultimate use is — meat, fiber, or work. This is the point where the market crashes, mainly because the breeders haven't developed an adequate end market.

The end market is important when you're looking at any alternative enterprise, but especially so if you're considering an exotic species that sells at an astronomical premium above its salvage value.

The *salvage value* is what you can get out of an animal that must be put down — say, it breaks a leg. I don't know about other parts of the world, but I know that no one eats llama in the United States, so the only salvage value for a llama is its fiber or hide. They produce about 8 pounds (3.6 kg) of fiber every other year. That's not much to salvage.

Now, before everyone who raises llamas begins screaming, let me say that I like llamas and that they do have some value — as pets, as guard animals for sheep and goats, and as pack animals. If you have a job to support your farm and want to keep some llamas around for pets — go for it. Or if you're raising a flock of sheep and are considering llamas for guard animals — that's great. I just want to warn new and aspiring farmers of a real hazard with exotic animals — they have wiped out many bank accounts over the years.

Let's go back to the idea of end markets (and this time I'll pick on the emu crowd): A mature emu weighs about 100 pounds (45.4 kg), and yields about 25 pounds (11.3 kg) of meat. Seeing as they don't sell emu meat in Safeway, or Albertson's, or Wynn-Dixie, you will have to take personal responsibility for marketing this 25 pounds (11.3 kg) of meat. Even if you're a great salesperson and are able to sell all of it for $7 per pound ($15.43 per kg), you'll receive $175. Out of that $175 must come your operating expenses, like feed, and depreciation on capital expenses, like fencing and buildings. When you look at salvage value, it is hard to justify spending many thousands of dollars on breeding stock that have such low salvage values — yet people did it for a few years.

The breeders who are getting into elk, deer, and bison have a stronger end market: There is a demand for this type of meat, particularly in upscale restaurants; there are many by-products from these animals; and there are opportunities for on-farm hunting. But again, if you're looking at any exotic species to create a profitable enterprise you must evaluate the salvage value of the animals, the start-up costs (fencing and other facilities, as well as the cost of breeding stock), and your ability to market the end product yourself.

Exotic species definitely have a place in the agricultural landscape, but you must *research, research, research,* before you invest! And research is not simply reading through the glossy little brochures put out by the breeders' group, because they benefit by keeping the bubble afloat. Eventually, each exotic bubble breaks, and you don't want to be in the fallout when it does. Even if people are willing to pay $500 for an egg, ask yourself, how long it can go on. Obviously, once the breeding market is flooded no one is going to scramble up $500 eggs for breakfast! The planning exercises in part IV will help you with your evaluation.

have been lost or are in danger of being lost are considered dual-purpose breeds that can produce both milk and meat, or eggs and meat, or wool and meat well. These are traits that still work on a small-scale farm. Many of these minor breeds are regionally acclimated, or do well on a primarily forage diet. They tend to be excellent mothers, making the farmer's job easier.

We became interested in minor breeds early in our farming operation. Our milking herd included not only Holsteins but also Milking Shorthorns, Jerseys, and some crossbred cattle. One of our top-producing cows was actually a cross between a Black Angus and a Jersey. Belle didn't know that black cows weren't supposed to be dairy cows! Our beef herd included Scotch Highlands, and our sheep flock primarily consisted of Karakuls, a fat-tailed breed of colored sheep that originated on the steppes of Russia. A walk through our henhouse would have revealed many minor breeds: Black Jersey Giants, Barred Rocks, and Rhode Island Reds, to name a few. One of our favorite pet animals on the farm was a Broad-Breasted Bronze turkey named Tom — we're original with names! Tom was also a

FARMER PROFILE

Greg and Lei Gunthorpe

In 1991, Greg and Lei Gunthorpe purchased a 100-acre farm across the road from his folks' farm. Like his father, grandfather, and great-grandfather before him, Greg's main enterprise is pasture-raised pigs.

"My dad has been raising hogs on pasture at that farm since 1951, and the neighbors still think we're all crazy. And that's in spite of the fact that most of them have gone out of business raising hogs in confinement during the same period. There used to be hundreds of small hog farmers in LaGrange County [Indiana], and now there're only about three others left."

Greg and Lei tried confinement hogs for a little while, because there was already a confinement barn on the place they bought. "I tried it for about four years. Two of those years I did okay, but in the other two I got sick of the death loss."

According to Greg, when you're raising pasture pigs most death loss comes in the first 48 hours: either babies that are rolled on by the sow, or runts that don't have the stamina to survive outdoors. But with confinement, many of the deaths fell on pigs that were well along. "When you lose a 200-pound pig, you lose a lot of money."

Greg has good luck with spring and fall farrowing on pasture. He averages eight to ten finished pigs per litter, with little or no purchased feed.

Greg's rotation includes soybeans, corn, wheat, and pasture. The only crop he harvests is the beans; the pigs harvest the rest right out of the field.

Over the years, the Gunthorpes began to notice that the breeding stock they were buying didn't really fit their system. "Modern breeding programs had developed strains of pigs that only work well in confinement. These pigs have lost the ability to forage, and to gain on a high-forage diet." Tamworth hogs, a minor breed, became a crucial link in Greg's breeding program. "The Tamworths are hardy and rugged beasts. They haven't lost their foraging traits."

It was a little tough to find a minor breed, like the Tamworths, at first. Greg had to drive four hours to find the nearest breeder, and there were several places he could have stopped in between to pick up commercial boars, but the drive was worth the effort.

Another benefit of keeping a minor breed going, Greg and Lei have found, is that it has helped their marketing. Although most of their hogs are still marketed conventionally, they do have one restaurant customer in Chicago who purchases one whole hog every week of the year; they also sell some hogs to a natural foods store in San Francisco. "These customers like the idea that they are helping to keep a minor breed going when they purchase from us. It has helped us define a marketing niche."

But Greg emphasizes that matching animals to a low-input system is his primary reason for opting for a minor breed. "I can't overstate the importance of how the animal's genetics need to match the land and the system."

favorite with visitors who came to the farm, and he got his picture in the newspaper several times.

Minor breeds may not work as well on a highly industrialized farm, but they can work very well on small-scale farms where forage is the primary feed, animals are expected to harvest most of that feed themselves, and the farmer doesn't want to build expensive and specialized housing. Our Karakuls didn't produce the large litters of lambs that most commercial sheep producers favor, but they were perfect for us. First-time ewes always had just one lamb — which works out well, because who needs a teenage mother with three or four babies to look after? Older ewes usually produced twins, and sometimes triplets. The ewes that did have triplets were capable of raising all three lambs themselves. The only bottle lamb we ever had — Junior — had to be hand-raised because his mother had severe mastitis in one half of her bag, so she could only feed one of her twins.

Finding minor breeds can take a little effort, but for small-scale farmers this effort is worth it. They aren't usually available at a local sale barn, so you'll probably have to look for individual producers who own the breed you're interested in. The marketing of minor breeds, like the marketing of exotic species, can be a real challenge. The conventional marketplace, including sale barns and packers, often discounts minor breeds. If you do seek to work with minor breeds, part III should help you think of ways to market them.

One good source of information on minor breeds is the American Livestock Breeds Conservancy (ALBC) of Pittsboro, North Carolina. The ALBC is a nonprofit organization dedicated to helping preserve these minor breeds. It offers a number of publications, including a minor breed census and lists of breeders of breeders around the country. (See appendix E, Resources.)

Breeding and Genetics

In 1865, an Austrian monk, Gregor Mendel, presented a paper before the Society for the Natural Sciences in Brunn, Austria. The paper's title was "Experiments in Plant Hybridization." Mendel's work received little interest at the time, but around the turn of the century other scientists independently verified Mendel's results, and breeding began its transition from a haphazard — but successful — art form to a science. Mendel may have been the father of genetics, but the science didn't come to be known by that name until 1905, when William Bateson, a scientist who applied Mendel's work to chickens, coined the term.

Developing an understanding of basic genetics helps when you're raising animals, and also comes easier when you're applying it on a farm than it does when you're trying to understand it in a high school biology class. As Ken said when he saw me reviewing some papers on Mendel for this chapter, "I never 'got' genetics in school, but once I started to raise animals it began to make sense."

Biology 101

If you struggled with the concept in school, or avoided it at all costs, a condensed biology lesson may be helpful. Mendel worked with peas and demonstrated that observable traits — such as flower position, pod color, and stem length — could be passed from parents to their offspring. He surmised that there was some type of a hereditary unit — now called a gene — that passed information along from both parents. Mendel hypothesized that each gene could come in different forms. In the pea experiment, for example, there were two forms for pod color — green and yellow. These gene forms are called *alleles*, and the offspring actually receives one allele from each parent for each of its genes.

Mendel also demonstrated that some of these alleles were dominant and others were recessive. In the case of Mendel's peas, the green allele was dominant (Figure 5.3). If a plant or an animal has received a *dominant* allele from one parent and a recessive allele from the other parent, it will always show the dominant trait. *Recessive* traits only show up when the recessive allele has been received from both parents. Our German shepherd, Kima, is solid white. White is a recessive color in German shepherds (black patterned with tan is the dominant color for the breed), so Kima received the white allele from both of her parents.

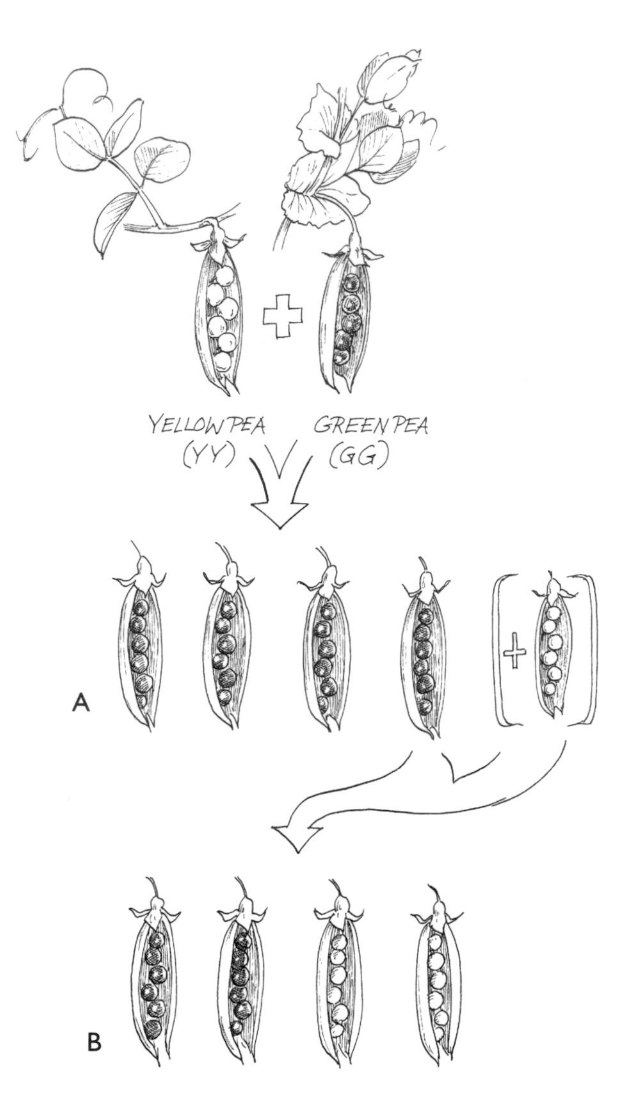

Figure 5.3. Mendel's experiment. **(A.)** All offspring will show the green trait, as G is the dominant allele. **(B.)** If one of these offspring are crossed with the yellow pea (YY); one-half would show the green trait, and one-half would show the yellow.

Some traits are passed along on a single gene, but most traits are actually passed along on more than one gene. As an example, scientists have so far determined seven genes that are responsible for color in horses, and they think there are probably still more to be found. When multiple genes are responsible for a single trait, one gene may be dominant, or *epistatic*, over the others. An epistatic gene simply obscures the expression of other genes. In horses, the white gene is such a gene; if its dominant allele is present, the horse lacks any pigment in its skin and hair, like an albino.

Geneticists use letters to represent the alleles for each gene. Convention has established that a capital letter is used for the dominant allele and a lowercase letter is used for the recessive allele. For the white gene in horses, the letter used is *W*. There are three possible combinations:

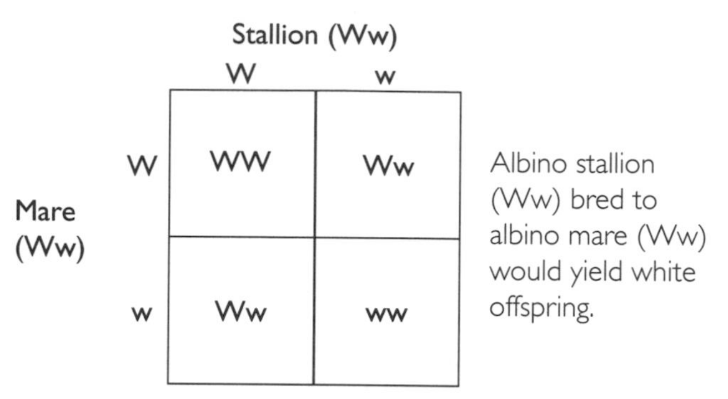

1. *WW*. A horse with a *WW* set of alleles received the dominant allele *W* from each parent; however, this particular allele combination is lethal before birth.
2. *Ww*. This set shows a horse that received the dominant *W* allele from one parent and the recessive *w* from another parent. Any horse with this combination exhibits albino characteristics. In the mating of two albino horses, approximately one-fourth the time their offspring would receive the lethal *WW* form, one-quarter would receive the *ww* combination, and one-half would receive the *Ww* combination (Table 5.1).
3. *ww*. This final combination of white alleles in horses is also the most common. It means the horse received the recessive w from each parent.

Table 5.2 describes the best current knowledge of color inheritance in horses.

You may have heard of the Human Genome Project. A *genome* is simply the full complement of genetic material for any living thing. All living creatures, from single-celled organisms to humans, have a unique genome that repeats itself in every single cell. Take a brain cell, a blood cell, a skin cell, or any of the other seven trillion cells in your body; in each one, basically the exact same set of genetic data is found. The Human Genome Project's goal is to map all 100,000 human genes (which are located on 23 chromosomes). Recently, veterinary geneticists have begun genome-mapping projects on most major livestock species. Mapping genes is another technological two-edged sword; it may be of great benefit, but it could be highly abused. The ethical debate over gene mapping and genetic engineering is one that we small-scale farmers need to follow.

Traits such as hair color and eye color — or in the case of Mendel's peas, pod color — are completely based on heredity; they're traits that are in the genes. But some traits, including temperament, intelligence, and size, are based in part on heredity and in part on environment, and are affected by the attention an animal received after birth. Proper handling and training can make a naturally high-strung animal pleasant to work with: improper handling and training can make a monster out of a naturally docile and calm animal. An animal can have the genetic potential for quick growth, but if it isn't receiving adequate nutrition or if it is carrying an excessive parasite load, it won't realize its full growth potential.

Table 5.1

Genetics of Horse Colors

Gene	Allele Combinations	Action
W	WW	Lethal before birth
	Ww	Horse typically lacks any pigment in hair, skin, eyes
	ww	Horse is fully pigmented
G	GG	Horse shows "silvering" with age, but is born a nongray color
	Gg	Same as GG
	gg	Horse does not show "silvering" with age
E	EE	Horse has ability to form black pigment in hair and in skin; black pigment in hair may be either points or evenly distributed
	Ee	Same as EE
	ee	Horse has black pigment in skin, but hair appears red
A	AA	If horse has black hair (EE, Ee), then the AA gene gives it a pattern of black tips, or points; the A gene has no effect on red horses (ee)
	Aa	Same as AA
	aa	If horse has black hair (EE, Ee), then it is uniformly distributed over the body
$C(C^{cr})$	CC	Horse is fully pigmented
	CC^{cr}	Red pigment diluted to yellow; black-pigmented horses are unaffected
	$C^{cr}C^{cr}$	Both red- and black-pigmented horses are diluted to pale cream; skin and eye color also dilute
D	DD	Horse shows diluted body colors and has dark points, including dorsal stripes, shoulder stripes, and leg barring
	Dd	Same as DD
	dd	Horse has undiluted color
TO	TOTO	Horse shows white spotting pattern (paint); legs are usually white
	Toto	Same as TOTO
	toto	No spotting pattern

Source: Modified from Ann T. Bowling, "Coat Color Genetics: Positive Horse Identification," *Equine Practice* (1979, Nov–Dec). <http://www.vgl.ucdavis.edu/~lvmillon/coats2.html>

Table 5.2
Color Definition in Horses

Genetic Formula	Color
W	White
G	Gray
E, A, CC, dd, gg, ww, toto	Bay
E, aa, CC, dd, gg, ww, toto	Black
ee, CC, dd, gg, ww, toto	Red
E, A, CC^{cr}, dd, gg, ww, toto	Buckskin
ee, CC^{cr}, dd, gg, ww, toto	Palomino
$C^{cr}C^{cr}$	Cremello
E, A, CC, D, gg, ww, toto	Buckskin dun
E, aa, CC, D, gg, ww, toto	Mouse dun
ee, CC, D, gg, ww, toto	Red dun
E, A, CC, dd, gg, ww, TO	Bay tobiano
ee, CC, D, gg, ww, TO	Red dun tobiano

Note: See Table 5.1 for key to abbreviations.
Source: Modified from Ann T. Bowling, "Coat Color Genetics: Positive Horse Identification," *Equine Practice* (1979, Nov–Dec). <http://www.vgl.ucdavis.edu/~lvmillon/coats2.html>

Emotions and Senses

Speak of an animal's emotions and some people get kind of testy (especially people with PhDs who study animal behavior): Referring to an animal's emotions is viewed as being anthropomorphic (attributing human characteristics to nonhuman beings). In his book *When Elephants Weep*, Jeffrey Moussaieff Masson tells the story of a trainer who says that animals don't get angry. I don't know what that trainer trains, but I do know that I've seen animals angry, happy, scared, contented, and loving. Shauna, our mare, was plain mad at us when we sold two of her colts, and she displayed that anger for weeks. Calves playing "king of the hill" on a pile of dirt in our barnyard were obviously having great fun and were quite happy, at least until Leona decided that her calf, Les, should be the winner of the game. She began bashing the other calves angrily off the hill until Les was on top.

Then there are the people who say this or that type of animal is stupid — not just that a single animal isn't so bright, but that the entire species is dumb. Sheep, cows, and turkeys are especially vulnerable to this assessment, but even horses are occasionally accused of stupidity, which is far from the truth — horses are intelligent animals. Animals respond to certain stimuli differently than we do, and react based on long-standing genetic programming. These responses and reactions are sometimes taken as signs of stupidity. Learning to understand these patterns of behavior, and learning to "read" an animal's emotional state, makes working with animals much easier and less dangerous for you and the animal.

The emotion that animals are usually evincing when people accuse them of being stupid (or dangerous) is fear. I once heard an old rancher say that a critter remembers the worst thing that ever happened to it and the most recent bad thing that has happened to it. Now scientists have proven him right. Fear memories, they've discovered, are stored, basically forever, in a portion of the animal's brain that isn't very well developed. An animal that has had an extremely bad fright — say, on entering a trailer — will continue to be afraid of trailers until brain centers that are more highly developed learn to block the fear. Overcoming an animal's fears can be a time-consuming process, and requires patience on the part of handlers and trainers.

Understanding how your animals use their senses (sight, sound, smell, taste, and touch) will help you understand why fear reactions can happen so quickly. Remember that livestock species are primarily prey animals, so their senses have developed to help them find food and shelter, for navigation, and, most importantly, to avoid predators.

Vision

As prey species, livestock have a wide field of vision (Figure 5.4). This allows them to see predators moving in on them from almost any direction, and is made possible by the biological design of the animals' eyes: large relative to their head size and located toward the sides of the face. Predators' eyes are smaller in proportion to head size and located to the front of the face, providing predators with a narrower field of vision, but the centered eye placement enhances depth perception through bifocal vision.

Bifocal vision occurs when both eyes can independently focus on the same thing at the same time. To put these differences in perspective, a prey animal can see a predator well on the horizon in almost any direction, but it has trouble seeing a stationary bug sitting on a leaf. Most predators don't see as well at a long distance, and they can only see in the direction they're facing; still, they can make out even slight movements within their field of vision and can differentiate three dimensions well. Horses and pigs have more bifocal vision than cattle or sheep, because their eyes are located more forward on their heads than are the eyes of cattle and sheep (Figure 5.5).

Despite their wide field of vision, livestock species have a very narrow blind spot right in front of their noses, and a slightly larger blind spot directly behind them. When you work with animals that are secured (tied or stanchioned), you learn (sometimes the painful way) to talk to them before approaching them directly from behind. One day Ken walked up behind one of our milk cows, Libby, to take the milking unit off her while she was in her stanchion. She hadn't heard him coming, so when he kneeled to remove the

Figure 5.4. Cows (and other prey species) have monofocal vision; that is, they only can see something with one eye at a time. Because their eyes are situated more to the sides of their heads than to the front, their depth perception is not very good. Still, though they may lack depth perception, their blind spots are minimal, which allows them to see almost all the way around themselves. In fact, the field of vision for cows is almost 330 degrees. Most prey species have a small blind spot right in front of their noses and a slightly larger blind spot behind themselves.

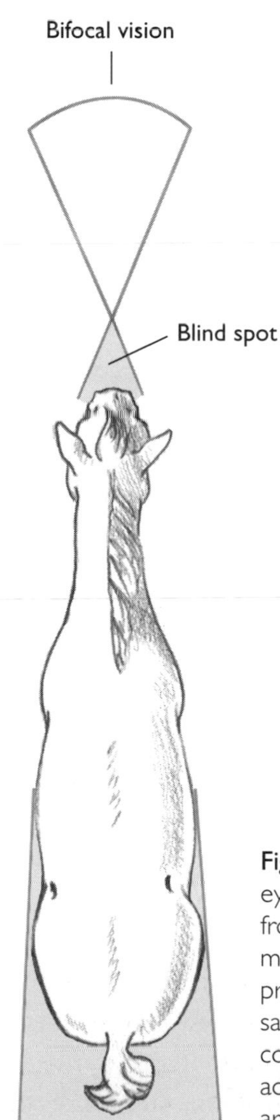

Figure 5.5. Horses, with eyes situated more to the front than cows' eyes but more offset than predators' eyes, share the same large field of vision as cows, but they have the added advantage of a small area of bifocal vision.

unit she kicked out; Ken went flying and she stomped his glasses. Libby was a fairly even-tempered girl, but she reacted out of fear.

Visual frights are responsible for the majority of difficulties that arise when working animals. If an animal is acting skittish, look around the area for items that may be causing it stress. Shadows or reflections off water; something out of its regular place, even something as simple as a jacket hung over a shovel handle; bright-colored objects; or an object flapping in the wind can spook an animal. The play of light can also affect animals. They are most comfortable in a diffusely lit area; extremely bright lights will cause them to balk.

Hearing

In most land animals, hearing is controlled by a three-part ear system. The external ear acts as a gathering device and canal for sound waves, moving them into the middle ear. Sound waves are basically just air particles, jostling each other at various pressures and speeds.

The size and shape of the external ear was, in many ways, a product of the environment in which the animal developed. Asses have the largest external ears of any of the domestic equines, and horses have the smallest. Mules' ears, as a cross between asses' and horses', are between the two. Asses evolved in a desert environment, which forced the animals to spread out over large geographical areas to find adequate feed. Big areas required big voices, and ears that were capable of hearing them. This was the only way they could find each other for companionship, support, or mating. In comparison, horses evolved in a grassy and lush savanna. In this environment, the land supported larger communities of animals. The horses could "talk" to each other without raising their voices, and they didn't need as large an external ear to hear each other.

The middle ear transmits the vibrations, which are created when sound waves strike the eardrum, into the inner ear. The inner ear acts as a messenger service, translating the vibrations for the brain and nervous system to interpret.

After sight, hearing is the second most critical sense for those of us who handle animals to understand. Low-pitched, rumbling sounds don't tend to bother livestock, but any kind of high-pitched whine will scare them. This is due to the range of sounds that their ears are most sensitive to: Human ears are most sensitive to sounds in the 1,000 to 3,000 Hz range, but livestock are most sensitive to the higher sounds of the 7,000 to 8,000 Hz range. The difference in sensitivities is due to the size and shape of the eardrum.

Although high-pitched, loud, or novel sounds can cause fear reactions, not all sounds are bad. Animals do well when a variety of sounds are part of their normal environment. We always played a radio in the barn while we were milking. We noticed that most of the time this didn't bother the animals at all, but sometimes a very loud commercial would come on and really get a reaction out of the girls. We switched over to a cheap CD player, and the cows did better. It didn't seem to matter what we played — Garth Brooks, Jimi Hendrix, or Mozart — so long as the volume was consistent and changes were made incrementally.

Touch and Smell

Livestock use their noses both for smelling things and for touching them. Notice how an animal approaches something new that's been placed in its environment: It lowers its head and approaches with its neck stuck out, first sniffing and then, slowly, touching the article with its nose. The nose has large bundles of nerve endings, making it the organ of choice for "feeling" new things.

The sense of smell is stronger in most animals than in humans and the other primates. A horse that spots something moving on the horizon will put its nose into the wind and begin sniffing the air. The sniffing action brings more air through the nasal cavity, increasing the likelihood that it will pick up the scent of what it sees.

Males of most species use their sense of smell to detect females that are coming into heat or are ready to breed. The male smells and licks the female's vulva; then he holds his head out and curls his lip up, in a behavior known as a Flehmen response. What he is sensing is a chemical attractant the female releases called a *pheromone*.

The sense of smell also allows mother animals to identify their own offspring. Mothers detect their own odor on a newborn animal, allowing them to identify

(cont'd on p. 54)

FARMER PROFILE

Kevin and Marcia Powell

Mulefoot hogs are an almost extinct minor breed. Developed in the Midwest in the middle of the nineteenth century, they were used by farmers along the Mississippi and Missouri Rivers to forage on river islands during the summer months. Today, to the best of Kevin Powell's knowledge, there are only seven breeding herds left — but Kevin is on a mission to change that.

Kevin grew up on a family farm near Strawberry Point, Iowa, and like many Iowa farm kids he attended Iowa State University's College of Agriculture. But unlike many of his peers, Kevin was accepted into the Biotechnology Scholars Program, which puts ten incoming freshmen onto an advanced track for their four years of school.

"The Biotechnology Scholars Program emphasizes studies of biotechnology and genetics. The more I learned about genetics, the more I became concerned about the loss of genetic diversity. I started to realize that we were losing genes that couldn't be replaced — and that someday a lost gene could be crucial to saving an entire segment of agriculture. My concern for both plant and animal genetic diversity grew proportionally to what I learned."

When Kevin returned to the farm from college, he and Marcia farmed full time. "We raised corn and alfalfa hay, hogs and cattle; but poor markets and high debt did us in. We tried to diversify with dairy goats, and the dairy goats paid pretty well, but the existing debt from our conventional operation was just too much."

Kevin gave up full-time farming and went to work as a quality engineer for the Square-D Company (manufacturers of electrical system components). "I still wanted to farm part time, so I took over 40 acres of my parents' farm. And I decided to do something about my concern for genetic diversity."

Kevin and Marcia contacted a group called the Institute for Agricultural Biodiversity in Decorah, Iowa. The institute ran a farm that maintained a number of minor breeds. "We went to visit in 1995. The director of the farm explained to us that they had a 'nucleus herd-lending program.'"

The lending program was designed to loan out breeding herds to farmers interested in helping maintain the breeds. The idea was that a farmer would keep a small breeding herd for a year or two then send it on to another farmer, but keep back some of the young stock for his or her own use. Kevin and Marcia took a breeding herd of Mulefoot hogs (eight sows and two boars), but during the time that they had the herd on loan the institute ran into financial trouble. "We ended up purchasing the herd," Kevin says.

Mulefoot hogs are a breed unique to the United States. They evolved out of a highly recessive trait in hogs: Instead of the normal two-toed, cloven hoof, they have a single toe, much like a mule or a horse. Farmers in the Midwest bred this trait up for the hogs they used on the islands, because they didn't tend to sink in the mud as much, and they had less trouble with foot rot.

The breeding population remained small but viable until the 1950s, when the U.S. Army Corps of Engineers made it illegal for farmers to drop the hogs on the islands in summer. "The corps was beginning to push recreational opportunities of the rivers, and they thought that allowing the hogs to forage on the river islands was detrimental to that goal. Large numbers of Mulefoots were sold for butcher after that decision was made."

Kevin explains that Mulefoot hogs should still have a place on small farms. "They are very gentle and calm animals. I raised commercial hogs for years, and these are much more pleasant to work around. They are also still good foragers, and they have very flavorful meat. Mulefoots grow a little slower than the modern breeds, but they do make comparable size."

Note: If you know of a breeder of Mulefoots or are interested in helping to preserve the breed by starting a small herd, see appendix E, Resources.

it as their own; later, they detect their own milk odor. One method of "grafting" an orphan onto a mother animal is to feed the orphan her milk for a few days prior to introducing it. Then place Mom, her own baby, and the orphan together in a small pen. Mom smells her own milk on the orphan and figures the little guy must be hers. Sheep farmers also use this approach to graft a lamb from a large litter onto a sheep that lost a lamb at birth; they skin the dead lamb and place its skin over the "extra" lamb. The ewe smells her own smell and allows the lamb to nurse. After two or three days, the skin is taken off, but by this time the mother smells her own milk.

Training and Handling

Training and handling really go together; every time you do one, you're doing the other. To be good at training and handling animals requires both skill and the patience of a saint. Calm and quiet movements on your part offer the best hope of success.

Problems

When problems arise in training and handling, they generally fall into one of three categories: the animal's disposition, the facilities, or the handler's interaction with the animal. Each type of problem can be overcome, but doing so takes time on your part.

Disposition. Disposition problems are more common with some breeds than others. Despite their placid, doelike eyes and small stature, Jersey cows are playful to the point of being a nuisance at times, and farmers and ranchers often call Black Angus cattle "Black Anguish" because of their high-strung natures. Even within a breed, some critters are just plain high strung and more difficult to handle than others. Regardless of the breed you are interested in, the best way to avoid disposition problems is to search out and keep calm individuals as your foundation breeding stock.

Facility. Facility problems are usually easy to correct, and are most often the result of a failure on our part to recognize distractions — and by that I mean the types of things that an *animal* senses as being wrong. Is there a strange, high-pitched noise coming from a piece of machinery? Are you trying to force animals to move from a well-lit area into a dark one? Is there something in the animal's field of vision that isn't normally there? Are floors too smooth and, therefore, slippery for a hooved creature? For all but the most trusting critters, anything out of the ordinary will cause a fear reaction. When a distraction is causing trouble, either correct it or allow the animals time to overcome their fear. Chapter 7 discusses facilities in greater detail.

Handler. Stressful interactions between a handler and an animal are probably the hardest problem to overcome. Handler problems require true self-evaluation, and none of us likes to think we could be at the heart of the problem. Yet most often we are. As Bud Williams (a real guru on animal handling and training) says about trying to control an animal, "*Believe* that she is responding to what *you* are doing *right now*."

Training Styles

Traditionally, people have tried to manage animals through fear and force. Such methods are counterproductive, though; they will cost you through increased health problems and injuries, reduced weight gain and productivity, and general aggravation on your part. Movies may depict cowboys whooping, hollering, and racing animals around, but that just stresses your herd, and you'll feel your herd's stress in your pocketbook at the end of the year.

Getting an animal to go where you want it to go, or do what you want it to do, is easiest if you can make it see the benefit of taking the action you desire. This is the good old coaxing school, and it works with critters just like it works with kids. Not only that, but it's often quicker, causes less stress, and is far less likely to result in injury (or death) to you or the animal.

About three days prior to our farm auction, the auctioneer stopped by. When he saw all our animals still spread far and wide over the whole farm, he freaked out. "How are you ever going to get all these animals contained in time for the auction?" he wanted to know. We told him not to worry or get excited. To his amazement, when he showed up the morning of the auction about 6:30 A.M. all but four wayward lambs were penned where they were supposed to be.

The lambs had bolted at the last minute that morning when a truck with no muffler drove up next to the pen we were working the sheep into. Once they bolted, we knew we wouldn't catch them that morning, so we didn't try again. All together, we penned 5 equines, 36 head of cattle, 45 head of sheep, and about 75 chickens during the evening before and the morning of the auction. With the exception of the lambs, it went very smoothly. We sold the incorrigible lambs to one of the farmers who bid on the sheep, and he picked them up two days later, penned in the barn with a minimum of fuss.

Working Tame Animals

If you're working with fairly tame and calm animals, simply bribing them with treats is the most effective form of coaxing. We strive to tame down all our animals so they respond to some form of treat: a carrot or apple for equines, a bucket of grain or alfalfa cubes for ruminants, or some table scraps for pigs. Just give them something special, in small rations so it's always a treat, and they'll follow you wherever you want to lead them. Ken calls this "training 'em to eat," and it's generally quick and easy to accomplish.

When we settled back in Colorado, our first purchase was a pair of spotted asses (donkeys and asses are basically the same beast, just called by different names; burros are a small breed of donkeys; mules are a sterile cross between a horse and an ass). We love any kind of equine, but we wanted asses because the larger ones are suitable for riding, and they make the best mountain pack animals. Duke was about five and very scared of everything and everyone; Jessie was about three and showed little fear. Duke got loose one evening from the small training corral where we were keeping him while we were taming him down, but by that point we'd successfully trained him to eat, so he followed me back in with carrot bribes. No hassles, no injured animals, no sweat! The entire operation took about 20 minutes.

Working Less Tame Animals

With less tame beasts that haven't been trained to eat yet, you must use some techniques that play on the animal's natural behavior. An animal's *flight zone* is its private space, and by learning how to move along the edge of its flight zone you can learn how to efficiently move it. Animals also have a *pressure zone*, which lies outside the flight zone. The objective is to move the

MORE TRAINING TIPS

- Training requires time and patience. If you don't have either, find someone else to be responsible for training.
- When introducing something new, let the animals investigate it at their own pace. If you're teaching a foal to load into a stock trailer, park the trailer in its pasture for a few days with some feed inside. If you are going to introduce a saddle pad, leave it hanging over a sawhorse in the pen for a day or two.
- Short lessons, given regularly, are the most effective. Half an hour per day, four days a week is far better than two hours, one day a week.
- Reward good behavior. Give animals a treat or a pat and kind words for good behavior.
- Stern words issued in a low, deep voice — not yelling or hitting — are the best for discipline.
- Lessons must be repeated frequently, until the task becomes completely familiar. Repetition solidifies behavior.
- Be consistent in everything you do.
- The smaller the training area, the more control you will have over an animal's movements and behavior. Circular arenas work best, because the critter can't get into a corner.
- In general, the younger an animal is when training begins, the quicker it learns. Still, the world-famous Royal Lipizzaner Stallions don't receive any training until they are three years old — so you don't have to be in a big rush.

animal by working in the pressure zone (Figure 5.6). A herd of animals also has a *group flight zone*, and moving properly at the edge of that zone will allow you to move the whole herd. Bud Williams, who has used his knowledge of flight-zone behavior to move everything from cattle and sheep to buffalo, reindeer, and elk, says there are four main principles to remember about moving animals:

1. They want to move in the direction they are already heading.
2. They want to follow other animals.
3. They want to see what is pressuring them (you).
4. They have very little patience.

In Bud's method, you are simulating a predator's stalking behavior. This, in turn, elicits avoidance behavior in the livestock. Let's look at what happens under this scenario in the wild: A lion locates a herd of animals. It begins walking in a slow circle around the herd, looking for weak, young, or old members. This circling action by the lion causes anxiety in the herd, which begins to bunch into a tighter group. As long as the lion's circle is far enough out from the animals' flight zone, they don't become frightened enough to break into flight — they just stay tightly bunched and focus on the lion's whereabouts. As the lion begins to increase its speed and move in on the herd, though, anxiety is replaced by fear, and the herd enters full flight phase. They run like hell! If the lion is lucky, a weak animal falls behind, and it has dinner. If it isn't lucky, the herd remains tight and it has to try again another day.

Once the herd has entered full, breaking flight, it will take at least one hour, and possibly more, before it begins to settle down again; when you're simulating a predator, then, you don't want to move the herd from anxiety behavior into fear behavior. Your goal is to maintain light pressure — just enough to cause anxiety, but not enough to cause fear. Your movements need to always be slow, steady, and calm, with no running, whip cracking, or yelling.

These techniques take practice and are easiest to learn when you don't actually have to succeed at moving the animals someplace in particular. They don't work well with tame animals, because tame animals know you aren't a predator.

Point of Balance

An animal's *point of balance* is at its shoulder. If you are in front of its point of balance, it will back up or turn away, and if you're behind its point of balance it will move forward (see Figure 5.6).

Gathering Loose Animals

If the herd you're going to work is spread out in a pasture, you first need to gather them into a loose bunch. Do this by walking in a very wide, slow arc behind the outermost perimeter of the animals. If a few stragglers

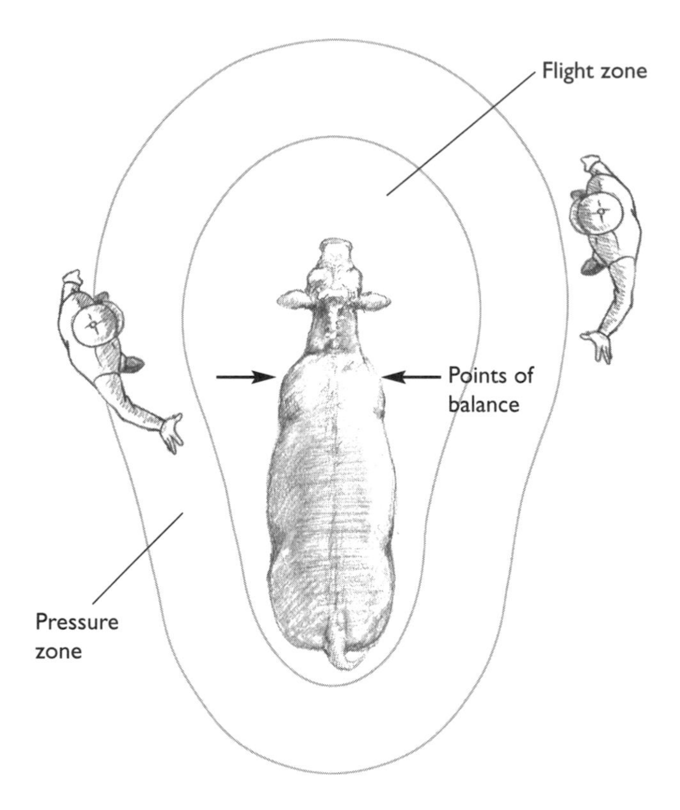

Figure 5.6. The *flight zone* is the animal's private space. Cross into that area, and the animal runs. In the *pressure zone*, the animal will move away but not run. Flight zone and pressure zone vary from one animal to the next, and vary depending on how calm or excited an individual animal is. The flight zone gets larger as an animal becomes more excited.

When working animals, try to stay out of their blind spots, and don't approach from the front. Approaching near their shoulder from an angle is the best way to safely interact with an animal.

are off someplace, don't worry about them; they'll join the rest of the herd soon enough. Your slow arcing movements will cause the animals to begin bunching up. If you are too far away, the entire group will simply turn to face you, and if you are in too close the animals at the rear will try to cut back around you or begin running. Depending on the number of animals you are working and the size of area they are spread out in, gathering can take from five minutes to half an hour to accomplish (Figure 5.7).

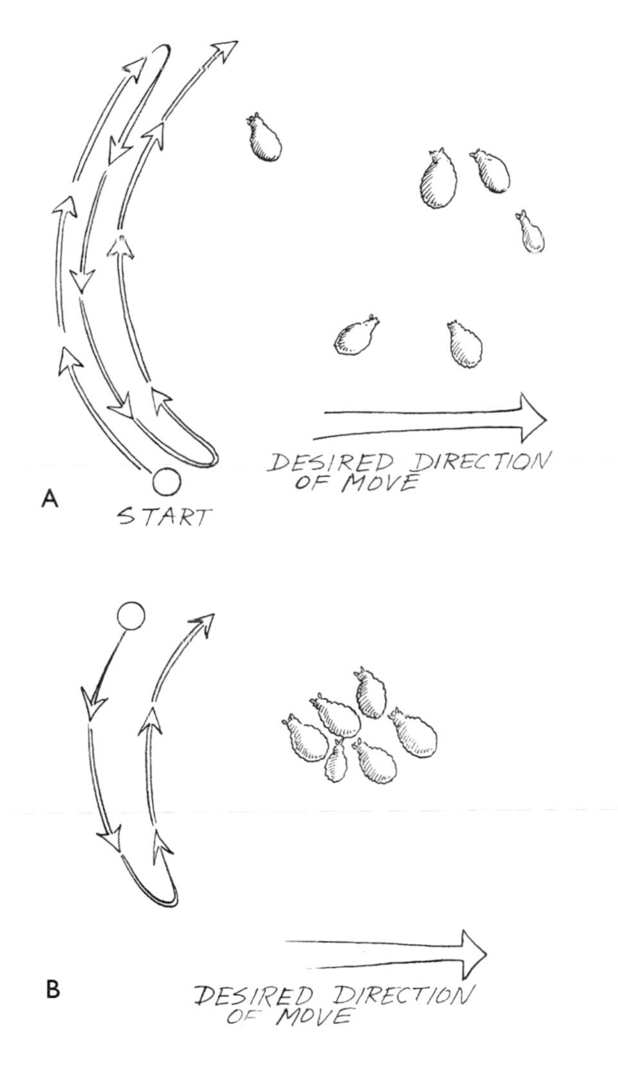

Figure 5.7. When animals are spread out in a pasture and you want to move them, first you must gather them into a group. **(A.)** Move back and forth in a wide arc behind the animals and on the far side of the direction in which you want to move them. **(B.)** The animals will gather in a group to study what is pressuring them (i.e., you). Now tighten the arc, which will turn and move the animals in the desired direction.

(Modified from Temple Grandin, Low-Stress Methods for Moving Cattle on Pastures, Paddocks, and Large Feedlot Pens. <http://grandin.com/B.Williams.html>)

Moving the Group Forward

When the group is bunched and you're ready to begin moving them, press in slightly. As soon as the group begins to move forward, quit pressing in on them and simply resume your arc behind the group. Too much pressure will cause them to begin running, and when they run the game is up. The continued arcing motion keeps you from staying too long in any one animal's blind spot. When the group's forward motion begins to slow down, press in slightly again.

Controlling Direction and Speed

Once the herd is moving forward easily, you can begin changing directions. If you extend the arc to the left, the herd will move right; if you extend the arc to the right, the herd will move left.

Moving parallel to the livestock in the same direction they are moving will tend to slow them down. Moving parallel to the livestock but in the opposite direction will tend to speed them up.

Remember — you are always trying to work just at the edge of the flight zone.

Purchasing Livestock

Purchasing livestock is a major step for beginning farmers. It's easy to be taken advantage of when you're starting out: You don't know about markets, you aren't a good judge of "flesh" or conformation, and you don't know the signs of age, illness, or bad temperament. Decisions need to be made before your first purchase: Do you want to buy registered animals or grade animals; do you want to buy young stock or more mature animals? Where will you buy, and how will you transport your animals? And probably the most important question for a beginning farmer: Is the asking price for the animal you're interested in a fair one?

Consider Your Goals

The price you're willing to pay has to be evaluated in light of your goals. Registered animals don't pay! That's right — the extra expense of purchasing and maintaining registered livestock won't pay in the long run, so if your primary goal is to make a living raising livestock, buy good-quality grade animals. But if your main goal for keeping livestock is to provide your kids with the opportunity to show livestock, say in 4H, then registered animals may be worth the extra money.

In the box about exotic species on page 45, I discuss salvage value. This is always an important number to calculate: It's the dollar value of an animal that has to be disposed of quickly. If you're direct-marketing, it's the value of the meat you'll be able to sell; if you market conventionally, it's the value of the animal on the market at the time you must sell. At the low end of the cattle cycle, the salvage price for a healthy cull cow may be as low as 35 cents per pound. It hurts badly when you have to sell that registered heifer you bought for $2,000, at the bottom of the market, and you're forced to accept whatever you can get.

If making a living is important in your goals, then stick with grade animals, and really study the market before you buy. Pay the going price, or possibly a slight premium, if you are dealing directly with a farmer who will stand behind his animals' health.

If you are interested in minor breeds, talk to the producer honestly. Tell him or her that you are really interested in working with minor breeds. Be willing to pay a small premium, but explain that you can't afford to pay far above the market value. Since most producers who are into minor breeds really want to see their breed continue, they will usually work with you. If they won't, go somewhere else.

Seek Good Advice

Learning to evaluate good flesh comes with time. See if you can get an experienced farmer to help you look at the animals you are thinking about buying. And read through chapter 8, Health & Reproduction; it will help you develop the observation skills you need.

When spending a large amount of your available money on livestock (say you're planning on purchasing a whole herd or flock of animals, or you're considering an expensive, registered breeding animal), hire your veterinarian to inspect the animals prior to purchase. Paying the vet to find out whether you should or shouldn't make the purchase may be one of the wisest investments you make.

Deal with Reputable Farmers

Initially, purchase your livestock directly from reputable farmers. When you look at their farms, don't look at the fanciness of their facilities — this is one case where the cover of the book may not tell you anything about the story. See if the animals appear healthy, and if the facilities are relatively clean. Ask them if they will stand behind the animals' health for at least three days. Some won't do this, and with good cause: They don't know that you will take good care of the animals. Many will, though.

Purchasing any livestock at a sale barn (livestock auction house) is a dangerous proposition, especially for beginners. Healthy animals go to the sale barn and come into contact with sick animals. Visiting the sale barn is a good way to study the market but not a good way to buy your animals, at least not until your doctoring skills are well developed.

Consider Young Animals

Young animals require a much smaller investment than mature animals, but you have to grow them out before you have a product. We began our dairy herd by purchasing weaned heifer calves and raising them up. This saved us money, allowed us to develop a good working relationship with the animals before it was time for them to be milked, and gave us time to work on buildings and fences. But we didn't have any real farm-generated income for almost two years! As with just about everything else, you need to evaluate the options based on your goals, and your pocketbook.

CHAPTER 6

Feeds & Feeding

Humankind has historically fostered and relied upon livestock grazing for a substantial portion of its livelihood because it is the only process capable of converting the energy in grassland vegetation into an energy source directly consumable by humans. Biochemical constraints determine that herbivores function as "energy brokers" between solar energy captured by plants in the photosynthetic process and its subsequent use by humans. The inability of humans to directly derive caloric value from the 19 billion metric tons of vegetation produced annually in tropical and temperate grasslands and savannas provides the ultimate justification for evaluating grazing as an ecological process.

— R. K. Heitschmidt and J. W. Stuth (ed.),
Grazing Management: An Ecological Perspective

THE SINGLE MOST IMPORTANT FUNCTION a farmer or rancher serves in the lives of his or her animals is providing for their nutritional needs. As Dr. Bruce Haynes, DVM, says in *Keeping Livestock Healthy*, "It goes without saying that a well-fed animal is more likely to be a healthy animal."

Water

Water is the major constituent that makes up all living organisms. At birth, water may account for 75 percent of the total body weight of a calf, but at two years a finished steer's body will only be 45 percent water by weight. Water provides structure to cells, it allows nutrients to be transported and broken down, it flushes toxins from the body, and it moderates the body's temperature. An animal could live without any food for a couple of weeks, but keep it from an adequate supply of water on a very hot day and it can succumb to heat stroke after only a few hours. Even during the winter, three waterless days can result in an animal's death. Water must be available every day!

Keeping Water Clean

One of the biggest challenges in the water department is keeping it clean. No animal wants to drink water that has been contaminated by manure or urine; yet if I had a dollar for every time we had to drain and clean a tank with manure or urine in it, I'd be a heck of a lot wealthier! Even if your water system is automatic, check your tank every day: The water should be clean, and there should be plenty of it. If you're going away, have a neighbor — preferably one who also keeps livestock — check the water.

The best tanks for small critters such as sheep or goats, or for just a couple of bigger critters, are probably the newer rubber pans. We use one that's about 25 gallons (95 L) in capacity. They can be set up to run with an automatic valve, so you don't have to keep refilling them, but if they become fouled they're quick and easy to dump and rinse. These tanks are great in freezing climates, because they can just be tipped upside down and stepped on — the ice simply pops out, even if the pan was frozen solid.

If you have a large herd of bigger animals, you'll need a larger stock tank, but again get the smallest one you can get by with so it will be easy to drain and clean. There are a number of brands out now that are made of rubber or hard plastic, so they won't rust out on you. If you do live in a severe freezing climate, the hard plastic units will break if they freeze solid, though some are designed to work with an external heater. Try to steer clear of the style of electric stock-tank heaters that are suspended in the water; these are expensive to run, and if an animal chews through the cord it creates a major hazard for both animals and people. Consider these alternatives for freezing climates:

1. Run a small trickle of water continuously — though this tends to make a major ice monster by the end of the winter.
2. Fill water tanks once a day during the coldest part of winter with just the amount of water that meets the animals' needs. The goal is for them to have all the water they need but to drink the tank almost dry each day. We successfully used this technique, filling the tank first thing every morning; our animals learned to come up and get a couple of good, long drinks as the tank was filling. With our size tank and our stock, we learned to leave the tank about half full after everybody got their first drink, so by late afternoon the tank was empty again.
3. Splurge on one of the newer insulated water-tank systems. Even in central Minnesota, where temperatures fall below −30°F (−34°C) for weeks at a time (our record was −49°F, or −45°C), on the thermometer), these units have a good reputation.

Galvanized tanks are readily available and come in a wide variety of sizes and styles, but they do tend to rust out. In Minnesota, our water was quite corrosive; galvanized tanks rusted after five years.

Pigs create their own problems with almost any water system. They love nothing more than to dump their water or roll in it; they get their water dirty almost as quickly as you clean it. We used one of the same rubber pans that I mentioned earlier for our pigs, but we set it inside a large steel container so that they couldn't tip it quite as easily. Some farmers set up water fountains for their pigs; others pour a concrete pad and tank (Figure 6.1). Poured-concrete tanks can work for any type of livestock, but if you do pour a tank and pad, make sure to roughen up the pad surface so it won't be slippery for hooved animals. Provide a way to drain the tank for cleaning — or you may be manning the bucket brigade more often than you'd like.

Composition of Feeds

Most livestock species are herbivores, so plants are their exclusive source of nutrition under natural conditions (though some commercial feed supplements may contain animal by-products, such as blood meal, dried whey, or meat and bonemeal). Pigs and poultry are an exception to the rule — like humans, they're omnivores, and at least a portion of their natural diet is composed of animal matter (which includes insects!).

Whether the source is a plant or an animal, there are certain basic properties that all feedstuffs share. They are made up of three major components: water, organic matter, and mineral matter or ash. If you took a sample of feed and ran it through a laboratory test to differentiate the three, you would weigh the initial sample, then bake it at 120 to 150°F (49 to 66°C). This first step would drive off the water, and the difference between the initial weight and the dried weight would give you the weight of the water. The amount of sample remaining would be the dry matter, and it is composed of both organic matter and mineral matter. The next step would be to burn this remaining sample at a very high temperature, which destroys

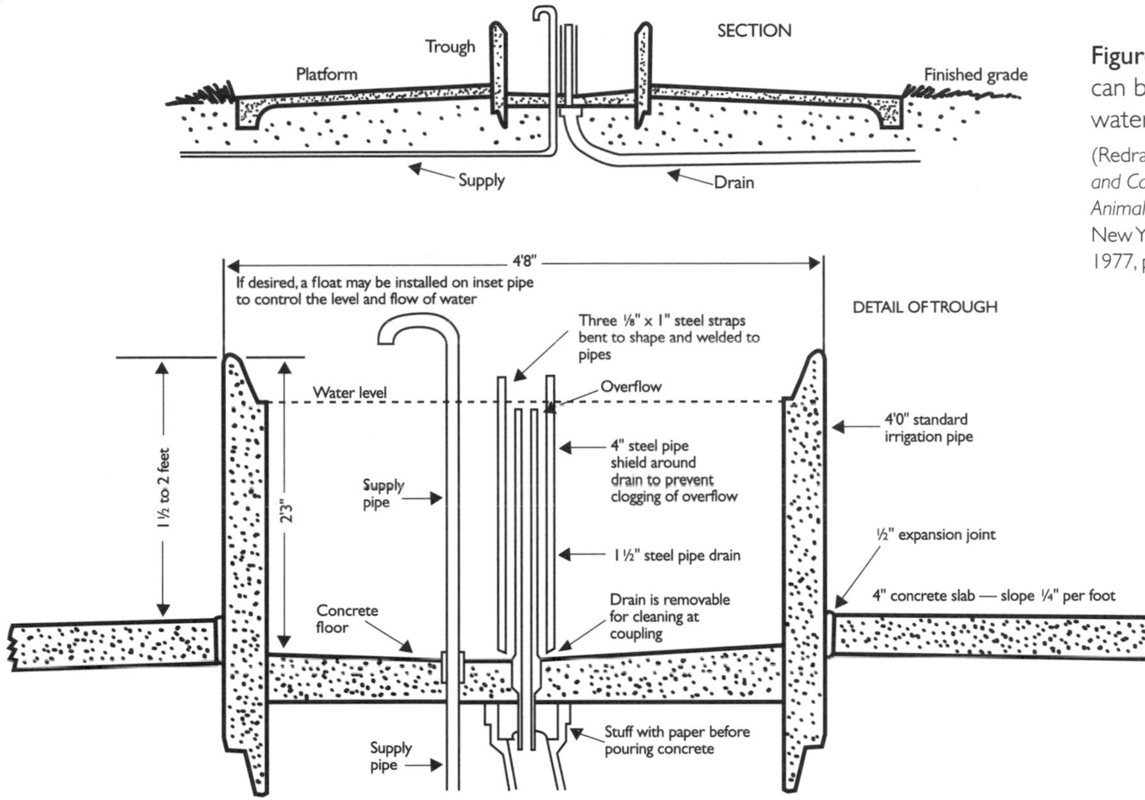

Figure 6.1. Concrete stock tanks can be used where a permanent water installation is desirable.

(Redrawn from Guy Lockwood, *Raising and Caring for Animals: A Handbook of Animal Husbandry and Veterinary Care.* New York: Charles Scribner & Sons, 1977, p. 37.)

TERMINOLOGY

▶ ***Balanced ration.*** A ration that provides all the nutrients, in the proper proportions (including energy, fiber, protein, vitamins, and minerals), for the animal's needs based on its age and its level of work.

▶ ***Concentrate.*** The grain or grains being fed as part of the ration.

▶ ***Dry matter.*** The mass of the ration or feedstuff if the water is "baked off." For example, a sample of mixed meadow hay might contain 85 percent dry matter, so your 60 pound (27.2 kg) bale of hay would actually weigh 51 pounds (23.1 kg) on a dry-matter basis (0.85 x 60 pounds).

▶ ***Energy.*** The part of the ration that is made up of sugars, fats and fatty acids, and starches used by the body for muscle and nerve activity, growth, fattening, and milk secretion.

▶ ***Feedstuff.*** Any food intended for livestock consumption.

▶ ***Fiber.*** The part of the ration that comes from cellulose and hemicellulose in plant matter; it is broken down in ruminants and horses to create additional sugars and fatty acids.

▶ ***Forage or roughage.*** The hay or pasture portion of the ration.

▶ ***Protein.*** The portion of the ration that contains amino acids, which are required by the body for cell formation, development, and maintenance, especially for muscle and blood cells.

▶ ***Ration.*** The combination of foods in a specific diet, for a specific animal or class of animals, at any given time. Includes everything the animal is receiving.

▶ ***Supplements.*** The vitamins, minerals, or protein being added to the ration.

▶ ***Total digestible nutrients (TDN).*** TDN is the portion of the ration that the animal actually is able to take advantage of. Feed reports, feed tags, and feed charts report the TDN of the feedstuff. If the TDN on the previous sample of hay was tested as 60 percent on a dry-matter basis, the bale would contain 30.6 pounds (13.9 kg) of digestible nutrients (0.6 x 51 pounds).

the organic matter. When the sample is weighed a final time, the remaining amount is equal to the mineral matter, or ash.

The *organic matter* can be grouped into four parts: the carbohydrates; the fats and fatty substances; the proteins or nitrogenous compounds; and the vitamins and minerals. All organic matter is principally made up of carbon, hydrogen, and oxygen, though other elements may be present (Figure 6.2).

Carbohydrates

Carbohydrates form about three-fourths of the dry matter in plants, so they are the most significant component of feed. The carbohydrate group can be broken down into sugars, starches, and fiber; the proportion of each varies according to the plant's age, environmental factors, and the type of plant.

The carbohydrates contain atoms of carbon (C) attached to molecules of water (H_2O). For example, a simple sugar molecule would contain six carbon atoms and six molecules of water ($C_6H_{12}O_6$). Starches are composed of groups of sugar molecules strung together. All animals easily digest both sugars and starches, so they provide a relatively high feed value. On the other hand, the fiber component is made up primarily of lignins and cellulose. The lignins are completely indigestible, and the cellulose, which accounts for about 50 percent of the organic carbon on earth, requires bacterial fermentation to break it down into usable sugars and starches. All animals are able to ferment a small amount of cellulose in their intestines, but only the ruminants, such as cattle and sheep, are able to convert the bulk of the cellulose in their diet to usable sugars and starches. A more detailed discussion on digestion by different species follows this section.

Fats and Fatty Substances

Like carbohydrates, fats and fatty substances are made up of carbon (C), hydrogen (H), and oxygen (O); however, the proportions of carbon and hydrogen are much greater than that of oxygen in a fat. For example, a common fat in plants is olein; its chemical formula is $C_{57}H_{104}O_6$, meaning that there are 57 carbon atoms, 104 hydrogen atoms, and 6 oxygen atoms in each molecule of olein. Despite our current fear of fat, it is an essential nutrient for all animals (especially young animals), including humans. Fat provides more than twice the energy than a carbohydrate provides, and it helps an animal maintain its body condition and temperature.

Proteins

Proteins are essential for the development of all cell walls. They're also critical in forming muscles, internal organs, blood cells, hair, horns, and bones. In most animals, protein accounts for 15 to 20 percent of the animal's weight (Table 6.1).

Unlike simple sugars, which may contain as few as 20 atoms, each molecule of protein is made up of thousands of atoms. In nature, as in construction, it's often easier to build a complex structure by using substructures, such as building blocks or prefabricated roof trusses. In the case of proteins, the building blocks nature has developed to simplify construction are called *amino acids*. There are many thousands of amino acids, but only about twenty are critical for protein construction. Words are built by varying the

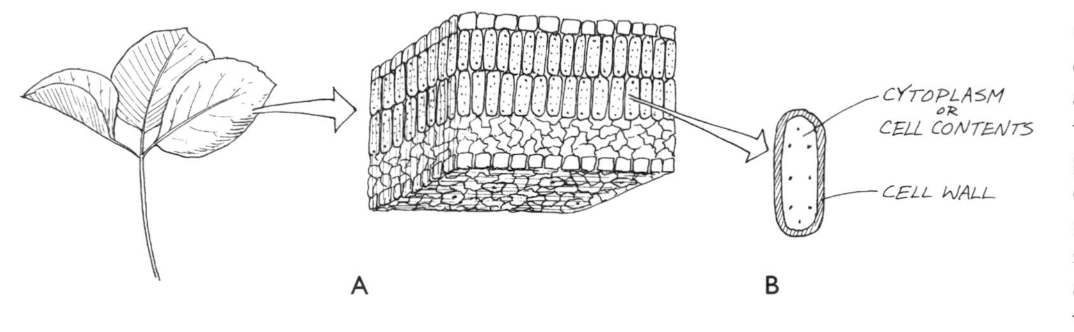

Figure 6.2. (A.) Chlorophyll-containing nutrient-rich cells are sandwiched between fibrous surface cells. **(B.)** The proportion of cell wall to cytoplasm is higher in older plants than in younger plants; in stem tissue than in leaf tissue; and in warm-season grasses than in cool-season grasses.

Table 6.1

Composition of the Bodies of Farm Animals

Animal, Weight	% Water	% Protein	% Fat	% Mineral
Calf, 100 lb (45.4 kg)	71.8	19.9	4.0	4.3
Fat steer, 1,200 lb (544 kg)	48.0	16.0	32.3	3.7
Dairy cow	56.8	17.2	20.6	5.0
Mature, horse	61.9	18.2	14.1	4.7
Lamb, 80 lb (36.3 kg)	50.9	17.4	24.9	4.2
Pig, 100 lb (45.4 kg)	66.8	14.9	16.2	3.1
Pullet, 0.5 lb (0.3 kg)	71.2	20.8	3.5	3.6
Pullet, 4 lb (1.8 kg)	55.8	19.2	20.0	3.1

combinations of the twenty-six letters of the alphabet, and similarly, proteins are built by varying the combinations of these twenty amino acids. As with fiber, the ruminants have a real edge when it comes to using all the amino acids and proteins that are in their diets.

Vitamins

Vitamins are organic in nature (they burn off in the laboratory sample with other organic matter), but unlike carbohydrates, fats, and proteins, there is no rhyme or reason to their structure: Each one is unique from the others in its chemical formulation.

Vitamins are required in very small quantities, but deficiencies of vitamins in the diet can result in a wide range of diseases, including rickets, anemia, and muscular dystrophy. At the same time, some vitamins, such as vitamin A, can be toxic if given in too high a quantity. Vitamin D deficiency is common in animals that are reared completely indoors; however, animals that spend at least some time regularly in the sun don't have this problem, because vitamin D is synthesized by the body when exposed to sunlight.

Minerals

The mineral, or ash, component of feed is what remains after the laboratory fires the feed sample at a high temperature. Like vitamins, most of the minerals aren't required in very large quantities, but deficiencies cause a wide range of health problems, and toxicity can occur when there are mineral excesses in the diet. The minerals include such elements as sodium, calcium, phosphorus, and selenium.

Mineral deficiencies (or excesses) usually occur where soil mineral imbalances exist. Plants that are grown in a soil that is either too low or too high in any given mineral will reflect the soil imbalance in their tissue. The best way to learn about soil mineral levels on your farm is to have soil and forage samples run (see appendix E, Resources). As an alternative, check with your local County Extension Agent, a reputable feed dealer, or your veterinarian for information on the general status of soil mineralization in your area. They can help you evaluate what types of mineral supplements will be best for your situation (your animals, your soil, and so on).

Mineral supplements are best fed free-choice. Animals are really quite efficient at controlling their intake of mineral supplements in order to meet their own needs. Our approach is to always have a free-choice plain white salt block and a free-choice trace-mineral block available. Another excellent source of vitamins and minerals that we put out is kelp meal. This dried sea plant provides a smorgasbord of vitamins, minerals, and amino acids that all animals seem to love.

Digestion

Some foods, such as sugar water, can be absorbed directly into the bloodstream, but most foods must be broken down into simpler elements and molecules prior to absorption. Digestion is the process by which this breakdown occurs, and it includes mechanical, chemical, and biological steps.

The first step in the process is always mechanical and involves the gathering, tearing, and grinding of food into smaller pieces with the mouth and teeth (or beak). Tearing and grinding food into smaller pieces allows it to pass through the esophagus, or gullet, to the stomach; it also increases the surface area of food particles, so that once they're in the stomach gastric juices can attack these particles more readily. The gastric juices, which consist of acids and enzymes, are responsible for most of the chemical breakdown of food. Fermentation is the biological process, and it involves the help of beneficial bacteria.

Animals may be classified according to the configuration of their digestive system. There are three major classes (Figure 6.3):

1. ***The monogastric fermenters,*** or single-stomached critters (e.g., pigs, poultry, humans).
2. ***The postgastric fermenters*** (e.g., horses, rabbits). Postgastric fermenters have only a single stomach, but they have a well-developed cecum — a fermentation chamber that's located between the small and large intestines. Although all animals have this organ, it is only well developed and highly effective in animals that are classified as postgastric fermenters.
3. ***The pregastric fermenters,*** or ruminants (e.g., cattle, sheep, goats). The ruminants have four stomachs, including the rumen, or first stomach, which acts as a large fermentation chamber. There is also a smaller class of pregastric fermenters that are known as *pseudoruminants* (llamas and alpacas). The pseudoruminants have three stomachs instead of four, but their digestive process is very similar to that of the true ruminants.

When food passes out of the stomach, or stomachs, it enters the intestinal tract. In all the species, digestion continues as the food winds its way through the intestines. The vast majority of the nutrients are actually absorbed into the bloodstream during the trip through the intestines, particularly the small intestine. That portion of an animal's diet that isn't absorbed into the bloodstream as a source of nutrients is waste, and is passed out of the body as manure.

Animals with a monogastric digestive system have some disadvantages when it comes to digesting their food: They can only ferment a very small amount of the fiber that's in their diet, and they can only synthesize a few of the many amino acids that their bodies require. This means that they get very little feed value out of hay or straw — though green grass, when it's young and vegetative, can provide a good deal of feed value to them. It also means that

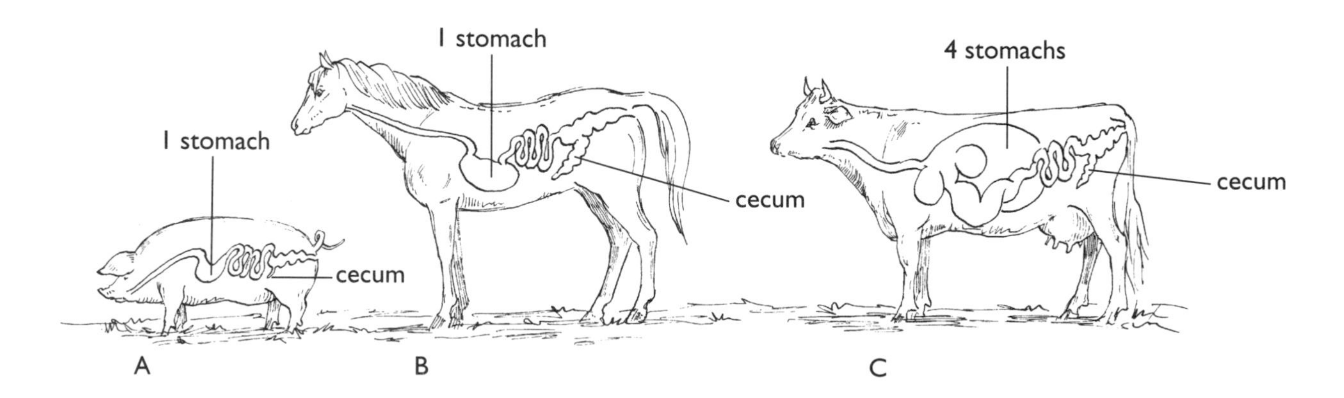

Figure 6.3. Mammals have three distinct stomach systems: the monogastric, the postgastric fermenter, and the pregastric fermenter. **(A.)** Pigs are monogastric, with a single stomach and a small, underdeveloped cecum. **(B.)** Horses are postgastric fermenters, with a single stomach and a highly developed *cecum,* or fermentation organ between the small and large intestines. **(C.)** The pregastric fermenters, like cattle, have four stomachs and a fairly well-developed cecum. Pregastric fermenters are also called *ruminants,* after the rumen, one of their four stomachs.

their diet must have a wider variety of protein sources, including animal proteins, to meet their amino acid needs.

When it comes to fiber digestion, the ruminants are the clear winners! The rumen allows food to be fermented very early in the process; by the time the waste leaves the body, almost all the cellulose fiber has been broken down into usable sugars and starches. Through bacterial and enzymatic action in their rumens, the ruminants are also able to synthesize all of their necessary amino acids from whatever nitrogenous compounds are present in the plants they eat.

The horse's cecum performs a similar function to a cow's or sheep's rumen; however, since the cecum comes fairly late in the digestive process, it is less efficient at both fiber digestion and amino acid synthesis. Even though ruminants and horses are physiologically capable of meeting most of their nutritional needs on a limited (grass) diet, there are times when their diet will need to be beefed up. Any animal that's at work has higher nutritional requirements; its rations must supply a significant level of both available energy and protein. I will discuss work and diet in more detail later on.

Ruminant Digestion

Ruminant digestion is an interesting process in its own right. A ruminant's mouth doesn't contain upper teeth. Instead, the top of the mouth is an extremely hard palate. A ruminant grabs a piece of food — say, a mouthful of grass — between its lower teeth and hard palate, and tears it off. The food is lightly chewed; just enough to moisten it a little and form a ball before it's swallowed. The ball of food, or bolus, travels down the esophagus and enters the rumen. Heavier food items such as whole grains (or stones, or pieces of hardware) usually bypass the rumen altogether and enter the reticulum, or second stomach, directly. Grain that bypasses the rumen generally remains intact, which is why you'll see some whole grain kernels in the manure if you're feeding whole grain. There are two ways to mitigate this: Feed ground or cracked grain, or feed the whole grains with hay or other light fibrous feeds, so it is captured in the rumen (Figure 6.4).

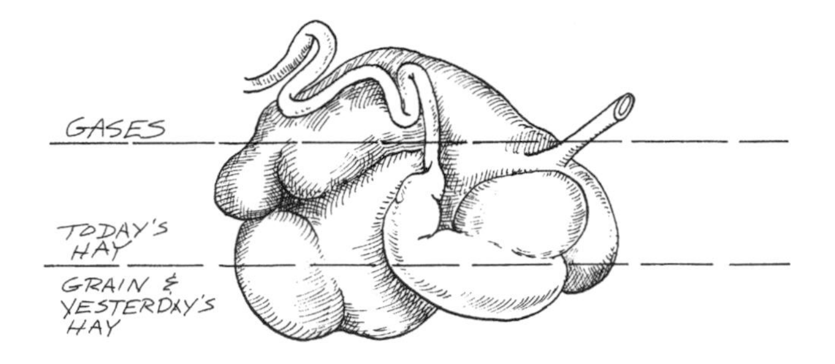

Figure 6.4. The rumen acts as a large digester, where bacteria and other microorganisms break down food. The top of the rumen is filled with gas, the middle contains recently eaten hay, which floats on the bottom slurry of yesterday's hay, grain, and fluid.

The *rumen* and *reticulum* are in fact an interconnected pair of stomachs that work together. They are sometimes referred to as the reticulorumen. The bacteria that are responsible for fermentation in the rumen live and reproduce in the reticulum.

Actually, saying that bacteria are responsible for fermentation is oversimplifying things. The rumen is populated by bacteria, protozoans, yeast, and fungi. For each gallon (3.8 L) of rumen capacity, there's up to 200 trillion bacteria and 4 billion protozoans. The quantities of yeast and fungi are more variable, but still number in the millions under normal circumstances. Now multiply these figures by the normal 25 to 30 gallons (95 to 114 L) of rumen content in a cow — or 3 to 5 gallons (11.4 to 18.9 L) in a sheep — and you'll get a good idea of just how large is the workforce that ferments food in the rumen. An important objective of a ruminant feeding program is to maintain an environment that is good for these microorganisms, so they can do their jobs well. A key approach to keeping happy microbes is to make any dietary changes slowly — say, over the course of two weeks — so the microbes can adjust to the change.

The rumen is like a boiling pot of water: There is always movement in it. But the movement takes place within layers. The top layer consists of gases (primarily methane, which is a by-product of the bacteria's life cycle). The bottom layer consists of a fluid mixture of grains and completely saturated roughage (yesterday's hay). The middle layer floats on the bottom layer, and is made up of newly eaten roughage. The

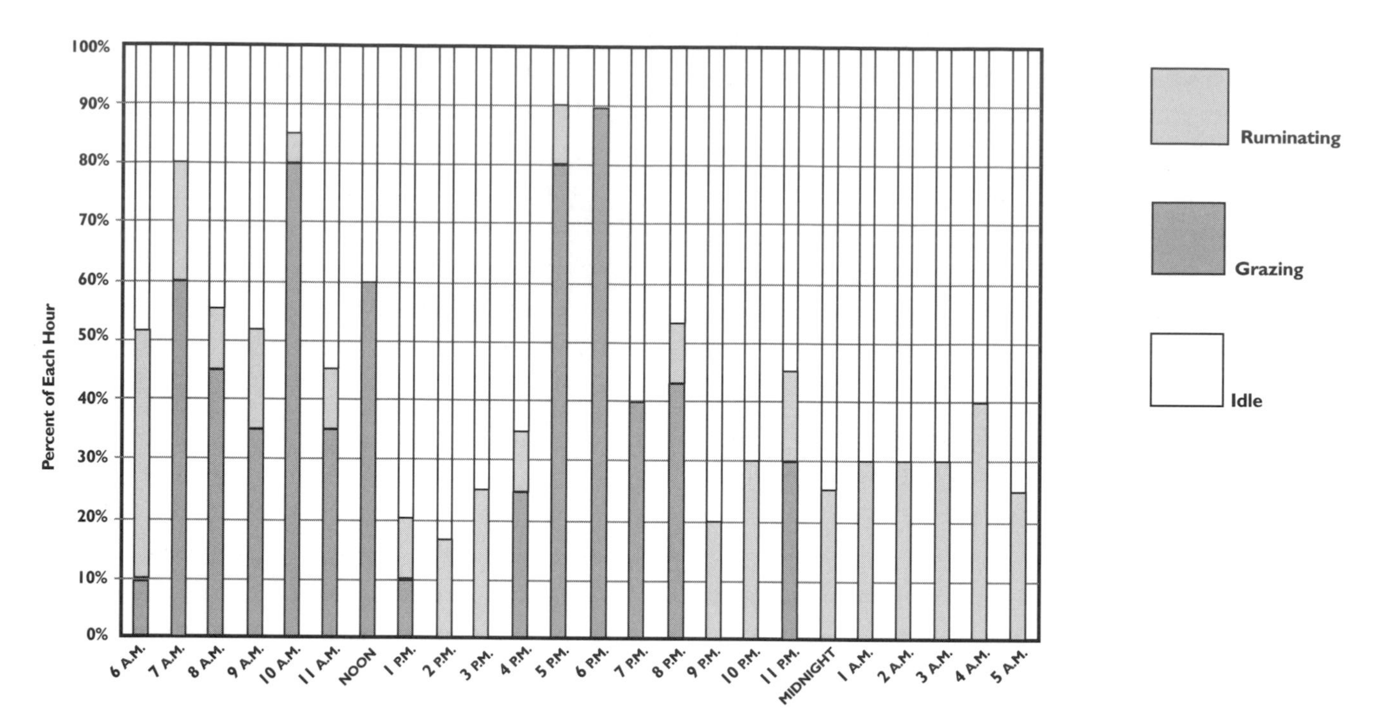

Figure 6.5. Ruminants have to spend a good part of each day chewing cud, or ruminanting. This chart shows the proportions of each hour during a day that cows eat, chew cud, and rest. For example, between 6 A.M. and 7 A.M., a cow grazes for 10% (6 minutes) of the hour, ruminates for about 42% (25 minutes) of the hour, and is "idle" for the remaining time (29 minutes). *Idle,* in this case, means time spent not eating or chewing cud. When idle, a cow could be sleeping, playing, grooming, or traveling.
(Modified from André Voison, *Grass Productivity.* Covelo, CA: Island Press, 1988, p. 70.)

rumen contracts every 1 to 3 minutes during the day, which helps keep everything in motion. These contractions work the heaviest materials into the reticulum, so they can continue their trip through the digestive tract.

Ruminants spend up to 6 hours per day eating, and up to 8 hours per day chewing cud, or ruminating (Figure 6.5). *Cud* is simply a bolus of food that has been floating in the rumen for some time, and is then forced back up the esophagus and into the mouth to be chewed again. Cud chewing serves two purposes: It provides some additional mechanical breakdown of food, and it provides for the introduction of large quantities of saliva into the rumen. During the course of a day, a mature cow produces up to 50 gallons (18.9 L) of saliva while ruminating, which helps neutralize acid in the rumen and maintain the rumen environment for the microbes that live there.

One side effect of fermentation in a ruminant is that a significant amount of gas is produced. In fact, a mature cow can produce up to 20 cubic feet (0.57 m^3) of gas per hour, which is enough gas to fill a balloon the size of a large chest freezer! This gas must be passed through regular belching, or eructation. Occasionally, sudden changes in feed will cause an animal to bloat. This is simply caused by the animal's inability to get rid of the gas it produces through belching, but it's far from a simple problem: It is a life-threatening condition. (See more about bloat in chapter 8.)

Food Requirements

Food needs vary. They change with age; they change according to the amount of work an animal is doing; they change with the changing seasons. One ration isn't always suitable for all animals, and the true husbandman must develop an eye for the condition of his or her animals. Feeds need to be adjusted according to the ever-changing needs and conditions of the stock.

Maintenance rations represent the food required to support the animal when it is doing no work, and producing no product (milk, meat, or fiber). A maintenance ration basically supports all minimum body functions, such as respiration and cardiovascular function. It maintains the animal's weight at a steady state: no gain, no loss. An animal kept on a maintenance ration wouldn't put on growth, reproduce, or do any form of work. Without an adequate maintenance ration, an animal slowly starves to death, eating away its own fat, muscle, and tissue until there isn't anything left to eat away. For most farm animals, up to one-half of their ration is strictly used to meet their maintenance requirement.

Providing adequate feed for growth, reproduction, and work is the goal of every farmer. And *adequate* is a key concept in my mind — you want to feed enough to keep your animals healthy and strong, but you don't want to overfeed them. Overfeeding not only costs too

FARMER PROFILE

Lanie Fondiler

Lanie Fondiler calls her farm in Westfield, Vermont, the Lazy Lady Farm, but her enterprise hardly fits her moniker. Since coming here in 1987, she has created a working farm operation, supporting herself off 35 acres, using goats and sheep.

"There was just a little cabin here when I bought the place," she says, "and the land was beat. The 12 tillable acres had been in corn for many years." The rest of Lanie's land is woods and brush.

Her first year, Lanie worked up 1g acres of garden and sold produce. That year she had to work a job in winter to support herself.

The following year, Lanie purchased her first livestock: seven ewes and their lambs. The second winter she used her wool to weave rugs for sale.

"I lived in France for a couple of years, which gave me the idea for doing goat cheese." Today Lanie's operation has both sheep and goats, with rugs, cheese, and specialty sausages as her main products.

Lanie spends about 60 hours per week milking the goats and making cheese, and another 30 at the farmers' market. She does all the other work of maintaining her farm, including tending her garden, caring for the animals, and weaving her rugs, in her spare time!

Her sheep are grazed through a series of paddocks. The goats have two paddocks near the barn, but they spend most of their time around the barn. As she says, "They're barn potatoes, and they'd be couch potatoes if I gave them one." To give them some exercise and let them eat some free browse, she has a few friends who stop by during the week to take the goats on walks through the woods and along the road ditches. These trips help cut down her feed bills.

Lanie sells most of her products at three farmers' markets: one in Montpelier, the state capital; one in Waterbury, which is a suburban bedroom community; and one in Stowe, a resort community that sees mainly tourists. She says, "Each market is different, with its own atmosphere and its own style of clients. The Montpelier market has lots of people who are involved with politics — lots of activists — these people have strong opinions and like to talk about issues. The Waterbury people are kind of low key, and small talk is the order of the day. Stowe is hardest for me, because the tourists are sometimes difficult to deal with. They'll try some cheese, and say things like, 'Well, it certainly doesn't taste like the cheese I had in France last year.' Now I've learned to just smile, and try to not let those type of people bother me, but at times it's real tough. I'm trying to learn how to schmooze with them, and I guess it is getting easier, because I haven't come home from there as stressed as I used to.

"I'd say the most important thing for really small farmers who hope to make a living at it is that you have to really know yourself. Know what you like, and try to figure out a way you can capitalize on it. Whatever you want to raise on this scale, you're going to have to sell. The best way to do that is to put a face behind the product. I have a much easier time selling at the farmers' markets, where my customers get to know me, than I have when I've tried selling through stores. And really, try to determine ways to save on your expenses. A 'penny saved' really *is* a 'penny earned' on a really small farm."

much, but it isn't good for the animal. An overly fat breeding animal has trouble getting bred, and young animals that are fed too much or too quickly grow so fast that they can develop severe joint and bone problems; this is especially true for young horses.

Appendix A has tables of feeding standards for various classes of livestock; appendix B lists the composition of common feedstuffs. Use these to calculate rations.

Economy Feeding

Grass is some of the most economical feed that "money can buy," but chances are it's not the only feed you'll ever need. If you keep pigs or poultry, they require a wider variety of feeds. If your plans include milking an animal — or many animals — you will definitely have to supplement them. If you are working horses, they too will require more than just grass and hay.

Your local feed dealer will gladly supply you with premixed, bagged rations for any class of livestock you're raising. Heck, if you buy through a Purina dealer you can purchase Rat Chow, Pigeon Chow, Trout Chow, or "Chows" for just about any critter you can think of. These prepared feeds are convenient, and if you don't need a great deal of supplemental feed during the course of the year they are probably worth going with. In quantity, they do get expensive, though.

One easy way to save money on feed purchases is to buy directly from area farmers during the harvest. Whether you're buying hay or grain, you can often find some good deals if you purchase the crop out of the field. This option requires adequate storage (see chapter 7). We often found a local farmer would allow us to take his grain wagon home, unload it, and return it as part of the deal.

Lawns, gardens, and orchards are great sources of feed. Use portable electric fencing to graze the lawn, or if you're worried about what the neighbors think, put a bagger on your mower and toss the fresh, chopped grass clippings over the fence. Garden waste can also be tossed over the fence, and windfall apples are like the finest imported chocolates to your animals! Limit the serving size of these delicacies at any one time, or digestive upsets may occur.

Pigs and birds, with their omnivorous diets, can eat many of the same items that we do, so table scraps don't go to waste. Thrown into the pig or poultry pens, these little treats disappear quickly.

Sometimes, "waste" feeds can come very cheaply, or free. Mary, who owned a natural foods store and deli near us, would save us goodies. Cheese or lunch meat that wasn't quite fit to sell anymore (the odds and ends, or the slightly dried-out stuff) would go into the freezer for my next visit. Freezing prevents the food from going rancid, since the idea is to feed waste — not spoiled garbage! Other businesses that might provide usable food waste include processors, bakeries, groceries, and restaurants. The one caveat: Pick up your freebies on a very regular basis, so your supplier isn't inconvenienced.

Poisonous Plants and Feeds

It is tragic to lose an animal, or many animals, to food poisoning, but each year it happens to farmers around the country. Unfortunately, many poisoning agents have no antidotes, so prevention is truly the best medicine.

The first step to take so that this doesn't become a personal tragedy is to learn what poisonous plants grow in your area. This is another time your County Extension Agent should be able to help you; he or she will know what offensive plants grow locally, what they look like, if their toxicity is strictly seasonal or continuous, and to which classes of stock the plant is toxic.

Some crop plants are also poisonous at certain periods in their growth; for example, Sudan grass and sorghums can cause prussic acid poisoning if consumed while the plant is immature, or immediately following a frost. Grazing of these plants at a stage when prussic acid levels are high can result in very quick death (Table 6.2).

Animals often become ill due to spoiled feed, and it isn't always easy to tell that the feed is spoiled. Mold is the most common cause of spoiling in feed. Cattle tend to be less sensitive to mold than horses, sheep, or pigs, but all species can develop health problems. Though mold itself isn't generally fatal,

Table 6.2

Seasonal Toxicity of Plants

Name	Season of Toxicity	Animals Most Affected	Habitat	Effects	Comments
Bitter weed *Hymenoxys odorata*	Spring	Sheep, cattle	Flooded areas, overgrazed range	Vomiting, green nasal discharge, anorexia	Toxin is cumulative — avoid overgrazing
Water hemlock *Cicuta* spp.	Spring	All	Open, moist to wet areas	Salivation, muscular twitching, dilated pupils	Generally fatal, and extremely toxic
Larkspur *Delphinium* spp.	Spring, fall	All	Either cultivated or wild; wild usually in open foothills and meadow among aspen or poplar stands	Arched back, falling, constipation, bloat, vomiting	Moderately toxic, causes death in some cases; young plants and seeds are most toxic
Pokeweed, poke *Phytolacca americana*	Spring	Pigs most affected, also cattle, sheep, horses, humans	Disturbed areas with rich soil, pasture, waste areas	Vomiting, abdominal pain, bloody diarrhea	Mildly toxic in small doses, but may result in death when consumed in large quantities
Cocklebur *Xanthium* spp.	Spring, fall	Pigs most affected, but also all other animals	Fields, waste places, edges of ponds and rivers	Anorexia, depression, vomiting, weakness, muscle spasms	Extremely toxic
False hellebore, corn lily *Veratrum* spp.	Spring	Cattle, sheep, fowl	Low, moist woods and pastures; mountain valleys	Vomiting, excessive salivation, cardiac arrhythmia, muscle weakness, paralysis, coma, birth defects in offspring of dams who have consumed it	Moderately toxic
Buckeye, horse chestnut *Aesculus* spp.	Spring, summer	All grazing animals	Woods and thickets	Depression, incoordination, twitching, paralysis, death	Young shoots and seeds are highly toxic, otherwise moderately toxic
Oaks *Quercus* spp.	Spring, summer	All grazing animals	Deciduous woods	Anorexia, rumen stasis, constipation followed by black diarrhea, dry muzzle	Moderately toxic; diet must consist of over 50% buds and young leaves for an extended period
Mesquite *Prosopis glandulosa*	Summer, fall	Cattle and goats; sheep are resistant	Dry ranges in brittle areas	Chronic wasting, excessive salivation, facial tremors	Moderately toxic; animals must graze for extended periods; mixed-species grazing reduces losses
Yellow star thistle, yellow knapweed *Centaurea solstitialis*	Summer, fall	Horses	Waste areas and roadsides	Involuntary chewing movement, twitching of lips, inability to eat	Moderately toxic; horses only graze it when there is a lack of other forage; death results from extended period of consumption
White snakeroot *Eupatorium rugosum*	Summer, fall	Sheep, cattle, horses	Woods, cleared areas, waste areas, moist rich soils	Weakness, trembles, weight loss, constipation	Extremely toxic, often resulting in death; toxins may be passed to humans through milk from affected animals
Chokecherry, cherries, and peaches *Prunus* spp.	All seasons	All grazing animals	Waste areas, orchards, fencerows, dry slopes	Excitement leading to depression, incoordination, convulsions, bright pink mucous membranes; bloat	Extremely toxic; generally fatal in less than 30 minutes; animals that live two hours normally recover
Milkweed *Asclepias* spp.	All seasons	All	Waste areas, roadsides, around woods	Staggering, bloating, dilated pupils, rapid and weak pulse	Moderately toxic, but may result in death if sufficient amount is consumed
Jimsonweed, thorn apple *Datura stramonium*	All seasons	All	Fields, barn lots, trampled pastures, waste areas on rich bottom soils	Weak and rapid pulse, dilated pupils, coma	Extremely toxic; generally fatal within 48 hours

This is by no means a complete list. In fact, it is estimated that there are about three hundred plants in North America that are capable of causing poisoning in animals at some point during their growth cycle. Some are as common, and widely used in pasture, as ryegrass. In the case of ryegrass, the problem occurs in seed heads that become infected with bacteria or fungi. (Modified from *Merck Veterinary Manual,* 8th ed. Rathway, NJ: Merck, 1998.)

some molds produce mycotoxins as a by-product of their life cycle, and these can be fatal. In other cases, the plant that has mold growing on it produces a toxin (called a *phytotoxin*) in response to the mold.

Mold develops on moist feed. If hay is baled damp, or if it's rained on after it's baled, it becomes moldy. Grains that were stored before they dried down sufficiently can also become moldy. Feeding moldy feeds should be avoided; however, if you have some that isn't badly molded and you need to use it up, feed little bits along with some good feed. For mold, "dilution is the solution."

Either molds or fungi can produce mycotoxins and phytotoxin responses. They are common throughout the world, and can be found on both stored feeds, and feeds in the field. Fescue poisoning, a common problem in many southern and western states, is the result of a phytotoxin. Again, talk to your County Extension Agent or your veterinarian to learn if mycotoxins or phytotoxins are a possible problem for your operation.

Feeding Babies

The best feed for baby animals comes right from their mothers. But at times, farmers need to step in and hand-feed little ones. In commercial dairy production, almost all calves (or kids) are bottle babies.

When bottle-feeding, the rule of thumb is to provide 10 percent of the baby animal's body weight per day in whole milk, preferably from their own species, but for most babies goat's milk works well, and cow's milk will get you by in a pinch.

If milk is unavailable, there is always commercial milk replacer. These products are made for most classes of livestock (and even for dogs and cats). Calf milk replacer is readily available from feed stores, but replacers for other species may have to be special-ordered. Look for unmedicated milk replacer. As Dr. C. E. Spaulding says in *A Veterinary Guide for Animal Owners*, "There [are] not enough antibiotics in a pound of feed [medicated milk replacers] to prevent scours or other diseases, and the antibiotic fed daily can damage enough of the natural and necessary bacteria in the gut to cause scours." Also, look for milk replacer that lists milk as the first ingredient; the cheap brands are often made with no milk at all.

Don't feed babies more than they are supposed to have just because they act crazy when their milk is gone. If possible, feed them smaller milk rations often throughout the day, rather than in one or two big feedings. Don't feed out of a bucket, because they inhale

Figure 6.6. Nipple barrels allow you to feed many calves at one time. The nipple design is actually superior to bottle feeding for calf health, as the calves suck milk up through a tube, which is closer to a calf's natural sucking behavior. Calves need to be trained to the barrel, which usually takes about 3 days, but once they get the hang of it, you can't pry them off! Keep similar-sized animals grouped together on one barrel.

the milk too quickly — use a nipple appropriate to the animal's size. Some nipples are designed to be used on a bucket or barrel and require a suction hose (Figure 6.6); these are convenient (especially for feeding groups of babies) and offer some advantages, but they aren't available everywhere (see appendix E, Resources). Most nipples are designed for use with individual bottles, and you can purchase these in almost any farm supply store.

Start babies on dry food quickly. If they aren't on pasture, offer a small pile of hay for them to nibble at right away. Whole oats are an excellent starter food, though for the first few days you may have to teach them to eat the oats. After feeding their milk ration and while they're still smacking their lips together, place a small handful of oats in their mouth. They may dribble out more than they swallow in the first day or two, but then they'll start to eat the oats. Once they get the hang of eating the solid food, leave a little pan out for them to eat free-choice after their milk.

It is crucial that all newborns receive the colostrum from their mother, within the first 24 hours of life. *Colostrum* is the first milk a mammal secretes after it gives birth, and it jump-starts the baby's immune system (see chapter 8). If for some reason, as with an orphaned animal, the mother's colostrum isn't available, there are commercial colostrum products you can purchase from your vet, or you can feed colostrum you have saved (and frozen) from other animals in your herd. It's best if the colostrum comes from the same species, but in a real pinch you can use cow's colostrum for other species. Most dairy farmers keep some in their freezer, and will usually share in an emergency.

CHAPTER 7

Facilities

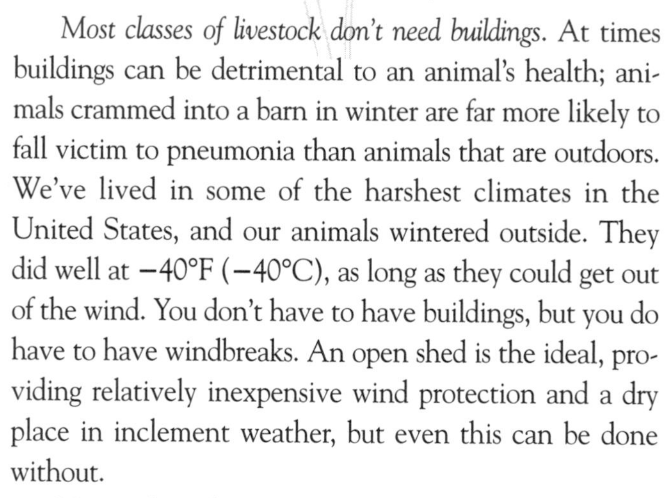

As my father always pointed out to me, farms and ranches that made a good profit were never flashy. They didn't have white fences. They were rather threadbare, austere, and rostered not one more asset than was absolutely necessary. . . . The problem for most farmers and ranchers is there is too much of their wealth invested in things that "rust, rot, and depreciate" and not enough invested in things that grow, add value, and reproduce themselves.

— Allan Nation, *The Stockman Grassfarmer* (September 1995)

LIVESTOCK FACILITIES include housing, fencing, storage areas, and so on. Good facilities can make your life a good deal easier, but there are lots of things that fall into the facilities category that you can live without!

Buildings

Buildings are one of the things that you can live without. Don't get me wrong, a nice barn can be a great asset, but you don't have to have a barn to keep livestock. In fact, for the small-scale farmer a conventional barn may just be a liability, because it's another building that must be maintained.

Most classes of livestock don't need buildings. At times buildings can be detrimental to an animal's health; animals crammed into a barn in winter are far more likely to fall victim to pneumonia than animals that are outdoors. We've lived in some of the harshest climates in the United States, and our animals wintered outside. They did well at −40°F (−40°C), as long as they could get out of the wind. You don't have to have buildings, but you do have to have windbreaks. An open shed is the ideal, providing relatively inexpensive wind protection and a dry place in inclement weather, but even this can be done without.

Now, what about the other classes of livestock — those that do need buildings? Small animals, like chickens, ducks, or rabbits will need some type of structure for protection from both the elements and predators, at least at night. For many wild creatures, it takes but a few nights to realize that a free dinner can be had at the farmstead.

Baby animals born in inclement weather must have protection from the elements; if you're hell-bent on lambing in January or February, or calving in February or March, plan on having a building!

Unless you get a barn as part of the deal with your farm, don't set your mind on the notion that a barn is the only type of building that will do. For poultry, a small traditional henhouse will work, or you can build a chickenmobile, which will let you move your chickens around

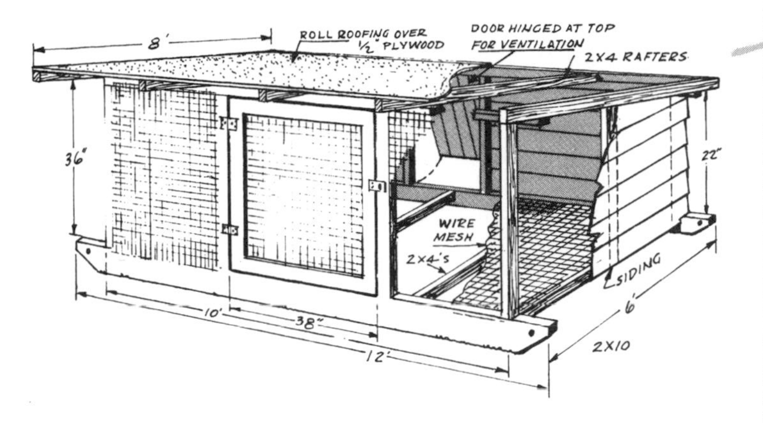

Figure 7.1. Portable chicken houses are a great way to move chickens around the farm, providing clean ground and access to new feed. Typically, chickens spread out in an area of about 150 feet (45 m) around their living quarters, so each time you move portable housing, allow at least 150 feet (45 m) from the last spot the house was parked.

your pastures for maximum productivity (Figure 7.1). Rabbits can be raised in individual hutches or in a pasture rabbit cage similar to the chicken-mobile.

Alternatives to conventional housing include straw bale structures, hoop houses, and small metal or plastic hutches; even tents or tepees may work.

At Little Wing Farm, an old dairy barn came with the land. This was the classic style of barn that you see throughout the Midwest; it was red with a large haymow above the main level, and a silo butted up against the side. The barn was still functional but needed lots of work, including replacing part of the metal roof, upgrading the wiring, and repairing the concrete floor. The milk room fixtures (compressor, vacuum pump, and stainless-steel bulk tank) were in working order, but the walls had rotted out.

Our original decision to milk cows was based, in part, on the fact that we had a dairy barn. We did some remodeling, fixing up the milk room to please the inspectors, and we built a three-cow, stanchioned milk parlor in the front of the barn. By making a parlor in front, we were able to bring the old barn up to grade-A standards without remodeling the entire place. We installed a pipeline for milking, so milk went from the cow to the bulk tank with no handling in between.

The back two-thirds of the barn remained much as it was when we first bought the place. We didn't regularly

THE CHICKENS AND ELAINE

All grade-A dairy barns have to be inspected twice a year (grade-B facilities are inspected once a year) by a state or federal inspector. For the first few years we milked cows, our inspector was Elaine.

Elaine was much feared in our neighborhood. You knew right away when she came through the area; her name could be heard on the lips of most patrons at the local café, and her name was generally accompanied by those colorful epithets that aren't spoken in church.

Elaine saw our chickens scratching around out in the back of the barn on one of her first visits and said, "Oh no, those chickens must go!"

"Well, Elaine," I said, "can you show me where it says in the regulations that my chickens can't live in the back of the barn? After all, I understand the regulations to say there can't be any other animals in the milking area, but that's up front in the parlor — not back here."

Elaine sputtered something about not having a copy of the regulations handy. Much to her chagrin, I pulled one out of my desk for her to refer to. She began leafing through it and muttering to herself. Finally, she said, "It's our interpretation that the chickens can't be back here."

Poor Elaine! She didn't know what to do when I said, "Elaine, you're paid to enforce the regulations, not interpret them. Show me where it says the chickens can't stay, or let it drop!"

The chickens stayed. I became a small legend at the café. And the moral of this story: If you are dealing with bureaucrats, get yourself a copy of the actual regulation that they are enforcing. What the regulation says you have to do, you have to do, but — that's all folks! There's more on this topic in chapter 10.

use it for any livestock, except the chickens, which were confined in it during the winter and slept in it during the warmer months. We would, from time to time, use it as a hospital area for sick or hurt animals, or to protect a mother with a newborn baby.

In contrast to Little Wing Farm, Bull Springs Ranch has no barn. A small old henhouse is the sole building that remains to attest to this place's agricultural past. Oh, and miles of ancient barbed wire also endure, testimony to some bygone rancher's labors.

Ventilation

If you have a barn or are building one, be sure to provide adequate ventilation. Drafts kill —ventilation *saves*. Windows that are up high on the wall and open in from the bottom provide excellent ventilation without drafts (Figure 7.2). In a large barn, ventilation flues along with the windows allow for good air movement. For small buildings, screened openings directly under the peak of the roof should do the trick.

Sanitation and Deep Bedding

When animals are housed indoors for any length of time, sanitation becomes critical. There are two basic approaches to indoor sanitation: Clean up all manure, urine, and bedding each day, and cover the floor with a fresh layer of lime and a little bit of bedding, or the method we prefer — deep bedding. With deep bedding, build up an initial layer of bedding (e.g., straw, wood shavings, sawdust, shredded paper or newsprint, dried leaves in fall) about 6 inches (15.2 cm) deep. With small critters like rabbits or chickens, every week or two the surface areas that are damp or full of too much manure (directly under the chickens' roost, for instance) will need to be shoveled out. Then add enough fresh bedding for the entire area to again be dry and clean. When using the deep-bedding method with large animals, you need to do the pitchfork work every day or two, but it usually only takes a few minutes; then add the dry bedding.

Once every year or two, you'll need to clean out all the bedding, down to the floor, and start over. This material is already partly composted! If you build a compost pile out of it and let it go one more year, you'll have the best organic soil amendment imaginable.

Deep bedding not only gives you a great compost product and cuts down on daily cleaning chores during the year, but it also keeps your animals clean, healthy, and happy. The deep bedding provides a nice soft, warm cushion for them to lounge around on. We had a first-calf heifer that gave birth to a very large calf in the barn. She became a "downer cow," unable to stand up initially because of the difficult birth. We called our veterinarian to get his advice, and when he heard she was in our barn on the deep bedding, he told us to just give her a day or two, rolling her over once or twice a day. Sure enough, the next morning she was up and about on her own. Had she been lying on a concrete floor, her condition would have been aggravated.

Another benefit of deep bedding is that it absorbs the urine very well, so less ammonia is generated. Ammonia fumes make the work environment unpleasant for you and cause a broad range of health problems

Figure 7.2. When animals are kept in a building, good ventilation is crucial to their health. Windows that are situated high on the wall and that are hinged at the bottom so they can open in and down are good. For very large structures, ventilation flues work well. The flue acts like a chimney, moving air up and out of the barn.

PEST CONTROL IN FACILITIES

I have a bad attitude toward flies, especially when they're buzzing around where I'm trying to work. I asked Roger Moon, an entomologist at the University of Minnesota College of Agriculture, to share with us some techniques for dealing with pests both in and out of the barn. He came to the farm for an afternoon and had plenty of nonchemical suggestions; some may help you also:

- Screen openings.
- Close doors and shade windows (old feed sacks work well both for shade and as a screening material); most bugs will not be interested in entering a darkened area.
- Use traps: sticky tapes or the jars with an attractant in them.
- Build physical traps in walls. These work because the flies head toward the light.
- Use sawdust or wood shavings for bedding material instead of straw. The wood by-products have a chemical compound in them that bugs don't like. The other good point about this option is that sawdust is actually far more absorbent than straw.
- Use natural organisms that control flies. Muscovy ducks and guinea fowl both eat flies and larvae. Chickens break up dung piles and eat larvae. Dung beetles and other beneficial insects can be purchased from hatcheries.

for your animals, including pneumonia and other respiratory problems. Ammonia fumes have also been known to cause blindness.

Sweden has some of the toughest animal welfare laws of any country in the world. To comply with these laws, Swedish farmers raising animals indoors *must* use deep bedding.

Windbreaks

As the temperature decreases, the energy an animal needs to survive increases but this demand is easily met with sufficient feed. However, when the temperature decreases and the wind speed increases, the additional energy requirements become more than the animal can easily make up. Windbreaks cut down the wind speed.

Windbreaks can be either natural or human-made. A thick grove of trees or a low willowy area provides excellent wind protection. A planted windbreak should contain plants of varying heights, from brush up to tall trees (Figure 7.3).

The sides of existing buildings can sometimes provide wind protection, but at other times they exaggerate the effects of the wind. The ideal human-made windbreak has spaces between the boards. These spaces slow the speed of the wind as it passes through them, allowing force to be significantly reduced (Figure 7.4). An L- or T-shaped break provides the best protection. Animals can move around the windbreak to always be on the lee side.

Handling Facilities

Unless you plan on keeping just one or two very petlike animals, you'll be well advised to develop some handling facilities early in your operation. These don't have to be fancy, and they don't have to be expensive, but the time and money you put into developing them will save you a great deal of grief, and possibly some money to boot.

The handling facility should be designed to make delivery and loading of animals easy on you, and easy on the animals. It also makes sorting animals simpler.

In Minnesota, we designed a system of gates and doors in our barn and silo room (which was attached to the side of the barn) for this purpose. It worked for moving animals in and out, and provided a controlled environment when animals were first dropped off, but it wasn't easy to move new animals through and didn't

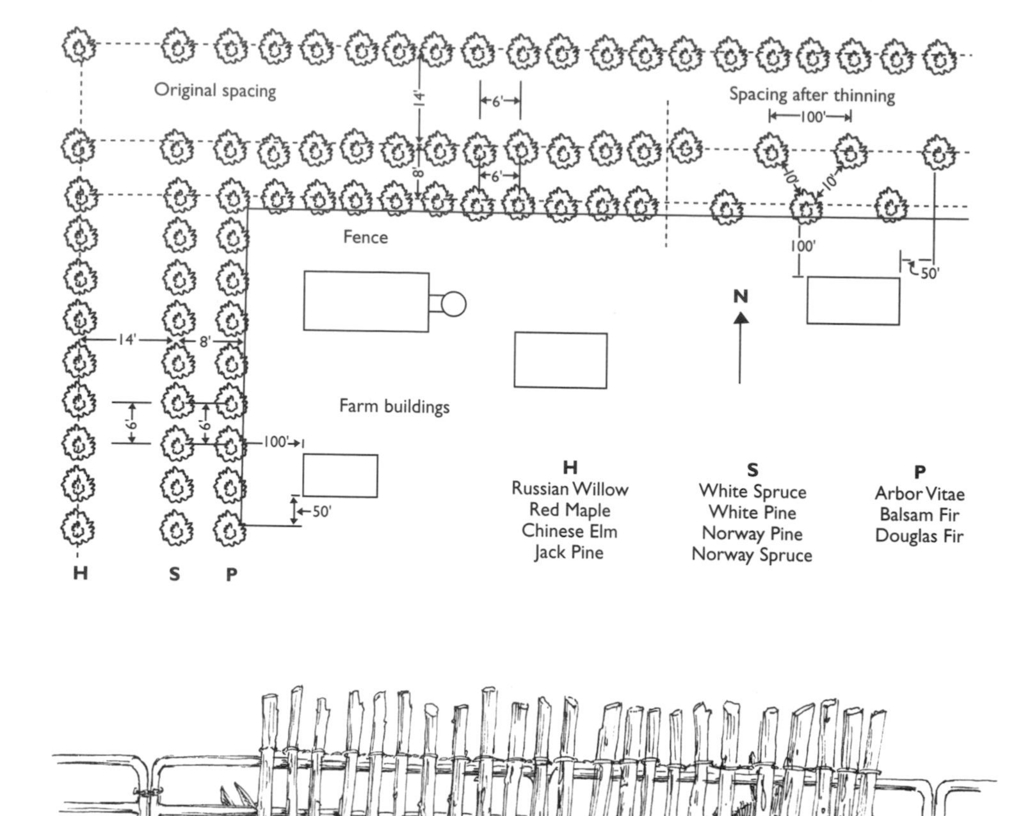

Figure 7.3. Planted windbreaks should be composed of several rows of mixed vegetation, including both brushy types of plants and tall evergreen species. County Extension Agents are a good source of information on which plants will grow well in your area.

(Redrawn from M. G. Kains, *Five Acres and Independence*. Greenberg, NY: Peter Smith, 1935.)

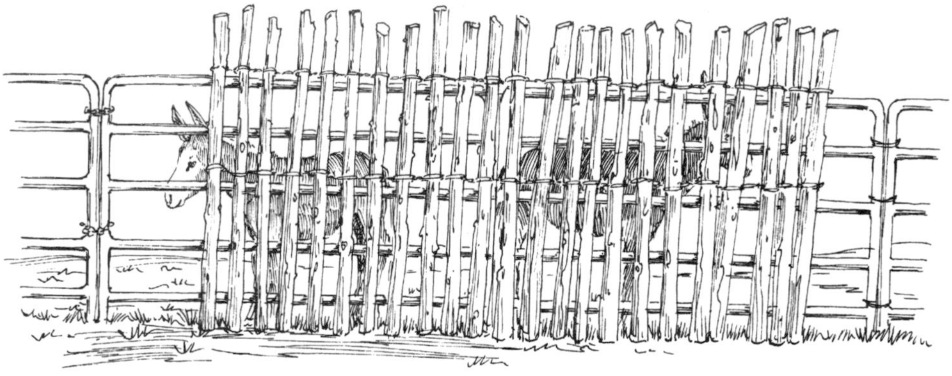

Figure 7.4. Man-made windbreaks can be used where no natural windbreaks are available. This windbreak is simply made with dead aspen (poplar) logs laced to stock panels. It's cheap and portable, so as we move stock to different paddocks, we can move the windbreak.

provide a separate holding area to isolate new arrivals. The next time around, we splurged and purchased stock panels and gates. The portability of these panels provides a great deal of flexibility while meeting the demands of a handling facility.

Handling facilities should be designed so that a truck or trailer can easily back up to a loading chute. These chutes should be just wide enough for large animals to fit in one at a time. The pen that feeds or receives animals from the chute should be curved (Figure 7.5).

Operations that move lots of animals in and out, such as stocker operations (all animals are brought in when the grass begins growing and sold as the grass becomes dormant), benefit from solid chutes and pen walls. If the animals can't see things going on around them, they calm down quickly. Such operations may also benefit from a scale incorporated into the loading chute, so animals can be weighed coming and going.

If only a few animals are brought in or out each year, slatted sides are fine. The chute itself should be constructed of steel or heavy wooden boards, but pens can be made of just about any type of fencing material. Just make sure it's well constructed, because when animals first arrive, they're scared and excitable — a combination that's hard on fences.

When new animals arrive, let them stay in the handling pen for a few days with food and water supplied right there. This gives you some time to observe them for signs of illness, which often shows up quickly after the stress of moving. If illness does appear, the chute comes in handy for performing veterinary proce-

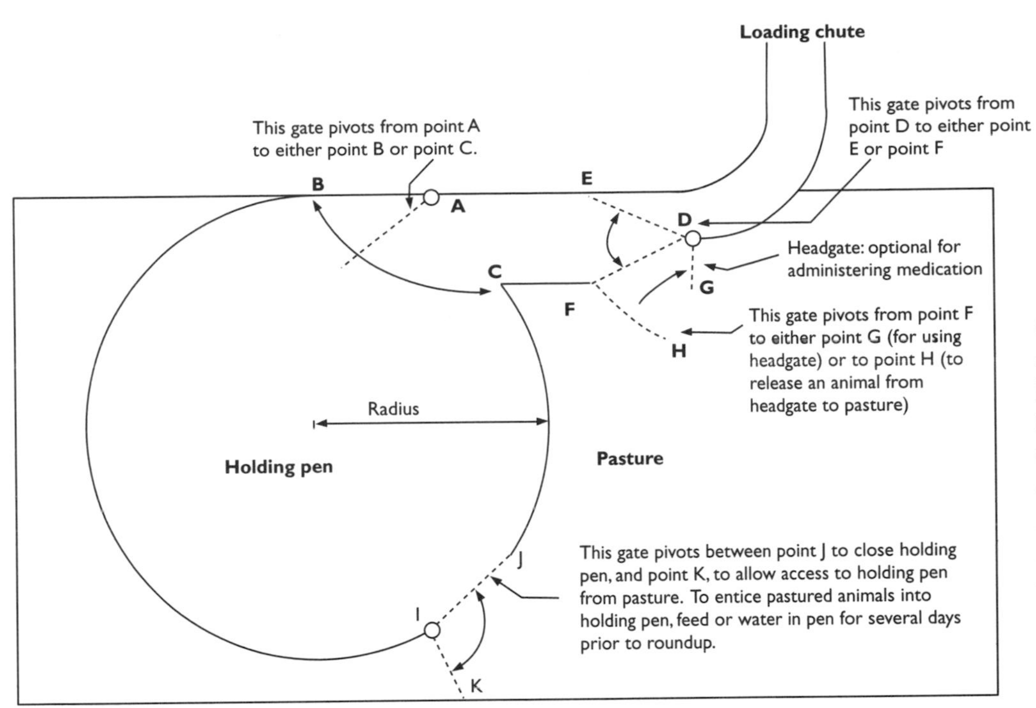

Figure 7.5. A curved holding pen can serve many purposes. It can be used as a training ring for horses, it can be used to isolate new or sick animals, and it of course serves well for shipping and receiving animals. The curved design keeps frightened animals from bunching up in corners and injuring each other. If lots of new animals come and go during the year, then make the sides solid.

dures, and the pen allows you to keep the illness isolated from your existing stock. Our Minnesota barn didn't provide us with an isolation facility, and we brought disease onto our farm with a load of calves we bought. Isolation of that one pen of calves would have saved us thousands of dollars and some terrible stress.

SIZING HOLDING PENS

The size of a holding pen depends in part on what type of critters you'll primarily be holding, and how many you expect to hold at one time. Large animals (cows, horses, buffalo) need at least 20 square feet (1.9 m^2) per animal, and smaller animals (sheep, goats, pigs, llamas) should have 7.5 square feet (0.7 m^2) each. Mother animals with babies should have an additional 5 and 2 square feet (0.5 m^2 and 0.2 m^2), respectively. The formula for the area of a circle is 3.1417 x radius^2. Using this formula, the area of a holding pen with a radius of 15 feet (4.6 m) would be 707 square feet (65.7 m^2) and would hold 35 cows, 28 cows with calves, 94 ewes, or 71 ewes with their lambs.

Sacrifice Areas

To a grass farmer, a sacrifice area isn't a place to take animals for a ritualistic slaughter; it's a place where the grass is going to get too beat up to ever really grow well. It is the spot where you feed hay during winter storms, or where you feed during the spring months if the fields are so wet that grazing them would destroy the sod.

With some planning, the sacrifice area can be designed to serve multiple goals. One paddock that contains the windbreak or an open-sided shed could serve as the sacrifice area. For a farm with horses, a training ring could serve as a sacrifice area. If new animals aren't added to an operation very often, the handling facilities can provide the sacrifice area.

Storage

Having critters around probably means having feed around. And feed is one item that needs to be stored well, or it spoils. Small, square bales of hay stacked outside, especially in a humid climate, become junk in short order. Grain, too, needs to be kept dry, and should be stored so as not to attract varmints. Rats have a habit of moving in where grain is easily accessible.

TOOLS YOU NEED

Some tools are absolutely necessary for small-scale livestock farming, including pitchfork, flat shovel, pointed shovel, wheelbarrow, hammer, pliers or wire cutters, buckets, posthole digger, and fence-post driver.

Beyond this group, there are lots of tools that come in handy but that you can do without. Too often beginners go to the farm or ranch supply store and buy everything in sight; unless you have lots and lots of cash, ask yourself if you really have to have an item before you buy it.

Storing hay to maintain its quality can be done in several different ways; the main thing to remember is that you want it covered. Tarps work and they're cheap, but they aren't a good long-term option. They tear, or blow away, unless very well secured in the first place; and sunlight eventually rots them, so they only last a season or two. One thing that can help is weighing tarps down well with old tires or blocks, or using tent stakes to tie them down (Figure 7.6). When using tarps, don't tightly enclose your whole pile of hay, or any moisture that is in the bales will cause them to rot in no time. Hay needs to have some air movement in and around the pile. Small hay piles benefit from being stacked on wooden pallets; this keeps soil moisture from ruining the bottom bales.

If you're dealing with large quantities of hay, a pole shed with just a roof, or a roof and one wall that blocks the predominant wind, protects hay well yet lets the air move around it. These structures are also relatively inexpensive to build (Figure 7.7). Hoop houses are also good for hay storage. Hay that is being stored in any type of enclosed structure must be adequately dry, or it may start on fire. In the Midwest, there are barn fires every summer and fall from hay being put in the mow a little too damp; as the hay cures, it heats up, and sometimes it can heat up enough to spontaneously combust!

Livestock kept with grass-farming strategies don't generally need grain in such large quantities that bulk storage space is necessary. The best way we've found to store the small amounts of grain we feed is in garbage cans or 55-gallon (208-L) drums. (Metal or plastic is fine, but make sure the drum contained food-grade materials before: You don't want to poison your stock.)

As with hay, grain must be well dried before it goes into storage. Dampness breeds mold (and possibly mycotoxins) and fire. Grain or premixed feeds that are being purchased from a reputable feed dealer should always be adequately dry, and most of the time area farmers whom you buy from directly won't be selling you wet feed. But when you buy grain at harvesttime, make sure the grain has been tested for moisture. Most feed stores can provide this service, or you can test your own sample. (For instructions, see page 177.)

Supplies and tools can usually be stored in a corner somewhere, but it helps if there is some kind of method to your storage. Vet supplies are a necessary evil (chap-

Figure 7.6. Tires are good for holding down hay tarps. Tarps should not completely enclose haystacks. By just tarping the top, the stack is able to breathe.

Figure 7.7. Where large quantities of hay need to be stored, an open-sided pole shed works best. Access is readily available, and the haystack is protected from most moisture while still being able to breathe.

ter 8 discusses the basic supplies you'll need to have on hand) and need to be stored with some type of organization. Some medications need to be kept in the refrigerator, but the rest of your vet supplies can be kept in one of those plastic boxes with the tight-fitting lids that are sold in department stores. These boxes keep your supplies neat, dry, and all in one place, whether that's in a barn, garage, basement, or a closet in the house.

FARMER PROFILE

The Van Der Pols

Jim and LeeAnn Van Der Pols both came from farming backgrounds. In 1977, they bought their own farm, 160 acres near Kerkhoven, Minnesota. Like many of their neighbors, Jim and LeeAnn raised corn, soybeans, and feeder pigs in confinement barns. But over the years, the work grew unpleasant and the operation wasn't quite profitable enough. Conventional wisdom said, "Get big or get out," but they decided there had to be an alternative.

Today, Jim and LeeAnn farm their 160 acres and rent an additional 160 from Jim's mom. They raise 1,200 pigs, farrow to finish, on pasture and in hoop houses. They've also diversified their livestock operation by adding 160 ewes and about 20 stocker calves to their farm. The variety of animals helps them control parasites, keep more of their land in permanent pasture, diversify their income, and spread out their workload.

Their crop rotation has changed: Soybeans, the darling of the Midwest, no longer find a place in their rotation. Instead, small grains (oats and barley planted together) and alfalfa are rotated with pasture and corn. The sheep and cattle (both nontraditional animals in this area) "provide appetite to eat the hay." Alfalfa is important in the rotation: It cuts the expense of raising corn, which their pigs must have. By bringing the ruminants into their operation, though, they don't have to mechanically harvest second or third cuttings — the animals graze them off.

"Our work is pleasanter," Jim says, "we've been able to expand the number of pigs enough to incorporate our son, Josh, and his wife, Cindy, into the operation without expanding the land base; and our long-term profitability is better."

The sows now farrow on seasonal schedules, spring and fall. In spring, they farrow directly on the pastures and have access to portable huts that look like large steel culverts cut in half. These spring pigs are finished in the hoop houses during the winter. The fall farrowing takes place in the hoop houses, the pigs are sold as feeders (at about 40 pounds), and the hoop house is cleaned for the spring pigs to come in for finishing. Straw from the small-grain crops is used to deep-bed the hoop houses, keeping a clean and healthy environment for the growing pigs.

"Since we've gone to pasture and hoop-house operations, we don't have to dock tails or cut eyeteeth out of the baby pigs," LeeAnn adds. This saves both money and labor, and makes for happier piggies. "The pigs burn their nervous energy running, playing, or burrowing in the straw, instead of fighting and chewing on each other."

Conventional confinement buildings have problems with noise, dust, and odors. Also, the manure from these structures, stored in liquid slurry lagoons, is a management nightmare. When spread on the soil, it has a tendency to kill beneficial organisms, like earthworms and the smell is overwhelming. For Jim, the end to manure-handling problems is one of the best benefits of their current system. "Now we have better nutrient cycling. Our land is improving — it's like we move one thing into place, and all kinds of good things start to happen."

The first hoop house went in during 1994. "Hoop houses cost about one-fifth the amount that a comparable confinement barn costs," Jim says. For example, a hoop house that can feed 400 pigs costs slightly under $10,000, whereas a confinement setup would cost more than $50,000. Last year, when Josh and Cindy joined the operation, another hoop house went in, allowing them to reach the 1,200-pig mark.

"Now, with the kids involved, we're beginning to work on direct-marketing. We may never market 1,200 pigs directly, but we feel we need to market as many as we can. Marketing through the commodity system is too chancy."

CHAPTER 8

Health & Reproduction

The most important "drug" you can give your animals is good husbandry.

C. E. Spaulding, *A Veterinary Guide for Animal Owners*

A 3-DAY OLD CALF LIES ON ITS SIDE, listless, dull eyed, not eating. The head of its tail is covered with caked yellow manure. *Scours.*

A goat has been straining for hours, and still no kids have come. *Breech position.*

A horse is off feed for the second day. Its eyes roll in its head, and it grunts in obvious pain. *Colic.*

A ewe has been walking stiff legged for a few days. Flies, far more than normal, seem to be buzzing around her. *A wound has become home to maggots.*

A cow is kicking at its side. It gets up, it lies down, it gets up, and it lies down again. It isn't interested in food. *Hardware disease.*

Although the best care, the most love and compassion, and the highest-quality feed significantly reduce the incidence of health problems, these things can still happen — animals get sick, they sometimes have trouble breeding, or problems occur while they're having their babies. For the small-scale livestock farmer, observation — not a peripheral glance, but true, studied observation — is one of the most crucial skills to develop: Your best chance of minimizing health problems is early detection.

Professional Help

The first thing a livestock farmer needs to do is develop a professional relationship with a veterinarian. Looking for a vet should be like looking for a family physician. You can expect, based on their diplomas and state licensing, that all vets are minimally qualified to provide service, but a vet needs to be more than minimally qualified — he or she needs to be someone you feel very comfortable with. Good vets don't only doctor your animals; they also answer your questions, they make good management suggestions that take into account the type of operation you're striving for, and they care about you and your animals' well-being.

Ask other farmers in the area whom they use as a vet, and what their opinion is of him or her. Have prospective vets out for something minor such as performing vaccinations, or take a dog or cat into their office for its annual shots. Feel out how their personality meshes with yours. Find out if they make emergency farm calls whenever you need them — which may happen to be midnight on Christmas! If they're unavailable, do they have an arrangement with another vet to

CAVEAT

Before you read on, I want to inject a word of caution: There is a phase that most medical, pharmaceutical, and nursing students pass through in which they become convinced that they're suffering from some terrible fatal illness (which they just so happen to be studying at the time). Sometimes they "suffer" multiple illnesses in the course of a few months. Then they finally figure out that they're healthy — as long as they don't worry themselves to death over all the things that they could possibly die from.

Well, reading veterinary books, or even this animal health chapter, can create similar reactions in livestock owners. You start to think raising animals is hopeless because they're bound to contract all the terrible illnesses you read about. During our tenure as husbandmen to a vast menagerie, we've dealt with our share of health problems, including those I listed above, but for each health problem dealt with we've raised hundreds of healthy, happy animals without incident. Take good care of your animals and your land, and they will take good care of you — generally with few health problems to weigh you down.

take their emergency calls? Will they accept and return client phone calls if you need their advice or opinion?

We lucked out with a group of six veterinarians who shared a practice in Minnesota. Each was professional, hardworking, and easy to talk to. They came promptly, day or night, for emergency calls. And they didn't pooh-pooh our interest in alternative practices and organic production. In fact, from time to time they'd ask us what we had been doing for a particular animal or situation; they didn't use alternative practices, such as homeopathic preparations or acupuncture, but they were interested in what they saw working.

When shopping for a veterinarian, remember that they have to deal with lots of different kinds of patients. One day, they might take care of horses, a parrot, a llama, and dozens of cows. The next day may bring an elk, a herd of sheep, some pigs, and a ferret. Obviously, no one person can be a complete expert on so many different critters, but they should be willing to admit when something is baffling them, and do the research that's needed to figure out what the problem is.

You'll find that throughout this chapter, I advocate using these fine professionals. In the early years of raising livestock, your veterinarian should be your teacher. The first few years, you should plan on using vets frequently. Their help and advice, especially when you're starting out, is worth every penny you pay for it! As your experience level increases, the need to call your vet will decrease.

Causes of Health Problems

Illness is generally the result of either a chemical or a biological agent, though sometimes it is simply the result of improper diet or a change in diet. As with people, injuries are the result of accidents. Falls and cuts are the two most common injuries livestock suffer.

Chemical Agents

Chemical agents cause poisoning. Some chemical agents, such as poisonous plants, may be biological by their nature, but it is the chemical in the plant that is toxic. Although chemical poisoning can happen with livestock, it is less frequently the cause of illness than a biological agent. To avoid chemically induced illness:

1. Learn about the poisonous plants that grow in your area of the country.

2. Store feed properly, and never store any kind of chemical products — including cleaning products — near your feed or where animals may gain access to it.

3. If you're using any chemical products around your farm, make sure to follow the manufacturer's instructions carefully, including directions for proper disposal of containers. A farmer we read about had a number of cows poisoned from eating hay. It turned out that some empty chemical bags had blown out into his windrows of hay as they were drying. The bags got baled with the hay. Certainly, this tragic experience could have been prevented by proper disposal of the bags.

4. If animals are kept indoors in an old building containing lead-based paint, lead poisoning may be a problem.

5. When grazing, keep animals away from fields that have recently been sprayed with fertilizers, pesticides, or herbicides. Field contamination from sprays may be the result of someone else's operation.

A neighboring farmer in Minnesota had the local co-op spray a field of red clover with Roundup (Monsanto, St. Louis, Missouri) prior to his fall tillage. Unfortunately, on the day that the co-op arrived our stock was grazing right across the fence, and the wind was blowing pretty hard. When we realized what was happening, we went out and brought our stock in close to the barn, but the damage was already done. Three cows showed signs of a toxic reaction. Our vet contacted a veterinary toxicologist at the College of Veterinary Medicine in St. Paul, who assured us that the Roundup shouldn't cause a fatal reaction. She told us to make sure the affected animals had access to copious amounts of water. We called both the neighbor and the co-op manager and threatened that if it ever happened again, we would sue for chemical trespass. They apologized and swore it wouldn't happen again.

We were far luckier than a farmer down the road, who had several cows die after the local electric utility sprayed along the edges of his field to kill the brush under the power lines. He was paid for the loss of his cows, but some losses money simply doesn't cover.

Biological Agents

Biological agents are the most common cause of illness. Complex organisms such as birds and mammals regularly act as hosts to a rather large menagerie of microorganism guests. In humans, it's estimated that 100 billion microorganisms routinely share our bodies. These regular guests are called *normal flora*, and for the most part are harmless to their hosts. In some cases, normal flora — like the digestive bacteria that help break down food — are actually beneficial. But under certain circumstances, these normally benign "bugs" can cause disease. When a microorganism causes disease, whether a member of the normal flora or a recently introduced bug that is just passing through, it's called a *pathogen*.

When conditions are just right for them, pathogens proliferate to the point that their numbers simply overwhelm the animal, like weeds taking over a garden. Some also produce toxins as by-products of their bodily functions; the clostridium bacteria, whose toxins can cause tetanus, botulism, and black leg, are prime examples. These toxin-producing pathogens are capable of causing illness even when only a few are actually present. The biological agents include bacteria, viruses, yeast and other fungi, and worms and other parasites.

Bacteria

These single-celled organisms are a funny bunch: Some of them are so helpful, but some are so deadly (Figure 8.1). The beneficial bacteria are known as *saprophytes*, and include bacteria that regularly live in the digestive tract. When the saprophytes are where they're supposed to be and in the right quantity, they help the body keep humming right along. If introduced where they aren't supposed to be, though, watch out. They become pathogens.

Bacteria can be treated with antibiotics; however, not all bacteria respond equally to all antibiotics. If a bacteria responds to treatment by a particular antibiotic, it's said to be *sensitive* to the antibiotic; if it doesn't respond to the treatment it's *resistant*. When treating an animal for a bacterial infection, it's best to have your veterinarian perform a culture and sensitivity test

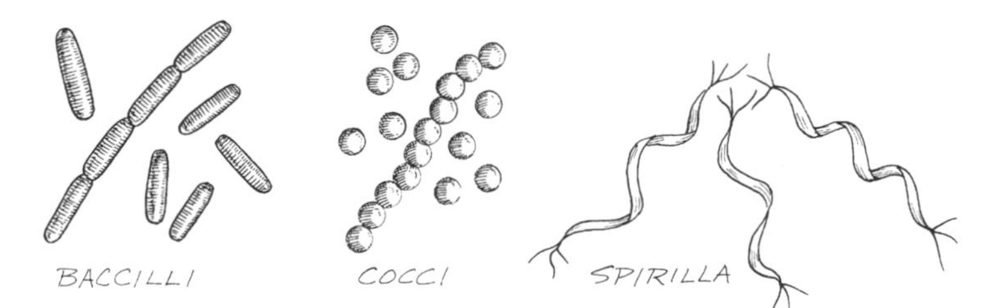

Figure 8.1. There are three major classes of bacteria: bacilli, cocci, and spirilla. These organisms are responsible for many diseases that affect humans, animals, and plants. At the same time, many bacteria are beneficial and even necessary, like those in the rumen of ruminant livestock.

if this is at all feasible (i.e., if you aren't dealing with an immediately life-threatening situation). This test will tell you exactly which antibiotic is most effective against the bacteria that are causing the illness.

Viruses

What can be said about a thing that is alive by some definitions of life, but isn't alive by other definitions? Like other types of cells, viruses are made up of proteins (Figure 8.2). And like other cells, viruses are able to reproduce; generally something that reproduces itself is considered to be alive. But viruses aren't independently living, breathing, reproducing organisms. Unlike any other type of cells, viruses can only reproduce inside a host organism's cells. Without a host cell inside an independently living organism, they do nothing.

Most cells, from the single-celled organisms such as bacteria to the individual cells that make up a tree, a whale, or an emu, are large on a microscopic scale. In fact, they can be seen under a high school biology class microscope. Viral "cells" are much smaller, and require an electron microscope to be seen. To put the scope of this size difference into some kind of scale, consider that one host cell can contain millions of viral cells. When the virus has reproduced to the carrying capacity of the host cell, the viral cells burst out of the host cell, killing it; they then go in search of new cells to infect, and start the reproductive process over again.

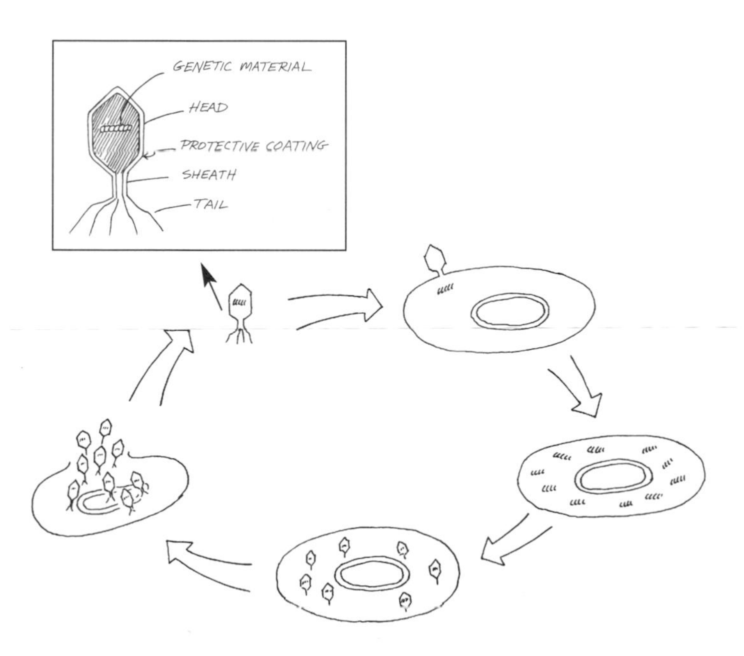

Figure 8.2. Unlike bacteria, there are no good viruses. Viral cells invade a host cell by penetrating the cell wall. They then deposit their genetic material into the host cell, where they use the host's genetic material to replicate. When as many new viral cells have replicated as the host cell is capable of accommodating, the viral cells burst the wall of the host cell and go out in search of new host cells to continue the process.

If you remember the discussion on genetics in chapter 5, living organisms have basic genetic material, which they ordinarily receive from their parents. Here, again viruses are different: They cheat and use genetic material from the host cell to reproduce their own genetic information, over and over again. This method of replication enables a single viral cell to enter the host cell and create millions of new viruses.

Unfortunately, viruses aren't affected — not even a little bit — by antibiotics. Once a viral infection has begun, the animal's immune system must combat the viral organism with antibodies, or the animal will die. Antibodies are like little warriors that are created in the bodies of all animals to combat alien proteins.

No drugs will cure a viral infection, but that doesn't mean drug therapy is never called for when dealing with a virus; drugs may be given to help alleviate certain symptoms (aspirin to reduce the fever, for example) or antibiotics may be called for to stave off secondary bacterial infections. Bovine viral diarrhea (BVD) is a good example of a viral disease that is highly contagious and causes high mortality levels (large numbers of exposed animals will die). But the virus itself isn't usually the cause of death; secondary bacterial pneumonia is, so as soon as BVD is diagnosed the animals are started on antibiotic therapy.

Viruses may not succumb to antibiotics, but many of the common livestock viruses can be prevented from causing illness through the use of vaccinations. Basically, what a vaccine does is teach the body to recognize the protein sequence of a given virus. After the body's immune system has recognized the virus as a foreign invader, it will quickly recognize it again. This preprogramming of the immune system allows antibodies to be instantly deployed when the virus first shows up, cutting down the immune system's response time to a point where the invading virus doesn't have much of a chance to begin reproducing.

Yeast and Other Fungi

Years ago, when I took biology in college, all living organisms were classified into two kingdoms: plants and animals. Members of the fungus family were considered part of the plant kingdom, but in recent years biologists have decided that two kingdoms aren't really adequate for classifying organisms, so they've gone to a five-kingdom system. Under the new system, the members of the fungus family have their very own kingdom. The fungus kingdom includes molds, mushrooms, yeasts, and lichens. Unlike bacteria and viruses, yeast and other fungi are multicelled organisms. These organisms normally don't cause problems in healthy animals, but when an animal's immune system is already compromised they can cause a variety of skin problems, respiratory problems, and mastitis (an infection in the udder, or milk-secreting gland, of a female animal). Often, these infections follow the use of antibiotics, because the balance of normal flora has been upset, providing an opportunity for the existing yeast to go crazy.

Worms and Other Parasites

Parasites aren't a single class of organisms. Instead they run the gamut from protozoans (single-celled members of the animal kingdom) to far more complex organisms such as worms and insects. In *Biology Today* (Del Mar, CA: CRM Books, 1972) a *parasite* is defined as: "An organism that lives on or in another organism and depends upon the host for its food." Most parasites are relatively benign, just part of the normal flora; many are a nuisance, such as biting flies; some cause serious and occasionally fatal illnesses.

Parasites are capable of attacking most parts of the body. In cows alone, there are almost one hundred known pathogenic parasites (those that are capable of actually causing illness). They can be found throughout the digestive system, on the skin, in the blood, throughout the respiratory system, in the eyes. Luckily, many of the worst parasites aren't found in the United States, and strong, healthy animals rotated on clean pastures are less likely to suffer from parasitic diseases. In this country, intestinal worms tend to be the biggest problem: Though rarely fatal, they reduce weight gain and milk production, and simply tax an animal's system. Medications are available for treating worms; however, before treating, have your vet run a stool sample for one or two animals. If worms are significant, the vet will find eggs; if eggs aren't found, then you don't need to treat the herd.

OF WORMS AND SHEEP

Before I leave this topic, I want to tell the story of our sheep flock. Common wisdom says you can't raise sheep without diligently worming them all at least twice per year, and many sheep farmers do it every other month. When we brought our original four ewes and a ram home, we wormed each of them and kept them in a small pen for a couple of days. Sure enough, the day after being given the wormer they were passing worms in their stools — lots of worms. We wormed them once more, about 14 days later, to make sure we got any little devils that popped out of eggs after the first worming. We kept our flock for 5 more years, increasing it to about thirty breeding animals, and *never* wormed any of them again. They were big, healthy, rambunctious sheep that never carried a significant load of worms. Now, we did have some factors working in our favor: Our farm had never had sheep on it before, so there wasn't a supply of sheep parasite eggs in the soil. Our farm had a weed called wormwood *(Artemisia absinthium)* growing on it, which sheep like to eat, and which acts as a natural anti-worming agent. Winters in Minnesota helped break parasite cycles. And finally, we were practicing multispecies grazing, which also helps break parasite cycles.

Natural Defenses

Earlier I mentioned antibodies, the little workhorses of the immune system, but the natural defenses that a body uses to block disease are complex and worthy of additional discussion. I looked at four types of pathogens, but within those four categories, there are thousands of individual species capable of causing illness. Yet in reality, thanks to the natural defenses, we rarely have to resort to medical intervention to fight these pathogens off.

The first defense a body puts up against invading pathogens is just good old physical barriers; a natural "close the door" approach. Skin, hair, and hooves prevent most organisms from getting into the body.

The second defense is simply washing away invaders with liquids. In the case of a cut, the liquid that flushes away germs is blood. Other bodily liquids, such as saliva, tears, and urine, are also capable of flushing invaders away.

Sometimes the normal flora acts as a defender, keeping down numbers of other microorganisms. For example, in most female mammals the normal bacteria that are present in the vaginal tract keep the yeast *Candida albicans* from overpopulating.

Enzymes that occur in bodily fluids such as saliva or tears also act as a defense against invaders. Notice how a dog will lick a wound clean (and lick, and lick, and lick). For any nonpuncture wounds, this licking action is a very successful strategy for preventing infection because of both the flushing action of the saliva and the action of the enzymes in the saliva.

If the pathogen has managed to breach all these systems, it is then the job of the immune system to fight it off. Some general immune system cells, particularly white blood cells, are constantly floating around in the bloodstream, waiting for the arrival of an alien organism. When they spot an invader, they cause an inflammatory response, which may be sufficient in its own right to kill off the invader. Signs of inflammatory response include swelling, localized heat, redness, and fever. The inflammatory response also provides a window of time in which antibodies can begin forming.

Antibodies take a week or two to develop after an invader has entered the system. If the invader has been introduced in the past, either naturally or through a vaccination, the body is capable of producing antibodies almost immediately. In some cases, this ability to remember and recognize an invader lasts for the remainder of the animal's life; however, in other cases it fades over time. If the same invader is reintroduced within a relatively short time, the immune response is much greater, and will last much longer, which is why some vaccinations are given in a series (Figure 8.3).

Antibodies can be formed to allergens, pathogens, and even cancer cells. Still, at times even antibodies aren't enough to return the body to a healthy state, and we have to turn to medical intervention.

A healthy and strong animal's natural defense system works much more effectively than an unhealthy or weak animal's possibly can. As farmers, one of the main things we can do to help our animals naturally defend themselves against illness is to minimize stress.

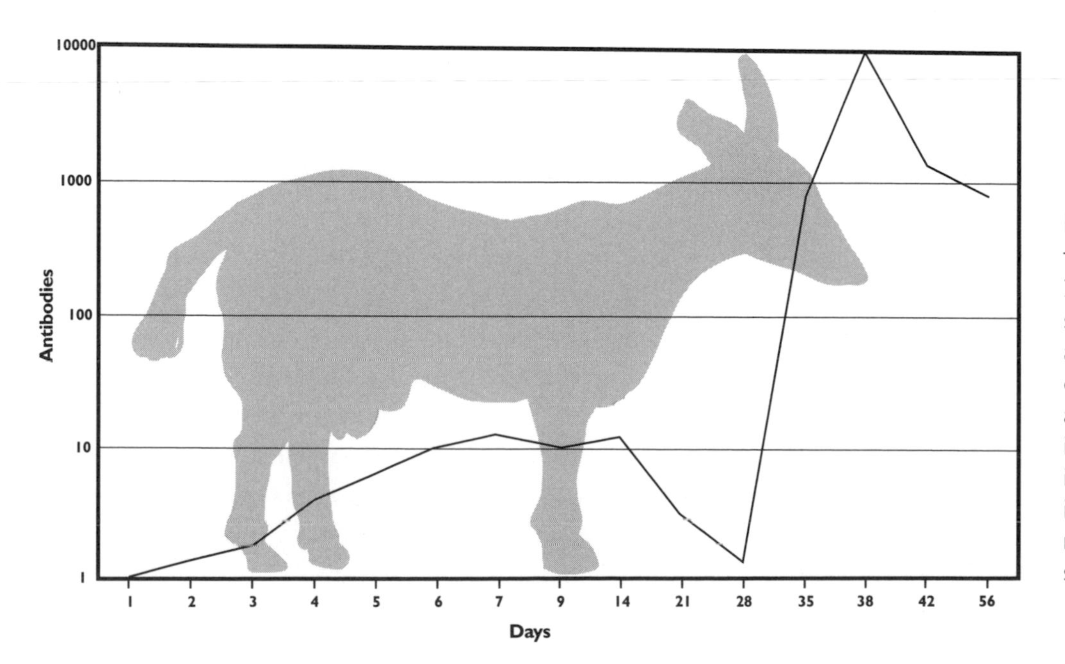

Figure 8.3. The animal is exposed to a vaccination on day 1. On day 28, the animal receives a booster shot. When first exposed, the animal begins to show low levels of antibodies, but when exposed a second time, the animal's immune system remembers the invader, and the response increases dramatically, both in numbers of antibodies and in speed of antibody production.

Colostrum

While we're on the topic of natural defenses, I want to talk about colostrum, the first milk that a mother mammal secretes when her baby is born. Colostrum is chock-full of all the different antibodies the mother has floating around in her own system. Babies are born without any antibodies of their own, and they come into a world full of possible pathogens, but during the first 24 hours of their lives they're able to absorb the antibodies in the colostrum directly through their digestive tract. These donated antibodies provide them with the protection they need while their own bodies begin the process of developing antibodies. Without a good dose of colostrum within the first 24 hours of life, babies inevitably stay sickly or die. (See chapter 6 for more about colostrum.)

The Power of Observation

Some illnesses and injuries cause readily apparent discomfort, but many don't. So if the power of observation is one of the best tools available to the small-scale farmer when caring for animals, what is it you're trying to observe? (These same observations should also be made when you are shopping for animals.)

1. Energy level. First and foremost, watch for listlessness. Healthy animals are "bright eyed and bushy tailed," as the old saying goes. They are active, moving around freely. Their heads should be held up, their ears and eyes should be responsive to their environments. Their appetites should be good, and they should drink

FARMER PROFILE

Linda Phillips and Susan Gladin

Controlling internal parasites in sheep can be a real challenge, particularly in the humid southeastern United States, but Linda Phillips and her partner, Susan Gladin, have turned this challenge into a small, home-based business.

Linda wasn't raised on a farm, though her family kept horses while she was growing up. "I guess those horses were the start of a lifelong love of animals," Linda says. Linda also loved to knit, crochet, and do hand spinning, so today, she and her family raise sheep, goats, and rabbits on their 13 acres, as well as keeping two horses. "Though the animals are really my project, Tim and the kids pitch in when I need help."

The family started out on a piece of land that was mainly wooded. The woods were nice but didn't serve their expanding livestock herd very well, so in 1998 they sold that piece and moved to the 13 acres, which was an open but worn-out piece of farm ground. "We renovated our land to new, improved pasture when we first purchased it, so for a while we couldn't let our animals out to graze. Luckily, other farmers in the area had places that could use some animal impact, so they let us graze for free.

"We moved the sheep and the goats through these plots using Electroplastic Net [a portable, electric woven-wire fence]. The animals would clean up the area of brush and weeds, as well as trim the grass. It's worked well for everyone involved. We got some free feed, and the neighbors got low-cost brush control."

When Linda first got into raising sheep, she struggled with internal parasites. "I used a commercial product to control the parasites, but I would still lose sheep to worms. I'd worm the flock, and 3 or 4 weeks later lose a sheep. Granted, I didn't really know what I was doing yet — like, I didn't know to worm them once, and then worm them again exactly 14 days later — but it was still horrible and expensive."

Linda began an extensive search for better ways to control the parasites. She read all the information she could find, within both the conventional system and the alternative. She studied the life cycles of the parasites, and she began having manure samples tested for fecal egg counts. In discussions with staff at the University of North Carolina, she learned that 1,000 eggs per gram of stool is the point at which worming is recommended, though most vets will recommend worming if any eggs are seen in a stool sample.

Through her research, Linda learned that by managing her grazing she could reduce worm loads. "Sheep will eat right in the area where they just passed manure, so they pick the worms right back up. By moving them often, they are eating off cleaner ground." But she also learned that in and of itself, a grazing program won't solve the problem. "The thing with parasite control, particularly alternative control, is that you have to have a real program. In and of themselves, no one of plenty of water. An animal that is just lying around, not eating, or not showing much interest in what's going on around it is probably ill. There are some exceptions to this rule. On really hot days, critters may just lie around in the shade looking pretty lethargic — but as the heat of day breaks, they'll get up and eat again. Still, even if they're lethargic from the heat, their eyes and ears should be responsive to what's going on around them.

Newborns can also be an exception to the listless rule. For the first week or two of its life, a newborn simply eats and sleeps, and its sleep tends to be very, very deep. Sometimes you'll see a newborn baby that you'll think has died, it's in such a deep sleep, but if you look more closely you'll realize it's simply in the black depths of newborn exhaustion. Coming into the world from the safety and warmth of the womb is hard work for a little thing.

One July afternoon, a wicked storm was blowing in. Our cows and their calves came and sought protection from the hard-blowing rain and hail in our open-sided shed. Layla, who had calved the day before, arrived with the rest of the herd, but her new baby wasn't with her. We mounted a search and finally found the little guy stashed away in some deep grass, sleeping the sleep of the dead. We carried him (all 70 pounds [31.8 kg] worth) up to the shed to join his mama. Layla (sufferer of temporary amnesia) suddenly remembered she had a calf and became quite excited when we brought him in.

these things will take care of everything."

Linda's search next led her to Susan Gladin, who was using a blend of herbs to control parasites; the basic recipe included wormwood and some other herbs. The biggest problem with the recipe was that it wasn't very palatable, so the animals didn't want to eat it.

"Susan and I did some more research into Chinese medicine and the preparation of traditional herbal remedies. We played with the recipe until we came up with the one we're now using." The current mixture is more palatable and requires fairly small doses that can be mixed with grain or minerals. The animals eat it without any trouble.

Next, Linda and Susan began looking for a way to make on-farm, inexpensive test kits for checking fecal samples. "Some veterinarians charge up to $50 to do one test. We knew that an inexpensive method would improve the control." They developed their own kits out of things you might find in any kitchen — like measuring spoons — but that you might not want to use for testing manure.

The herbal preparation and the test kits worked, and through word of mouth people were hearing about them and coming to purchase them. In 1996, Linda and Susan began marketing their products through their own company, Farmstead Health Supply.

In 1997, the University of North Carolina ran a study comparing their product to one of the top-selling commercial worming preparations. The sheep that the university used in the study were carrying egg loads of right around 1,000 when the study began. Upon treatment with the commercial preparation, this dropped to 180 eggs per gram, but then climbed back up during the test period to more than 1,000. The sheep treated with the herbal preparation dropped to only about 600 eggs per gram, but, unlike the commercially treated animals, their egg counts remained down around 600.

Linda says, "Some people put down alternatives. They say that they won't work, and that herbal medicine is just quackery. But we have relied on our products almost exclusively for over 3 years, and all my sheep are still out there in the pasture breathing. If they didn't work, I wouldn't have any sheep left."

Despite her reliance on herbal preparations, Linda indicates there are still times when the "knock-down" powers of the commercial preparations have a role. "When we moved to our new place, I wanted the animals to come to the land with as low an egg count as possible, so I wormed them with a commercial product twice, 14 days apart."

Linda's recommendation for any animal health program: "Really watch your animals. If an animal looks like it's getting behind, test for parasites. With sheep and goats, if they're getting a heavy parasite load, the skin under their eyelids begins to look really pale instead of pink."

In fairness to Layla, I have to say that most cows seem to suffer this memory loss from time to time. The calf is sound asleep when the herd moves off, and Mama moves with the herd. A little while later, you can watch Mama get a look on her face — hey, where's Junior? Suddenly, she trots off in a display of speed that would warm the heart of a Kentucky Derby fan, bellowing and snorting the whole time, until she finds her calf.

2. Hair and coat. This might be considered a vanity issue, especially if critters were like people, but it's not. Hair or wool should look shiny and healthy, and it should cover the body fairly evenly (unless you are looking at bison, which are always shaggy looking, or animals that are shedding out their winter coats in spring and early summer). Poor-quality coats can indicate nutritional illnesses, external parasites, or other systemic diseases. Also, the coat shouldn't be caked with manure.

Manure that's caked on the side of the body is often just a sign that the animal has been forced to lie in a manure pile, and may not be of great importance. But if the manure is caked around the tail-head and down the backs of the legs, it is a sure sign of diarrhea, or *scours*. (See more on scours on page 90.)

3. Discharge. Look for suspect discharges from the nose, mouth, ears, or eyes. Sometimes the nose or the eyes may have a little bit of watery discharge, and it isn't anything to worry about. But if a discharge is pussy looking, if there is crusty stuff built up around the muzzle or eyes, if there is excessive slobber or frothiness around the mouth, or finally, if there is any kind of discharge from the ears, the animal is not well.

4. Hydration. Look at the eyes to see if they appear "sunken." This is usually a good indication of dehydration. Dehydration often accompanies scours, or illnesses that cause a fever. If you can view the gums and the tongue, typically they should be light pink. If they are gray or white, chances are the animal is in shock, either from an injury or from dehydration accompanying an illness.

5. Breath sounds. Listen for any coughing or wheezing. Healthy animals breathe easily through their noses, not their mouths.

6. Mastitis. The last external check is done exclusively on milking animals. Though most common in dairy cows, mastitis can occur in any female animal that is producing milk in her udder — even a pregnant animal that hasn't given birth yet. On rare occasions, a young female that hasn't even bred yet can develop mastitis. A healthy udder should be warm, but not hot; pink, but not red; and soft, but not hard. The milk should flow smoothly and, except for colostrum, should be very liquid with no clots or lumps in it. Colostrum is almost like pudding, it's so thick, but it shouldn't have any lumps in it after the first few squirts.

If after making these external observations, you suspect an animal isn't well, remove it to a quiet and secluded location. This will allow further evaluation and treatment, and hopefully reduce the likelihood of an illness being passed to healthy herdmates. If the animal is tractable, check its body temperature. Animals' normal temperatures run in a slightly wider range than ours do, but if your animal has an elevated or a subnormal temperature according to Table 8.1, it's definitely time to make some decisions. Figure 8.4 shows some criteria we use in deciding whether or not it's time to call the veterinarian. There is, of course, a certain degree of flexibility in applying this, and as your nursing experience and comfort level increase you may wait longer to call. (Some vets or other farmers may disagree with this system, but it has worked for us.)

Table 8.1

Normal Temperature Ranges

Animal	Temperature (°F)	(°C)
Cow	100.4–102.8	38.0–39.4
Horse	99.1–100.8	37.3–38.3
Pig	101.6–103.6	38.1–39.8
Sheep	100.9–103.8	38.3–39.9
Goat	101.7–105.3	38.8–40.8

Source: Modified from N. Bruce Haynes, *Keeping Livestock Healthy* (Pownal, VT: Storey, 1994), p. 122.

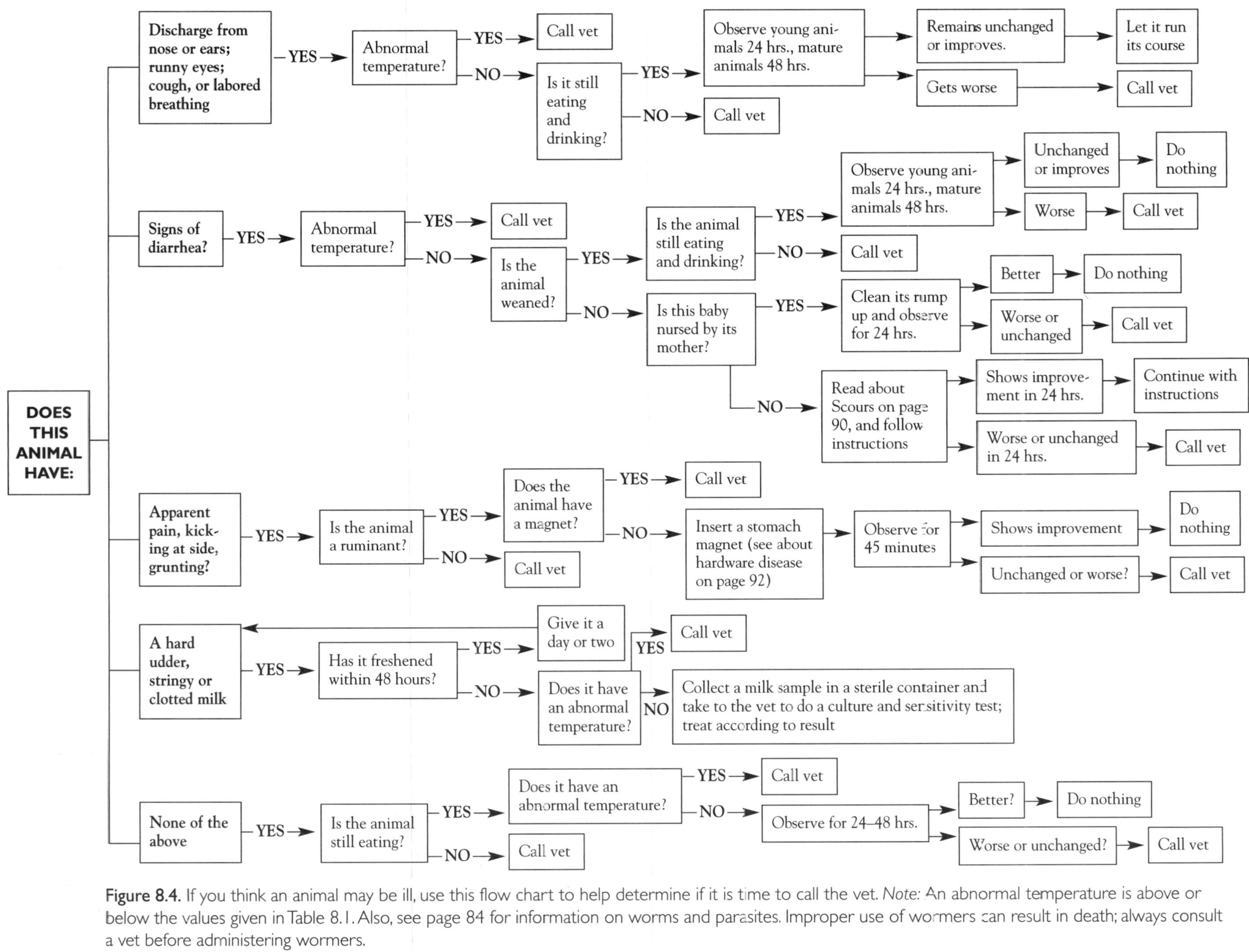

Figure 8.4. If you think an animal may be ill, use this flow chart to help determine if it is time to call the vet. *Note:* An abnormal temperature is above or below the values given in Table 8.1. Also, see page 84 for information on worms and parasites. Improper use of wormers can result in death; always consult a vet before administering wormers.

Common Illnesses

Let me say right up front that there are hundreds of different ailments that can cause problems and thousands of different causative agents for those ailments; I can't possibly cover them all in this short chapter. As you get more involved in animal agriculture, you'll need to get at least one or two books that specifically address animal health (see appendix E for a list of recommended books). We have half a dozen "vet books," and they are some of the most well-worn volumes on our bookshelf! The few illnesses that I'm going to mention here are some of the most common problems that farmers deal with, but even these can be the result of myriad causative agents. Read more, and talk to your professional veterinarian.

Scours

In adult animals, scours isn't usually fatal. Most often, adult diarrhea is the result of a change in diet or of consuming very lush pasture. Cases of diarrhea caused by diet change will clear up in two or three days, and don't have many other symptoms than the diarrhea itself. Lush-pasture diarrhea will continue for as long as the high-quality, moist feed lasts; but, like change-of-diet cases, it doesn't tend to have other symptoms associated with it. If you haven't adjusted the animal's diet, or it's not on lush pasture, the next most common cause of adult scours is excessive parasite loads. Parasitic scours typically isn't accompanied by a fever, but the animal will appear lethargic, and its coat may be dull. Diagnosis of parasitic scours requires a stool sample to be checked by your vet (unless you actually see worms in the stool). If an adult animal is both suffering from diarrhea and running a fever, it's probably time to call your veterinarian. The animal is either suffering from a viral or a bacterial case of scours.

Scours in baby animals — say, less than a month old — is always a very serious and life-threatening situation. Normal baby animal stools are yellowish and tend to be kind of gooey, like soft Silly Putty. Sometimes the stools may stick to the tail-head for the first day or two — during fly season this should be wiped away, if possible, to prevent screw flies from laying their eggs in the mess. The eggs develop into maggots, and the maggots don't stop eating when the manure is gone. In short order, they can do terrible damage to the baby animal, possibly even causing death.

With scours in very young animals, the stool becomes watery, or sometimes slimy, and if left untreated the baby will die within a few days. Scours is quite common in bottle babies — those being fed by humans instead of their mothers. The most prevalent cause of scours in bottle babies is overfeeding, especially of milk. The scours caused by overfeeding is the most easily cured kind, but without treatment it can take an otherwise healthy baby animal out in just a few days. (For information on feeding baby animals, see chapter 6.)

Other causes of scours in babies include the bad guys: bacteria, viruses, and parasites. One of the worst bad guys that we've had personal experience with (when we brought in a load of calves) is K-99 *Escherichia coli*. *E. coli* is a common and generally beneficial bacterium found in the gut of most animals. There are many strains of *E. coli*, but the strain that microbiologists have dubbed K-99 is highly contagious in baby animals during the first few days of life, and it's very deadly. Once we knew what the problem was, we were able to treat all the rest of the calves that were born that summer with a colostrum supplement of K-99 antibodies. K-99 scours sets in within the first 3 or 4 days of life. Common causes of scours in the 5- to 15-days-of-life range include rotavirus and coronavirus; at 2 to 6 weeks of life, salmonella species of bacteria are the primary culprits.

Treatment

Treatment of scours should be instituted as soon as the problem is recognized. The first thing to do is replace fluids and electrolytes. *Electrolytes* are basically the "salt" molecules that are normally found in the bloodstream and include such elements as calcium, potassium, sodium, and magnesium. Commercial electrolyte solutions are available at most feed stores, farm supply houses, or from your vet. We always made a homemade concoction (see box on page 92). Electrolyte therapy is good not only for scours but also for any illness that might cause dehydration. With an adult animal, simply provide a pan of water that has electrolytes mixed in.

If you're treating a baby animal, dilute its normal milk ration by half with water. Between the milk feedings, feed it a comparable ration of your electrolyte solution. For example, if an 80-pound (36.3 kg) calf is receiving 4 quarts (3.8 L) of milk per day (a whole-milk ration that weighs approximately 10 percent of its body weight) during two feedings, mix 1 quart (0.95 L) of whole milk with 1 quart of water at each feeding. Between the two daytime feedings, and again just before bed, feed it 2 quarts (1.9 L) of electrolyte. Don't feed the electrolyte with the milk, because the digestive process interferes with absorption of the electrolytes into the animal's system. Babies suffering from the overeating version of scours require no more treatment than this, but you should continue it for 2 to 4 days or until the stools return to normal. If you do suspect that the scours is being caused by a pathogen, antibiotics may be in order — check with your vet.

If the animal you are treating for scours is a ruminant, helping its normal flora return is crucial. The best way I know of, if you can get close up to a

VETERINARY SUPPLIES

Over time and as your experience increases, you will accumulate more vet supplies than I've listed here, but this is a good starting kit:

▸ **Antiseptic and sanitizing fluids.** The ones we find regular use for are alcohol, iodine, and peroxide. The alcohol is good for sanitizing thermometers and other supplies, and for cleansing skin prior to giving an injection. Iodine (7%) is a good general wound cleanser and works well for cleaning navels on newborns. The peroxide is the best thing to use for cleaning wounds in which maggots reside.

▸ **Aspirin boluses** (or the extra-large economy-size bottle of generic people aspirin), for relief of aches, pains, lameness and fever. If you use people aspirin, one tablet per hundred pounds of body weight works well (1 tablet per 45 kg).

▸ **Clean blankets, towels, and cloth "rags."** Old towels and blankets come in handy when drying off animals (like newborns) and when warming animals that are in shock. The rags have many uses, including cleaning wounds, cleaning caked manure from young animals' tail-heads, and other clean-up chores. The ideal rag is cut from old bath towels — 12-square-inches (30 cm^2) seems to be a good size. Thrift stores and yard sales are a good source for used blankets and towels.

▸ **Needles and syringes.** Keep several disposable needles (size 18 gauge–1.5" is the most versatile) and several syringes (6 cc, 20 cc, and 60 cc) on hand at all times. Syringes come in handy for feeding very weak animals (like newborns). To feed with a syringe, slowly dribble milk, electrolyte, or colostrum into the mouth.

▸ **Cow magnet.** If you plan to keep cattle, keep several cow magnets on hand. Cow magnets are about the size of an adult's pinky, rounded on both ends, and have no sharp edges.

▸ **Stomach tubes.** About 6 feet (183 cm) of soft rubber tubing works well. If you are using a tube to force-feed fluids, make sure that you insert it in the animal's stomach and not its lungs, or you'll drown the aniimal. To confirm that the tube is in the stomach, blow into the end of it — if you are on target, you'll see the animal's side expand.

▸ **Thermometer.** Absolutely the most important item in a vet box! Purchase a rectal veterinary thermometer with a ring top and tie about 2 feet (60 cm) of string to it. While taking an animal's temperature, either hold the end of the string or wrap it several times around the critter's tail. This prevents "losing" the thermometer inside the animal, or having the thermometer fall out, only to be crushed by a hoof.

HOMEMADE ELECTROLYTE SOLUTION

To 1 gallon (3.8 L) of warm water, add and mix well:

- 4 tablespoons (59 mL) of corn syrup or dextrose
- 2 teaspoons (10 mL) of table salt
- 2 teaspoons (10 mL) of baking soda

healthy herdmate that is chewing its cud, is to reach into its mouth and grab out the bolus of cud before the animal knows what hit it (yes, it's gross the first time you do it); then insert the bolus as far as possible down the throat of the ill animal, so that it swallows the bolus. The bolus from the healthy animal is full of good bacteria, and acts to recharge the ill animal's system.

Bloat

Limited to ruminants, bloat is a hazard when you're using a grazing system. Bloat is caused when excessive quantities of gas become trapped in the rumen; in extreme cases, it can be deadly within an hour or two. It is usually the result of eating lush, leguminous pasture, and is aggravated by moisture from dew or rain. It is most common on alfalfa, slightly less common on clover, and doesn't happen on bird's-foot trefoil pastures. Pastures with a high percentage of grass compared to legumes are the least likely to cause bloat, but even these can do it in early spring.

The most prominent symptom of bloat is a bulge on the animal's left side, just below the spine and in front of the hip bone. This area usually appears caved in, but in a bloating animal it sticks out. Bloating animals also quit eating, and quit belching.

Ruminants on pasture need to be watched for the first signs of bloat, especially in spring. To avoid it in the first place, limit access to lush pasture first thing in the morning, or right after rain, until the animals are well acclimated to the pasture. Feed some hay prior to turning them onto pasture, and leave them on pastures for short periods of time — 45 minutes to an hour is good to start.

Treatment

We never had a single case of bloat after our animals were acclimated to their pastures, but we did have one or two cases at the start of each grazing season. For cows, we'd administer a mixture of 1 cup (236 mL) cooking oil, 1 cup (236 mL) water, and 3 tablespoons (44 mL) baking soda, mixed well. (A squirt water bottle — the kind bike riders and hikers use — works well, just *dribble* the contents into the animal's mouth over a few minutes. They don't get it all, but they get enough.) Sheep and goats are much less likely to suffer bloat, but if they do, administer about one-fourth of the above mixture. After the animal drinks its "medicine," tie a stick in its mouth — sort of like a bit. This gets the tongue working, which helps kick-start the belching process.

As soon as the animal begins belching, you can watch its side go back down. With this quick fix, we never lost an animal and never had to move to more extreme treatments. In serious cases, a stomach tube can be passed down the animal's throat and into the rumen; the gas escapes out the tube. Stomach tubes can be purchased, but in a pinch use 10 feet (3.1 m) of garden hose, with any sharp edges filed smooth.

The last resort, and it should be used only in life-or-death situations, is to cut right through the animal's side and into its rumen. Vets carry a two-part tool for this purpose, called a trocar and cannula. In lieu of the trocar and cannula, a sterilized knife (boil in water, or soak in bleach, for about 5 minutes) might save the animal's life. In either case, the animal will need to be placed on antibiotics, because infection is bound to follow cutting into the rumen.

Hardware Disease

Unless you purchase a piece of completely bare land that has never had any buildings on it, chances are that you will, at some point, come up with hardware disease if you have any cows. Even baby calves can suffer from it. Sometimes it can happen in other ruminants, but it's most prevalent in cattle.

Hardware disease is caused when the animal eats a sharp piece of metal, such as a nail or a small hunk of wire. The piece becomes trapped in the reticulum, and can puncture the wall. Symptoms include obvious pain, kicking at the side, a slight rise in temperature, and getting up and lying down repeatedly. If left untreated, death may result.

Treatment

The cure, at least, is simple. Insert a cow magnet in the cow's stomach to "catch" the hardware. Some cattlemen insert magnets as a matter of course into all their animals; we simply kept magnets on hand, in our vet supplies, and inserted them when an animal showed the signs. If the problem is indeed hardware disease, the animal recovers almost immediately when the magnet is inserted.

Inserting a magnet can be done by hand or with the help of a bolus, or balling, gun. Much like feeding a pill to a dog, the goal is to get the magnet all the way down the back of the animal's throat so that it swallows the magnet. When done by hand on a full-grown cow, you must insert your arm into the cow's mouth, halfway up to the elbow! The magnet remains in the animal's reticulum for the rest of its life and attracts and holds onto any pieces of metal the animal swallows.

Pneumonia and Other Respiratory Disorders

Respiratory illnesses can occur in all species and are often caused by some of the normal flora that have gotten out of control when an animal is stressed. Stress caused by poor management (such as drafts or ammonia fumes in buildings, and poor nutrition) or transportation of animals is often the underlying cause of respiratory illnesses. It is very common in young animals of all species.

Treatment

If mature animals have no fever and are still eating well, we simply keep an eye on them, but respiratory illness in young animals is again far more serious. Keep them warm, administer electrolytes, and call the vet if the problem persists for more than 24 hours or seems to be getting worse.

If your animals are going to come into contact with other animals — for example, if you plan to show them, or if you'll be bringing new animals in and out of your herd — many contagious respiratory diseases can be prevented through vaccination.

Alternative Health Practices

This topic might be controversial in some circles. Many members of the medical and veterinary communities scoff at such practices as homeopathy, acupuncture/acupressure, and herbalism as nothing more than quackery. Personally, I don't agree; we've had good luck with alternative practices, and they have been in use in other parts of the world for considerably longer than our modern medicine has. I concur that these aren't a simple replacement for the practices that are considered conventional at this time and place in history, but many alternatives deserve consideration.

When we first began milking cows, we used antibiotics as a regular tool for mastitis control, but we had a few high-producing cows that suffered chronic, low-level infections. We would treat them with an antibiotic, and the infection would partially clear up, but in no time flat their somatic cell count (SCC) would start to climb again. *Somatic cell count* is a laboratory test routinely run on milk shipped to a dairy processor; dairy farmers are paid a premium for milk with a low SCC, and docked for milk with a high SCC. *Somatic cells* are a type of white blood cell that are part of the immune system's initial response to infection. The higher the SCC, the worse the infection is. Our interest in alternative practices really grew out of our desire to maintain a very low SCC without the constant use of antibiotics.

Learning more about alternative practices will take some time, and some additional research, but may be worth it for you. Also, some veterinarians are beginning to practice these alternatives. (Check appendix E, Resources, for information on finding one in your area.)

Homeopathy

Our earliest forays into alternative medicine began with *homeopathy*. Homeopathy is a medical practice that uses special preparations of natural substances to stimulate an immune response. These preparations contain minute quantities of a plant, animal, or mineral substance. For example, trace amounts of bee venom go into the preparation known as "apis mellifica," and trace amounts of wild hops are used to prepare "bryonia."

The principle behind homeopathy is that "like treats like." So a preparation like apis mellifica would be used for a bee sting, or "alium cepa" (red onion) would be used for a runny nose.

One of the first homeopathic preparations we began using in our quest to eliminate antibiotics was homeopathic sulfur, and our success rate with it was good enough to convince us that alternatives weren't simply quackery. Over the next few years, with lots of further study, we began using a wide variety of preparations for the various problems that crop up from time to time, both for the animals and for ourselves.

Acupressure

I never got the hang of acupressure myself, but I saw an experienced veterinarian use it; after watching him there was no question in my mind that it wasn't quackery. He was doing a demonstration of the technique on a 3-year-old dairy cow. Prior to the demonstration, he'd given us a chart of acupressure points on a cow. When he reached one of the points for the lungs, there was a definite and strong reaction. He'd never seen this cow before — or met the farmer whose cow it was. He turned to the farmer and said, "This cow suffered with a severe case of pneumonia recently, didn't she?" The farmer, who was in a state of shock, confirmed that she'd had a bad bout of pneumonia the previous winter!

Reproduction

For most of us, one of the greatest joys of raising livestock comes from the miracle of reproduction; witnessing the birth of a calf, seeing the first bumbling steps of a colt, or spying a clutch of chicks peaking out from under their mother's ruffled feathers is part of what draws us to animal agriculture in the first place. But the reproductive processes — breeding, gestation, and delivery — can also be a source of trouble.

Unless you are taking up an esoteric form of animal agriculture such as worm farming, you are dealing with animals that always reproduce through a sexual process. Sexual reproduction requires a male to supply sperm and a female to supply an egg.

Males produce sperm in the testicles. In mammals, the testicles hang outside the body in the scrotum, but in birds they are internal — a fact that makes the sexing of young birds a real challenge! The scrotum is designed to help regulate the temperature of the testes, dropping lower to reduce temperature during warm weather, and pulling in closer to the body during cold weather to maintain a higher temperature. Generally, sperm require temperatures that are 4° to 5°F (–16° to –15°C) cooler than normal body temperature to survive (Figure 8.5).

With each ejaculation, males send billions of sperm cells in search of an egg. Boars lead the pack, with 50 billion sperm per ejaculation. Only one sperm is actually required to fertilize each egg. Although billions of sperm cells are released with each ejaculation, many don't survive long enough to meet the egg. For example, bovine sperm remain viable for only about 12 hours, but chicken sperm can remain viable in the hen's reproductive tract for up to 2 months.

Sperm cells come in two varieties, called X and Y. The X type is responsible for female offspring, the Y type for male. The mother's egg is sexually neutral, so she plays no part in determining the offspring's sex.

Unlike males, which regularly produce new sperm cells from puberty until very old age, females have their full complement of eggs at puberty. Eggs are produced and then stored for life in the ovaries (Figure 8.6). All species contain two ovaries, but a quirk of nature

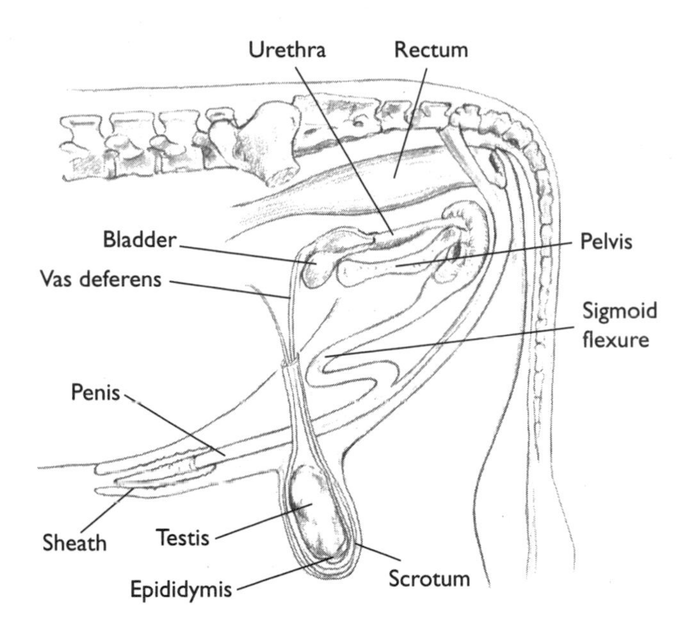

Figure 8.5. For a male animal to perform in a breeding capacity, his reproductive organs must be normal, healthy, and fully developed. Though most are, problems crop up occasionally. Retained testicles is possibly one of the most common problems in male animals. If the testicles do not drop down into the scrotum, the temperature in the testicles remains too hot, thereby reducing sperm viability.

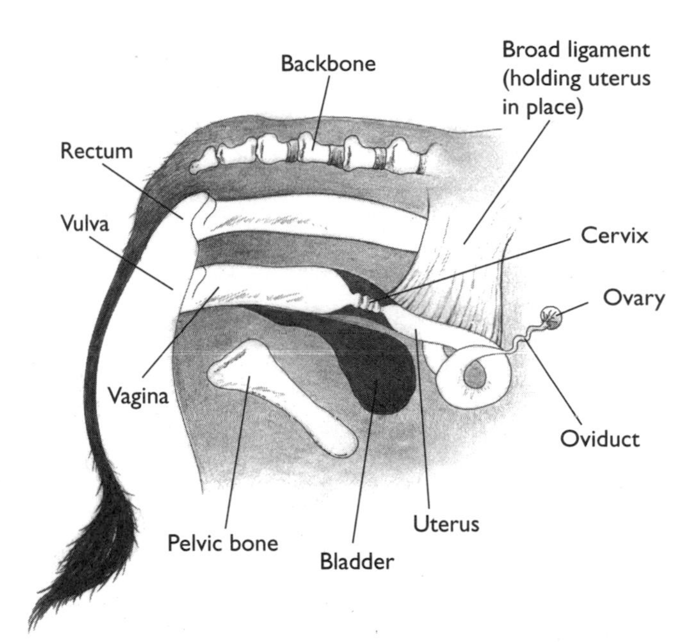

Figure 8.6. As with males, females require healthy reproductive organs for breeding. One of the most common causes of failure to breed in females is scar tissue in the vagina, cervix, or uterus, resulting from infection following a prior birth. It is important to check females after they've given birth for any pussy discharges that might indicate some internal infection is present. Slight cases often can be treated simply with a sanitizing douche.

provides poultry with only the left ovary as a developed, working unit.

Estrus Cycle

The whole process whereby eggs are released from the ovaries is referred to as the *estrus cycle*. This complicated process begins at puberty and extends into fairly old age. The cycle is controlled by the ebb and flow of four major hormones: follicle-stimulating hormone, estrogen, luteinizing hormone, and progesterone. The length of the cycle varies from species to species, and in most species the female will only allow the male to breed her for a short period during the cycle, called heat or the estrus period. Males sense the estrus period through smell.

Normally cows and mares release one egg per cycle, so twins are unusual, and triplets are a real rarity. Ewes typically drop between one and three eggs per cycle, though some breeds are capable of dropping up to six. As in the male of the species, sows are the big producers, dropping as many as twenty eggs per cycle. Chickens can drop about twenty-eight eggs per month (Table 8.2).

Multiple births are most often the result of dropping multiple eggs. Occasionally, however, one fertilized egg splits in two at the beginning of development. This is the anomaly that causes identical twins.

The estrus cycle in some species, including horses, sheep, goats, and chickens, is seasonally cyclic. Their cycles are controlled by the number of hours of daylight, and during part of the year they do not come into heat. Cows and pigs cycle regularly all year long.

When eggs leave the ovaries, they travel down the fallopian tubes. As the egg travels down the tube, it will become fertilized if sperm is present, and settle into the uterus for development. (Birds are the exception, with development taking place outside the body.) The period of development is called gestation or, in the case of birds, incubation.

Table 8.2
Reproductive Information

Species	Ovulation Rate	Gestation Period	Estrus Period	Estrus Cycle	Male Sperm Count per Ejaculation
Cow	1 egg/estrus	275–285 days	13–17 hours	21 days	7–15 billion
Horse	1 egg/estrus	330–345 days	90–170 hours	22 days	5–10 billion
Pig	10–20 eggs/estrus	112–115 days	48–72 hours	21 days	40–50 billion
Sheep	1–3 eggs/estrus	112–115 days	24–36 hours	17 days	1.6–3.6 billion
Chicken	28 eggs/month	26–29 days	na	na	6 billion

Fertilization

Fertilization may either take place au naturel — through copulation between a male and a female — or through artificial insemination (AI). AI is now a common practice for most species of livestock. Dairy cows and turkeys are almost all bred artificially in commercial agriculture. The use of AI is also increasing dramatically in the pork industry. Other species are still bred largely the old-fashioned way: Boy meets girl. AI does have some good points including ability for all farmers to have access to high-quality sires and the elimination of the need for keeping male breeding stock on site; also, the price isn't exorbitant.

The fertilized egg is called a *zygote*, and although it starts out as two unique cells — an egg and a sperm — it is considered a single cell in its own right. It begins to split into additional cells almost immediately. The zygote "plants" itself into the wall of the uterus, at which point it's called an *embryo*. In a truly remarkable process, embryonic cells continue to split into more cells, differentiating into the various types of cells that are ultimately required for a fully developed organism, including blood cells, muscle cells, and skin cells. The "instructions" for how to correctly split and differentiate are supplied in the chains of genetic material that each parent provides.

Infertility

Infertility problems can occur in both males and females, but are more common in females. Though we never had a single case of male infertility, when it does occur it is a much bigger problem than female infertility, because one male is responsible for breeding many females. If the prize ram you paid dearly for is a dud, there are no lambs next year! If a female is infertile, she won't have any lambs, but the rest of the flock will. Vets can test a male's fertility for you.

One of the main causes of infertility in females is an infectious process in the uterus following the birth of last year's babies. The infection occurs most often when the animal has failed to pass the afterbirth. It's a good idea to check for the afterbirth, though at times you won't find it, because many animals eat it. This habit probably evolved as a method of hiding the evidence of birth from predators. Even if the mother doesn't eat it, other animals may (our dogs spent most of the spring looking around the pastures for some of this fine stuff). If you haven't seen the afterbirth, keep an eye on the mother for discharges over the next week or two; clear to slightly turbid mucous discharges with some blood are quite normal, but if the discharge looks pussy have the vet come out and check the animal.

Delivery

At the end of gestation, the fully formed embryo is ready to pop out into the world as an infant. The last few days prior to delivery, Mama begins to show signs that the big event is near. Her udder begins to swell with milk. The area around her tail-head begins to flatten out, and appears sunken. Mucus begins discharging from the vulva, and the vulva itself looks puffy. Often, during these last few days, mothers act kind of funny:

They often go off by themselves, avoiding people and other members of the herd; they may "talk" more than normal; or they may act quite restless, lying down and getting back up frequently. Close to the time of birth, a sack of fluid may show up and break.

Labor can last just a few minutes, or hours. First-time mothers generally take longer for labor than those that have delivered babies in the past do. The most common delivery position is front feet first, followed by the nose (Figure 8.7). After the shoulders pass through the cervix (the strong muscle that protects the uterus from the outside world), the baby just about shoots out, but getting to that point can take time. Rear presentations, or *breech* births, are far less common than frontal deliveries, but they can happen.

Troubled deliveries are really few and far between if the mother is in good health. Over all the years and all the hundreds of babies born to our animals, we had to assist only a few times. As a rule of thumb, any animal that has struggled in labor for over an hour, or that is obviously weak and tired, needs assistance.

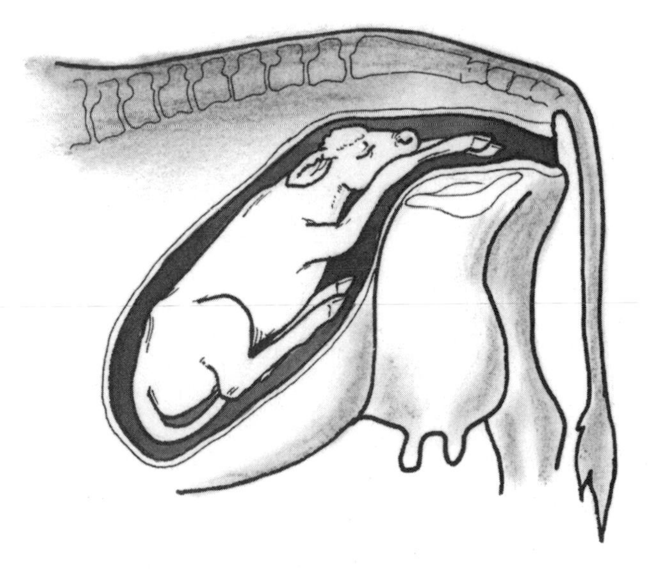

Figure 8.7. In a normal presentation birth, the front feet come first and facing up, followed by nose and face, shoulders, and then the rest of the body. Most births are normal presentation and require no assistance, though sometimes an extremely large baby will get wedged at the shoulders and require some pulling. Once the shoulders pass the cervix, the birth happens quickly.

Call a vet in the first time or two that you run into a problem delivery. After you've worked through the procedures with an experienced guide, you'll be able to do what's needed on your own most of the time, but that first exposure definitely benefits from guidance. I'd be cautious about asking other farmers for help and advice in this particular department; many of the old-timers we met were too quick to pull babies, used too much and the wrong kind of force, and caused more problems than they prevented. Most vets have at least one or two horror stories to confirm this observation.

Once the infant comes into the world, Mom will lick it down to clean off the mucus it's coated in. Sometimes a first-time mom is confused about what's going on. In this case, make sure the baby's air passage (the area around its nose and mouth) is cleaned up so it can breathe. If it isn't breathing, give it a good strong slap on the side.

One trick we used on more than one occasion to get a new mama "in the groove" was to bring the dogs over near the baby. Even if she isn't quite sure what the baby is or what she's supposed to do with it, if Mom senses a predator nearby her maternal instincts seem to kick in! (Just make sure you have enough control of your dogs that they can't attack the infant.)

Nursing

Baby animals need to begin nursing soon after birth for two reasons: to get the colostrum, and to get the energy. Watch Mother and Baby from a distance for about 45 minutes. If Junior hasn't found the teat by then, it's time to get involved. First, make sure that Mom's teats are open by hand-stripping a little milk out of each. Then nudge the baby into position and hold its mouth against the teat, working the mouth like an air pump. Once or twice over the years, in the case of especially weak babies, we had to strip a little milk into a nipple jar and feed the first serving that way. This usually got a baby's energy up to the point where it could work at feeding itself.

Restraint

If you can't control an animal, taking care of its problems is quite difficult. We work to get all our breeding animals tame enough so that we can move them easily into controlled areas. They can then be worked into smaller pens, which make restraint easier. This is something to begin early.

If you are dealing with fairly wild animals, you'll need an extensive handling system that includes catch pens, chutes, and headgates. These can be expensive, so consider the cost if you plan on running stockers or species like deer, elk, or bison!

At times you can exert enough control in a confined area, such as a stall, to do what needs to be done without additional restraints, but often you'll need more control. Halters, ropes, nose rings, and twitches are all good tools for controlling animals. Again, though, I can't do justice to the topic of restraint in this short chapter. The vet books listed in the appendix E, Resources, have excellent sections on restraint, as well as on administering medications. Check them out, or visit with your veterinarian.

FARMER PROFILE

Herman Beck-Chenoweth and Linda Lee

Locust Grove Farm, Inc., of Creola, Ohio, is a unique operation. When Herman and Linda started the farm, they had very specific ideas about what they wanted to do — they wanted to develop a sustainable living program, and they wanted to involve others in that program. Today, their farm is incorporated as a nonprofit organization and operated as an intentional community.

"We believe that the bottom line isn't the bottom line," Herman explains. "If we can improve our piece of ground by fostering healthy natural cycles — like the nutrient and water cycles — and if we can take care of most of our needs for food and energy, and make a little money to boot, then we are profitable. Making the most money possible but destroying our resources isn't profitable, even if traditional accounting says it is."

The farm is about equally divided among tillable ground, pasture ground, and woodlands. Goals of the community include being self-sufficient in food and using minimal energy. The community also markets excess produce, meat, and eggs, and runs a small publishing operation.

The community members began processing chickens for sale in 1992. "Using fairly inexpensive facilities ($250 worth of equipment), two people can process 30 birds per hour. The bottleneck beyond that is the evisceration step."

Locust Grove community members raise and slaughter about 900 chickens and 100 turkeys per year for market purposes, which is the maximum amount allowed to be sold, under Ohio law, to meet the on-farm processing exemption. The rest of the birds they raise are for community use.

The group is developing a larger-scale facility capable of processing up to 20,000 birds per year. "Since we are limited to selling 1,000 birds per year, our goal is to use this larger-scale facility as a teaching center.

"The publishing operation — Back Forty Books — grew out of the fact that we wanted to do a book and a newsletter on free-range poultry production." The community also offers quarterly workshops about on-farm processing of poultry. "Our workshop is for people who want to process poultry on-farm for resale."

PART III

MARKETING

CHAPTER 9

Finding a Niche

Successful direct marketers think like consumers and adapt production and products to consumers' wants. These marketers judiciously invest time, money, and creativity in pricing, advertising, displaying, and packaging their products in the best possible light.

— John Cottingham et al., *Direct Marketing of Farm Produce and Home Goods*

THE QUESTION OFTEN ARISES, especially among new and aspiring farmers: "Can I make money at farming?" The answer is maybe. Keep costs as low as possible, maximize income, learn to live very cheaply — and you just may make it.

Income can be increased in the conventional marketing system by providing a product that perfectly meets the demands of the conventional marketplace, by dealing in large quantities or by retaining ownership through the finishing phase of the animals' lives. But to really maximize income, the small-scale farmer or rancher needs to develop some alternative marketing strategies.

Most farmers and ranchers, if asked to list the reasons they chose to pursue their occupations, would come up with a long and varied list — working with nature, being their own boss, raising their children in the country, and carrying on a family tradition are a few of the common reasons given. But it's a rare farmer who would consider the opportunity to be a salesman as the driving motivation behind his or her choice. Farmers don't want to spend their time with a phone glued to their ear, or making presentations in stores. Catering to customers and maintaining more records than they already have to doesn't sound appealing: After all, it's not what farming is about!

Ken and I are with the majority on this. Neither one of us ever had a burning desire to be a salesman, but we quickly figured out that we needed to market at least part of our crop as a consumer-ready product. On average, conventional marketing strategies leave the farmer with far less than half the money that the consumer spends on meat and dairy products. Through direct-marketing, however, we were able to recapture part of that revenue, and put the extra money into our bank account.

Barb and Kerry Buchmayer (see the story on page 140) have summed up the situation well in their farm philosophy statement, which they developed as part of their planning:

We believe the key to thriving in agriculture is to take a high-value product, add more value to the product, and deliver it as close to the consumer as possible. In our case, we plan on starting with organic milk, bottling it to add more value, and selling it to retail stores or food co-ops. We believe the days of producing a commodity, taking it to market, asking what the buyers will give for it, and securing a profit are over. Profit may be possible if enough units are produced, but the small, family farmer usually does not have the capacity to run enough animals or acres to obtain these vast scales of economy.

Conventional Markets

Before I get into alternative marketing strategies, let me take a quick look at the conventional marketplace. Food is big business, and the meat segment of the food economy is no exception. About 95 percent of the meat in the United States is slaughtered and processed by less than 10 percent of the packers and processors. For example, ConAgra, Inc., of Omaha, Nebraska, is the largest meat packer and processor in the United States. This one company had more than $14 billion worth of meat-related sales in 1996, and handled beef, pork, lamb, chicken, and turkey! Two pork processors, IBP (Dakota Dunes, South Dakota) and Smithfield Foods (Portsmouth, Virginia), account for well over half the hogs butchered every day. Each handles about 75,000 animals per day every day of the year.

There are two conventional marketing concepts that you should become familiar with. First is the concept of industry concentration. *Concentration* simply means that a small number of companies controls a large share of the market. As concentration increases in a particular segment of the economy, the top companies serving that segment can exert greater control over the prices of both raw materials coming in and final products going out. The second concept is that of vertical integration. *Vertical integration* means that a company controls its product up and down the line, from cradle to grave, so to speak. The poultry industry is a good example of a concentrated and vertically integrated market segment. Almost 100 percent of commercial broiler growers in the United States are contract operators to the top processors. These processors, such as Tyson Foods (Springdale, Arkansas), own the birds and supply the feed to growers. They slaughter the birds and process the meat into all kinds of products, from frozen chicken parts (legs and thighs) to fried chicken dinners. In this segment, the growers are controlled by such tight contracts that they end up selling their labor instead of an agricultural product. However, unlike an employee selling his labor, the farmer has invested a huge amount of capital just to get the job and bears an unreasonable risk for the payback.

For big corporations, profits increase the more they are able to process the product. When an animal is slaughtered, some of its meat is simply cut, frozen, and shipped to the grocery store; other meat is made into "value-added" products such as hot dogs, lunch meat, canned stew or soup, sausage, and jerky. Dairy processors also receive a higher return on value-added items like yogurt, cheese, and butter. Process the product some more — say, into boxed macaroni-and-cheese dinners or canned franks and beans — and the processor's profits jump another notch.

Profits also come from the by-product stream. By-products are sold for processing into edible fats, leather, pet food, fertilizer, paints, inks, and even pharmaceutical products. In 1996, slaughter facilities were paid about $8 for each raw beef hide: not an insignificant amount when you consider that the top packers butchered almost 23 million head. Again, more processing means more profits. Next time you toss a ball to your kids, think about the 100,000 hides per year that are required to produce the leather that is used to make sporting goods!

The Farmer's Disadvantage

Farmers and ranchers are at a disadvantage when markets become highly concentrated and vertically integrated. As concentration occurs, many small packers and sale barns simply shut down, further limiting options for farmers in an area. Also, the relationship between supply and demand no longer works quite the way it should; low supplies don't necessarily trigger high prices, and the spread between consumer price and farm price grows wider and wider.

Another disadvantage to a concentrated market is that as concentration increases, the market looks for standardization; if you're marketing something that doesn't fit the norm, you're highly penalized. These standards are, for the most part, set for ease of handling in an industrial setting, not for quality of final product. In other words, if you try to market a hog that isn't between 235 and 245 pounds (107 to 111 kg), you'll be penalized, because packers want completely uniform animals. It doesn't matter if your hogs taste better than those raised in a factory farm setting, or that they had a good life; all that matters is that they weigh 240 pounds (109 kg) and don't have excessive back-fat.

Futures Markets and Other Options

Large-scale livestock farmers try to gain back some control over the price they receive by playing the futures markets. These markets are operated by exchanges similar to the ones that stocks are bought and sold on. The Chicago Mercantile Exchange is the largest exchange for meat, dairy, and livestock futures. What's actually being bought and sold on the exchanges are contracts for future delivery of a specified amount of product. One live-cattle contract, for example, requires that 40,000 pounds (18,144 kg) of live cattle be delivered at a predetermined future date, for a set amount of money. Futures contracts can be bought and sold for all kinds of commodities, including agricultural commodities, metals, and currencies (Table 9.1).

Most of us can't operate in the quantities that are required to actively participate in the futures market. As smaller operators, our conventional marketing options include selling at sale barns; directly to packers; or to individual buyers who act as middlemen, putting together truckloads of animals for the packers. Even these options become difficult if you aren't located in a major livestock region of the country.

Table 9.1

LIVESTOCK FUTURES

Commodity	Size of Contract (pounds)	Size of Contract (kg)
Live cattle (for butcher)	40,000	18,144
Feeder cattle	50,000	22,680
Boneless beef (90% lean)	20,000	9,072
Boneless beef trimmings (50% lean)	20,000	9,072
Lean hogs	40,000	18,144
Frozen pork bellies	40,000	18,144
Fresh pork bellies	40,000	18,144
Fluid milk	200,000	90,720
Butter	40,000	18,144
Cheddar cheese	40,000	18,144

Alternative Marketing

There are plenty of ways to market outside the conventional system. First off, really look around your farm, and make an inventory of all the possible products or services you might be able to direct-market. As food for thought, consider marketing the following:

- Breeding stock
- "Pet" animals
- Compost
- Fiber, tanned hides, bleached skulls, and other by-products
- Services, such as grooming, training, and stabling
- Agritourism (on-farm bed-and-breakfasts, hunting, fishing, camping, and so on)
- Live animals for butcher (halves and wholes)
- Cut and wrapped frozen meat
- Eggs
- More highly processed products, such as cheese, yogurt, jerky, stew
- Other secondary products, such as wood and flowers

Remember, the first key to alternative marketing is to carefully evaluate all the possible products that can come from your operation. The section on developing a marketing plan (see page 110) will help you evaluate and plan.

As small-scale farmers, we have one significant marketing advantage over the major corporations: We can capitalize on niches within the marketplace, such as natural, organic, grass finished, and humanely raised. Consumers are looking for fresh, wholesome food that has been raised in an environmentally sound manner, and we small farmers are in a position to supply them with that product.

Alternative marketing requires a strong commitment on your part. Your customers aren't faceless masses who live a thousand miles away. They are people who count on you to provide them with a high-quality, and unique, product. Alternative marketing also requires following additional laws (see chapter 10), extra planning and record keeping, and, most of all, time. Advertising, or getting the word out to your potential customers, becomes a major concern for farmers pursuing alternative markets. Educating your customers about your product, why it is special, and why they should buy from you becomes an important part of your job. But the payoff can make the extra effort worthwhile.

Group Marketing

Some alternative marketing approaches are easily done on an individual basis, but many work well on a group basis. Group marketing provides a mechanism for individuals to reap the benefits of alternative marketing, while spreading out the risk and reducing the workload of the individual members of the group. Groups often have an easier time meeting the needs of institutional and commercial markets, such as restaurants, which generally seek a regular supply of product throughout the year.

Some groups can be informal — say, a few neighbors joining together to market a semi-load of animals. Others are quite formal, like the new "value-added cooperatives" that are forming around the country. These co-ops are legal corporations that are democratically controlled by their members. Oregon Country Beef (Brothers, Oregon; see page 114), a good example of a small cooperative marketing organization, today represents twenty-nine family ranches. The group markets natural beef to food co-ops, natural and specialty grocery stores, and a Japanese firm.

Group members of Oregon Country Beef not only market their beef cooperatively, they also actively manage its day-to-day operations. Various ranch families take responsibility for certain aspects of the co-op's operation, including marketing, accounting, coordinating sales, and finishing. (At this time all Oregon Country Beef cattle are finished in a member's feedlot, though the group is looking into grass finishing for at least part of its production.) One of the things that make this co-op unique is that Oregon Country Beef doesn't purchase animals from members, add value, and then pay the members based on profits; instead, members continue to own their individual animals until they are delivered to the final customer. This approach makes members quality conscious from start to finish.

Some group-marketing arrangements develop after one farmer or rancher begins working on a direct-marketing strategy and does so well with it that he or she has to begin contracting with other growers to meet the increasing demand for the product. Coleman Natural Beef (Saguache, Colorado) and Laura's Lean Beef (Lexington, Kentucky) are two examples.

Coleman Natural Beef began selling meat in 1979, when fourth-generation Colorado rancher Mel Coleman decided to experiment with direct-marketing of the family's naturally raised beef. Today, Coleman beef is raised by more than 400 ranchers in the West, and is certified organic by the Organic Crop Improvement Association (OCIA).

Laura Freeman began Laura's Lean Beef in the early 1980s. After working as a journalist for several years, she returned to her family's 200-year-old farm in the hills of Kentucky. Laura began experimenting with managed grazing, and researched lean-beef genetics. In 1984, she began marketing beef directly from her farm. Today, Laura's company markets beef in 1,600 stores, located in 23 states; several hundred farmers — from the Southeast to the Northwest — provide the meat.

Each farmer or rancher growing for these two successful group-marketing efforts signs a contract with the company, which specifies the company's feeding and management requirements.

To find out if there are groups operating in your area, or to find help for developing your own group, check appendix E, Resources.

Niches

The goal of niche marketing is to find a way to differentiate your product from the tons of other products that consumers look at every day. No matter what you're selling from your farm, if you're going to direct-market it you need to think about how it will best fit into a niche. As Allan Nation of *The Stockman Grassfarmer* magazine says, "The first rule of niche marketing is to make your product as different from the dominant market as possible" (June 1998, "Allan's Observations"). Some of the niches that grass farmers can capitalize on include natural, organic, pasture finished, green and humane, lean, and family farm.

"Natural"

The U.S. Department of Agriculture (USDA) allows the word *natural* to appear on the label of a food product that has been minimally processed and that contains no artificial colors, flavors, preservatives, and so on. In the case of animal products, "natural production" methods must be documented. The commonly accepted methods for obtaining approval of the word *natural* on a label include pasture raising of livestock, no routine use of antibiotics, and no use of growth hormones. The animal's ration should contain minimally processed feedstuffs from natural sources, like cracked grains. Feedstuffs such as chemical urea aren't acceptable. Feed doesn't have to be organic.

"Organic"

The organic market is the fastest-growing segment of the U.S. food economy. For meat and livestock-related products to bear an organic label, the farmer must be able to prove that his or her operation has been"certified" to meet organic standards. Certification may be provided by a state agency or by a private, nonprofit certifying organization (see appendix E, Resources), depending on where you live.

Standards that are generally accepted by the national private certifying organizations for organic meat, eggs, dairy products, and so forth include:

- One hundred percent of the animals' feed must be certified organic.
- Baby animals must either be born to certified adult animals or brought in and fed certified feed as soon as possible for the species in question (chicks within 48 hours, for example, and calves within 30 days).
- Animals must be treated "humanely" at all stages of life.
- Animals must be clearly identified and trackable from birth to slaughter.
- Products used on, in, or around the animals have to come from an approved list of substances. For example, chemical pesticides are strictly prohibited, but diatomaceous earth is allowed as an insect control and wormer. Antibiotics are generally prohibited, and animals that have received them for therapeutic reasons are to be marketed through the conventional system.

Although we farmed our land in compliance with the requirements for organic certification, we never went through the process. Certifying can cost a considerable amount of money, and our customers knew us, or came to know us, so they felt comfortable with buying our meat without certification. If we began marketing in a broader area or through off-farm outlets, perhaps certifying would have made sense. The expenditure to certify, like all other expenditures, needs to be evaluated in terms of what purpose it will serve and whether or not it will pay for itself.

"Pasture Finished"

Pasture-finished animals are not fattened on grain. Interestingly, the United States is one of the few countries in the world where animals are routinely fattened on grain prior to slaughter. But pasture-finished animals can be equally as tender as fattened animals, they tend to have a more distinct taste, and they're considerably leaner.

One of the top direct-marketing livestock farmers in the United States is Joel Salatin of Swoope, Virginia. Joel has coined the term "Salad-Bar Beef" to describe to consumers the product he is selling, for his cows graze a veritable salad bar in his pastures. (Joel has written a couple of excellent books that explore his farming and marketing techniques in more detail — see appendix E, Resources.)

There are a couple of potential problems that may crop up when you market pasture-finished animals. First, Americans have become used to lots of fat in their meat, and the fat has a white color. Animals finished on pasture have fat that is more yellow than white. This pigmentation is a result of the high levels of vitamin A, vitamin D, and beta-carotene found in fresh forages. The color sometimes throws people, so you need to educate them about it. The second problem is caused by the lower fat content of the meat. Fat can form in the muscle tissue itself, and also externally to the muscle. Pasture-finished animals have less internal fat, or marbling, and significantly less external fat. This leanness is great from a health standpoint, but it causes the meat to cook more quickly and to dry out when overcooked. Teaching consumers to use slow-cooking methods is important to the palatability of pasture-finished animals. But consumers who cook pasture-finished meat correctly will be back for more!

The first time we took a steer in to the butcher we worked with in Minnesota, he was amazed at how high the yield was. *Yield* represents the difference between live weight and carcass weight hanging on the rail. Typical beef yields are less than 60 percent; "good" yields normally run between 60 and 65 percent. Our animals consistently yielded between 67 and 69 percent. These high yields were the result of the low quantity of fat on our animals. (*Note:* This yield is not the same as the amount of meat you get back — check chapter 11, Butchering and Processing, to determine how much meat yield an animal will have.)

"Green and Humane"

Green and humane marketing can attract consumers who still want to eat meat, but who want to know that the animals they're eating have been raised in a healthy and happy environment, in harmony with the environment. We developed a clientele that included a number of ex-vegetarians. They'd given up meat because of the way it was normally raised but were happy to find a source of meat that met their social goals. Coleman's Natural Beef uses this tack, advertising "beef without guilt."

"Lean"

Lean marketing attracts many customers who still want to eat meat, but who have health concerns about the amount of fat in their diet. Laura's Lean Beef strives to meet the needs of this class of consumer and has gone through extensive testing. The effort has paid off, with the American Heart Association (AHA) awarding Laura's products a "Heart Check" — an award given by the AHA to food products with discernibly reduced fat.

Family Farm

Many consumers want to support small-scale family farmers. They like the idea of their money helping keep more small farmers on the land and are willing to go out of their way to spend for this purpose.

Oregon Country Beef tries to touch on several of these niches at once with its advertising (Figure 9.1).

Marketing Venues

Where you market your products and how you reach your customers depends on your resources and your personality. Marketing venues run the gamut from on-farm sales to marketing through major grocery chains, with a wide variety of options in between.

The Farm

Most direct-marketers work right from their farms. Our customers just dropped by — though we recommended they call first to make sure somebody would be there. When marketing from the farm, there is one concern you must be conscious of: If your customers

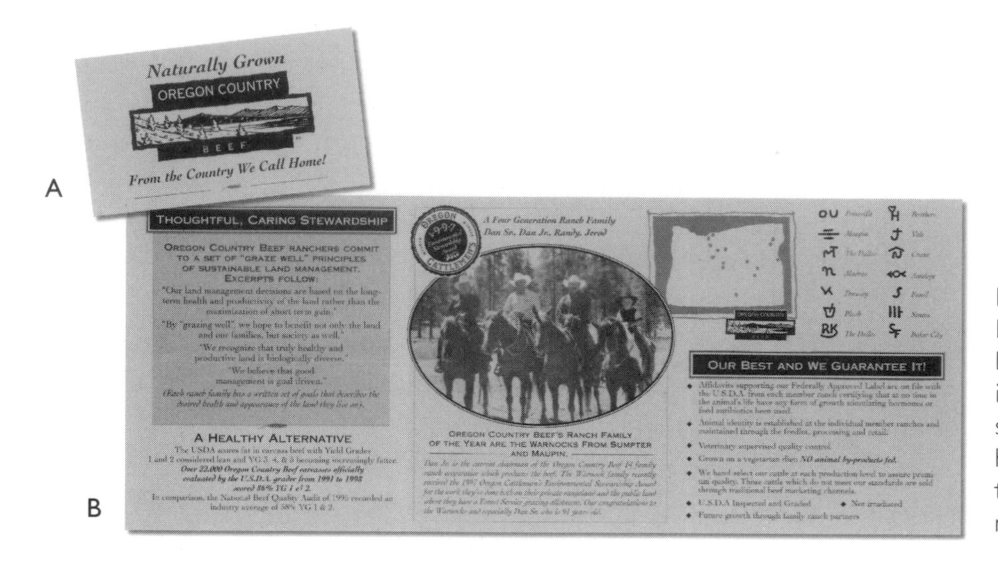

Figure 9.1. (A.) The Oregon Country Beef brochure folds out from a business card size; brochures are kept in display racks where the meat is sold. **(B.)** When opened, the brochure highlights the niche markets to which the group appeals — supporters of natural, green, lean, and family ranches.

can't find the farm, they can't buy from you. All written information (flyers, brochures, and so on) and advertisements should provide clear instructions on how to get to your place of business.

The Storefront

Direct-marketers Don Rathke and Connie Karstens, owners of The Lamb Shoppe, in Hutchinson, Minnesota, have built a storefront business on their farm, with regular hours. They have put in their own butchering facilities and a processing kitchen for creating value-added products right on their property.

Another approach to a storefront business is to move the store into town. Tommy Cashion, and his daughter Tommie-Lyn, of Woodbury, Georgia, run their storefront business in nearby Warm Springs. (See their story on p. 122.) The Cashions have always raised beef and goats on their ranch but felt they needed additional diversity on their place, so in 1994 they added red deer. Their in-town business consists of a pub and retail outlet that evolved out of their deer operation. The pub's menu features deer, and they sell a variety of deer-related products, including antler furniture, chandeliers, and lamps; wine gift baskets featuring venison sausage and jerky; and deer-hide rugs.

Farmers' Markets and Flea Markets

Farmers' markets and flea markets provide another outlet for meat and dairy products. When marketing meat or dairy products at a farmers' market, keeping things cold or frozen is a concern. Ron and Mindy Desens of Litchfield, Minnesota, regularly sell lamb and chicken at the St. Paul farmers' market. They have a chest freezer mounted in the back of their pickup that stays plugged in until it's time to head to the market; once at the market, they pay extra for a space with electrical outlets available.

Location. Farmers who use the farmers' market approach do best if they are located near a large, urban area, or a place where urban consumers are passing through regularly. From reports we've had and our own experience, rural farmers' markets don't do as well, because people in exclusively rural communities go to the farmers' market to find bargains. When you're direct-marketing, you want customers who are looking for quality and are willing to pay for it, not bargain hunters.

Presentation. Successful farmers' marketers also say that presentation is a major concern. Attractive displays draw people in. Meat is best packaged in Cryovac (vacuum-sealed packaging), which allows the consumer to view the product. Most butchers can supply this type of packaging for a small additional charge, but David Schafer, who markets at the Kansas City farmers' market, says this extra charge is well worth it. He also says, "Post anything that has ever been written about you and your farm. Customers like dealing with a 'famous farmer.'"

Samples. Another strategy that helps move product at a farmers' market is to give out samples. Heather Olson sells eggs, bacon, and breakfast sausage every week at the Litchfield, Minnesota, farmers' market. Heather takes a camp cookstove with her, and fries up samples for the crowd, drawing people over to her stand. She also brings along store-bought eggs: She breaks a store-bought egg in one dish and a farm-fresh egg in another. The old saying goes, "A picture paints a thousand words," and the beautiful golden color of her egg yolks next to the washed-out-looking store-bought eggs does just that.

Consumer-Supported Agriculture

Consumer-supported agriculture (CSA) operates on a subscription basis and most often involves market-gardening operations. However, farmers who run CSAs often offer meat and eggs — of their own, or from another area farmer — as part of the member's subscription package. Members pay at the beginning of the year and receive a share of the farm's production during the growing season. The up-front money helps the farmer plan for the year's needs and gives him or her money at the beginning of the year to purchase supplies. The members also bear some of the risk; in a poor crop year, members receive less than in a good crop year.

In one variation on the CSA theme, customers order their meat at the beginning of the year, so the farmer can grow basically what he has standing orders for. Joel Salatin markets pasture-raised chickens (and eggs), turkeys, rabbits, pork, and beef (mixed quarters, halves, or wholes) this way. Joel sends out an order form in January, with the customers specifying not only what type and quantity of meat they want, but also on which of several delivery dates that Joel offers they would like to receive their product. Chickens and rabbits are butchered and picked up at the farm. Beef and pork are butchered at a local packer, where Joel meets his customers on pickup day to complete the financial transaction. (The Salatins also offer federally inspected and labeled beef and pork for customers who don't want to deal with the large quantities of halves and wholes.)

Stores and Restaurants

Marketing to stores and restaurants takes a greater commitment of time on your part. These accounts must be serviced regularly. We tried to develop a working relationship with a small food co-op about an hour away. For us, this didn't work; the distance was too great for us to service them, given the small quantities of meat that they could turn over. Stores and restaurants also require a steady supply that may become difficult for individuals to meet. This is one area in which group efforts pay off especially well.

Image

Successful marketing, regardless of the venue you choose, is somewhat like improvisational theater. Your customers have a picture in their mind of what a farmer or rancher is supposed to look like, and it behooves you to play the part. Clean bib overalls and a straw hat, or a plaid shirt, jeans, and a cowboy hat fit the bill. Neither a scraggly hippie look nor young-businessman attire works well for your audience. If you're unsure of what look matches the mental picture that suburban and urban consumers have, pick up a Smith & Hawken or Eddie Bauer catalog and study the styles. That's country to city folks!

Developing a Clientele

Developing a clientele is one of the toughest jobs for the direct-marketer. Talk to any successful direct-marketer, and he or she will tell you that the key to success is developing a loyal customer base — folks that come back again and again.

Word of Mouth

Loyal customers are also your best advertisers: Word-of-mouth advertising grows a business better, and is cheaper, than any other form of advertising. We began marketing first to friends and acquaintances, but, largely through word of mouth, our customer base steadily increased.

There are two different terms that professional marketers use: advertising and promotions. *Advertising* is simply putting your business in the public's eye, primarily through the use of some "media." Everything from business cards to paid ads in commercial media (newspapers, Yellow Pages, radio, or television) falls into the category of advertising. T-shirts or baseball caps with your farm's name on them are also a form of advertising, as is providing free samples.

Promotions are a specialized form of advertising in which your business supports some community event or group. Sponsor a Little League team, and you are engaging in a promotion. Provide free, or at-cost, ground beef to the local Rotary Club for its annual community picnic, and you are doing a promotion.

Flyers

One of the most successful, and cheapest, advertising tools we found for getting our business in front of new customers was a computer-generated sign with our phone number on "tear-offs" at the bottom (Figure 9.2). We placed these on bulletin boards in laundromats, convenience stores, the library, and on other readily accessible public bulletin boards. Inevitably, each one of these signs garnered us at least one new customer.

Classified Ads

Newspaper classifieds didn't net us many meat sales, though we did sell some breeding stock and calves using classifieds. Some farmers do have success marketing meat through newspaper advertising. Stephen and Kay Castner of Kay's Home Farm Lean Meats in Cedarburg, Wisconsin, run a regular display ad in their local weekly paper.

Figure 9.2. Little Wing Farm brochure, with "tear-offs" at bottom, an inexpensive advertisement.

Internet

The Castners have also developed a Web page on the Internet. They've had it for a number of years, and although they don't feel it has had much impact on meat sales it has helped them market breeding stock. They raise purebred Galloway cattle and Targhee sheep on their farm, and contacts made through their Web page have netted breeding stock sales as far away as Kansas.

Ethnic and Special Markets

Ethnic markets sometimes provide a good clientele, but they may require special services. Sherry O'Donnell of Red Lake Falls, Minnesota, found an excellent market for her lambs among Muslims attending school in North Dakota. Muslims have religious ceremonies that are to be performed when the lambs are slaughtered, so they want to do their own butchering. Sherry set up an area for butchering right on her farm specifically to accommodate the needs of this clientele. Muslim families come out, spend the day, and butcher several lambs each for their freezers.

Farm Visits

Most successful marketers encourage their clients to visit the farm, even if sales don't occur there. Some host an annual barbecue to which all their regular customers are invited. Others just extend a fairly open invitation. No matter what approach you use, the goal is always the same: to create a proprietary interest in your operation among your customers (Figure 9.3).

Our customers had a standing offer to come out to the farm anytime, though we did encourage them to phone first and make sure we'd be home. Most included an outing to our farm anytime they had out-of-area visitors. Our customers loved the fact that they could bring their friends to a real farm where the kids could pet baby animals, or even some big ones. Some of these visitors became long-distance customers, purchasing meat from us whenever they came through the area, but even if the visitors never purchased anything, the "hosts" became more dedicated customers!

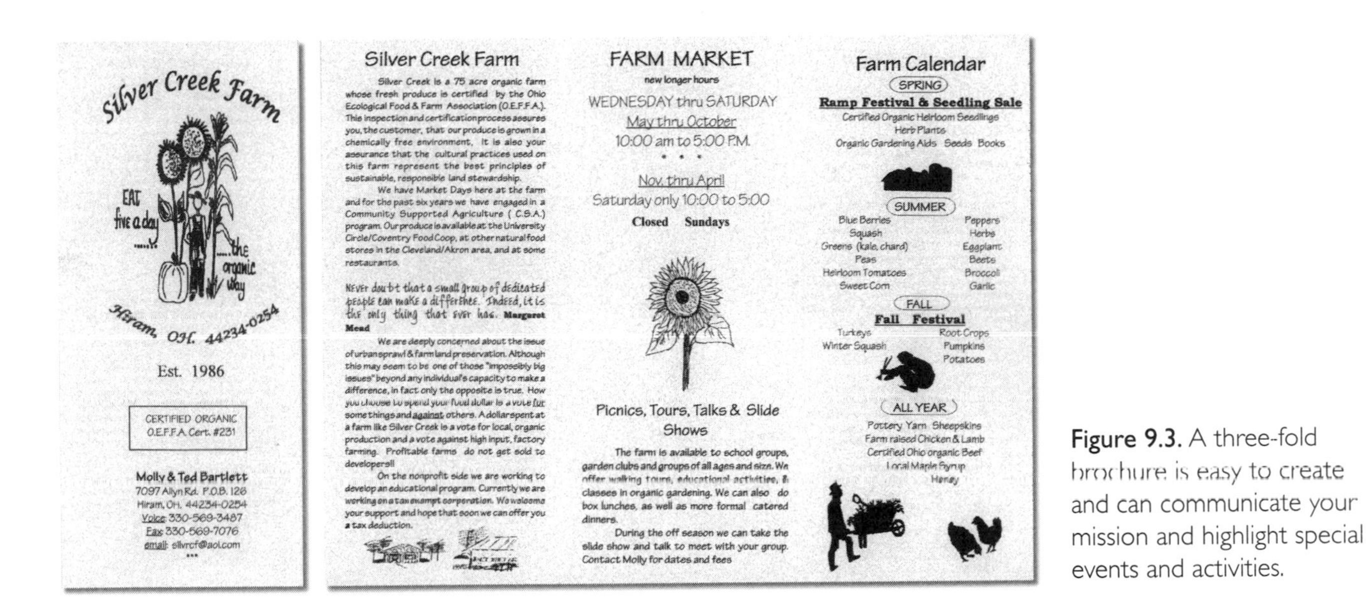

Figure 9.3. A three-fold brochure is easy to create and can communicate your mission and highlight special events and activities.

Community Outreach

Another strategy for developing a client base is doing promotions for civic groups. Offer to attend a meeting of the Rotary, a homemaker's club, or an environmental group. Prepare a slide show and presentation describing your farm's operation and the environmental benefits of grass farming. Have some sample products to give out after the presentation: meatballs in a Crock-Pot, say, or hard-boiled eggs. If your goal is to offer on-farm hunting or fishing, attend a sportsmen's club meeting.

Samples

Offering samples works not only at the farmers' market and when you're doing promotions for civic groups, but also in almost any public place where you can set up to do them. Our local natural foods store lacked the freezer space to carry our meat, but Mary, the owner, let me come in from time to time and set up a table to give out samples. The smell of cooking meat helped draw people into her business, and helped me find new customers: a win-win situation!

Home Delivery

Home-delivery services work for some farmers. Dick and Pam Bowne run Gemini Golden Guernsey Milk of Palisade, Minnesota. A home-delivery route for bottled milk from their herd of Guernsey cows is a primary part of their business. Guernsey milk has a great golden color and wonderful taste. "We wanted to be able to provide this quality of milk directly to an appreciative clientele," says Pam. To initially build that clientele, the Bownes went door to door in target neighborhoods, leaving a small sample bottle of milk and information on ordering. As well as running a home-delivery service, Pam and Dick market their milk and ice cream at stores in the Minneapolis–St. Paul area and at the Minneapolis farmers' market from mid-April through mid-November.

Mail Order

Mail order is the final strategy successful direct-marketers use. One problem with the mail-order approach is that shipping costs are quite high for meat or dairy products, but specialty meats and other farm products may have a high enough profit margin to do well from a mail-order catalog. Specialty meats include those coming from "exotic" animals, such as bison or ostrich, or certain processed products like jerky.

Maintaining a Clientele

Once you've identified clients, keeping them becomes your next challenge. Find some way to keep in touch with regular customers. Flyers and newsletters, handed out during presentation, or mailed to your customer list, work well (Figure 9.4). Tell people what's happening on

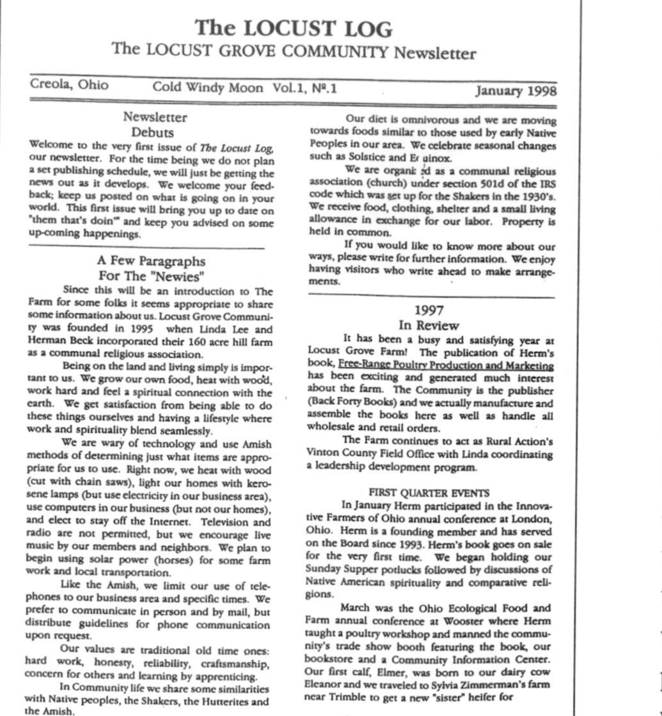

The LOCUST LOG

The LOCUST GROVE COMMUNITY Newsletter

Creola, Ohio — Cold Windy Moon Vol.1, №.1 — January 1998

Newsletter Debuts

Welcome to the very first issue of *The Locust Log,* our newsletter. For the time being we do not plan a set publishing schedule, we will just be getting the news out as it develops. We welcome your feedback; keep us posted on what is going on in your world. This first issue will bring you up to date on "them that's doin'" and keep you advised on some up-coming happenings.

A Few Paragraphs For The "Newies"

Since this will be an introduction to The Farm for some folks it seems appropriate to share some information about us. Locust Grove Community was founded in 1995 when Linda Lee and Herman Beck incorporated their 160 acre hill farm as a communal religious association.

Being on the land and living simply is important to us. We grow our own food, heat with wood, work hard and feel a spiritual connection with the earth. We get satisfaction from being able to do these things ourselves and having a lifestyle where work and spirituality blend seamlessly.

We are wary of technology and use Amish methods of determining just what items are appropriate for us to use. Right now, we heat with wood (cut with chain saws), light our homes with kerosene lamps (but use electricity in our business area), use computers in our business (but not our homes), and elect to stay off the Internet. Television and radio are not permitted, but we encourage live music by our members and neighbors. We plan to begin using solar power (horses) for some farm work and local transportation.

Like the Amish, we limit our use of telephones to our business area and specific times. We prefer to communicate in person and by mail, but distribute guidelines for phone communication upon request.

Our values are traditional old time ones: hard work, honesty, reliability, craftsmanship, concern for others and learning by apprenticing.

In Community life we share some similarities with Native peoples, the Shakers, the Hutterites and the Amish.

Our diet is omnivorous and we are moving towards foods similar to those used by early Native Peoples in our area. We celebrate seasonal changes such as Solstice and Equinox.

We are organized as a communal religious association (church) under section 501d of the IRS code which was set up for the Shakers in the 1930's. We receive food, clothing, shelter and a small living allowance in exchange for our labor. Property is held in common.

If you would like to know more about our ways, please write for further information. We enjoy having visitors who write ahead to make arrangements.

1997 In Review

It has been a busy and satisfying year at Locust Grove Farm! The publication of Herm's book, Free-Range Poultry Production and Marketing has been exciting and generated much interest about the farm. The Community is the publisher (Back Forty Books) and we actually manufacture and assemble the books here as well as handle all wholesale and retail orders.

The Farm continues to act as Rural Action's Vinton County Field Office with Linda coordinating a leadership development program.

FIRST QUARTER EVENTS

In January Herm participated in the Innovative Farmers of Ohio annual conference at London, Ohio. Herm is a founding member and has served on the Board since 1993. Herm's book goes on sale for the very first time. We began holding our Sunday Supper potlucks followed by discussions of Native American spirituality and comparative religions.

March was the Ohio Ecological Food and Farm annual conference at Wooster where Herm taught a poultry workshop and manned the community's trade show booth featuring the book, our bookstore and a Community Information Center. Our first calf, Elmer, was born to our dairy cow Eleanor and we traveled to Sylvia Zimmerman's farm near Trimble to get a new "sister" heifer for

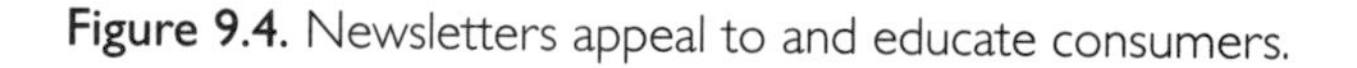

Figure 9.4. Newsletters appeal to and educate consumers.

your place, tell stories about your animals, or let them know about food issues and policies that affect them and you — like a factory farm trying to come into the area.

All correspondence with your customers should include information on ordering, product availability, new products, and price changes. Computers are useful for preparing newsletters, as well as maintaining mailing lists and preparing flyers and labels, but they aren't absolutely necessary. Hand-prepared materials work, but they must be neat and legible.

Developing a Marketing Plan

The first step into alternative marketing is planning. Take the time to work through a planning process, and your likelihood of success will be greatly increased. Planning for marketing is special, and separate from the general planning you'll work on in part IV, so I'll cover it here. Still, some of the information you'll need for your marketing plan — such as cost of production — will actually be developed as part of the whole-farm planning you'll do in part IV.

The goal of your marketing plan is to help you understand where your product fits into the market, who your customers are, what they want from you (their likes and dislikes), and who your competitors are. It should help you develop advertising and pricing strategies. It's primarily a tool for your own use, but if you need to seek outside financing to begin your marketing endeavors bankers will want to see it. Developing your plan is a form of mental exercise; work on it in family brainstorming sessions.

Brainstorm

Begin by identifying all the possible products you could sell from your farm. The list I presented earlier under Alternative Marketing (see page 102) can provide a starting point, but try to think of other items as well, and break the list down into as much detail as possible. At this point, your objective is to come up with the largest laundry list of ideas you can; some may seem crazy or off the wall, but list them anyway. After you generate the ideas, you can begin to evaluate them individually — and sometimes a crazy idea turns out to be not so crazy after all.

Evaluate: Research and Critique

Evaluation is really a two-part affair. The first part is doing some preliminary research, and the second part is comparing the items against each other.

Consider Legal Issues. Research comes first, because much of what you find out will be incorporated into your comparison. For each idea, explore what kind of legal concerns may come up (see chapter 10 for details on legal research). Do you need a license to establish the business? Are you going to need additional insurance?

Survey Potential Markets. Finding out what potential customers want and need is another step in the research process. You need to understand your customers' wants and needs even better than they do, and the best way to find out is to ask. Survey friends, family, and business acquaintances. Ask them what they like, how much they buy, what they're willing to pay, and so on. When possible, avoid questions that result in a yes or no response; instead pose questions that dig deeper into people's feelings — leave them open ended. Don't say, "Do you like beef?" Instead say, "How do you like your beef?" and, "Which cuts do you prefer?"

If you're considering selling at a farmers' market, attend one some Saturday with clipboard and questionnaire in hand to survey the crowd. Say, "Excuse me, folks, my name is Janet Jones, and my husband and I own a farm about 20 miles from here. We're thinking about starting a booth here at the farmers' market to sell our [insert product name]. Would you mind answering just a few questions?" Ask why they come, and how often. Ask what items they think are missing from the market now. If you are thinking about eggs and chickens, then ask if they'd be interested in purchasing eggs from chickens that didn't have to spend their lives in a square-foot cage. Most important, ask why they shop at the farmers' market. If the bulk of respondents say they do so to save lots of money, your prices will need to be lower than grocery store prices — or your sales pitch will need to be great! If regular responses include things like supporting family farms, or obtaining extra-fresh or natural food, you can probably charge a premium over grocery store prices.

The same type of survey can be used for businesses that you think may be interested in purchasing or carrying your products. If you survey businesses, have your questions well thought out ahead of time. A busy manager or chef may spare you a few minutes, if you can ask pointed questions, but he or she isn't going to spend large amounts of time with you. Surveying busy people is like fishing; your hook needs to be baited before you throw it in the water.

Understand the Competition. Understanding your competition is another important aspect of research. Who already sells what you are considering selling? What do they charge? Do they offer specials? Is there some way you can effectively differentiate your product from theirs? Study packaging and advertising.

At the farmers' market, study the booths that do a booming business. What's the difference between them and the booths that take home most of what they came with? How do the successful marketers display their wares? How do they interact with the crowd?

Matrix Analysis

Once you've done the preliminary research, you need to compare the ideas. An excellent approach to doing this is using a technique known as "matrix analysis." Don't let the fancy name throw you — the technique itself is easy. You are simply designing a graphic model that lets you easily assess each idea's strengths and weaknesses. Table 9.2 shows a model matrix.

When using matrix analysis, you judge each idea against various criteria, with each criteria having a "weight" or a certain number of assigned points. *You need to determine the criteria, and how to weight the responses within each criteria,* but the following list should give you some ideas. (Again, to come up with the criteria, brainstorm, and talk about what's important to your success and your quality of life.)

1. ***Current status.*** Is this something you are currently raising, something you could easily venture into, or something that would require lots of work to begin raising in the future? *Weight:* 10 points for an animal, crop, or situation you already have set up on the farm; 5 points for those that could be easily set up; 0 points for those that would be difficult to add.
2. ***Volume.*** Consider current and projected volume. How many head, dozen, acres, and so on, do you currently produce? Can production readily increase if demand warrants it? What is the maximum production you could realistically deal with? *Weight:* 10 points for items that you could begin marketing with current production volume; 7 points for those that would require only a slight increase.
3. ***Customers.*** Who might buy from you? Individual consumers, institutions, retailers, restaurants? *Weight:* If you are trying to start small, give higher weight to individual consumers. If you already raise 1,000 pigs per year that you market conventionally, then the opportunity to market to an institution might be worth more points.
4. ***Channels.*** Will customers come to the farm? Will you attend a farmers' market, or will you sell to a small cooperative? *Weight:* If you enjoy company, you might score customers at the farm high, but if you like privacy it may score 0.

Table 9.2

Example Marketing Matrix

Possible Enterprises	Fresh Eggs	Broilers	Turkeys	Grass-Finished Beef, Conventional	Grass-Finished Beef, as Wholes and Halves	Grass-Finished Beef, by the Pound (kg)	Firewood	Paid Hunting
Who are my customers? Directly to consumers = 10 points, value-added retailers or restaurants = 5 points, wholesale or conventional markets = 0	10	10	10	0	10	10	10	10
Is this something I'm already doing in my operation, or would it be a new endeavor? Current operation = 5 points, new operation = 0	5	5	0	5	0	0	0	0
How will I advertise? (cumultative points) Word of mouth = 10, promotions = 7, other free or low cost = 5, paid = 0	22	22	22	0	22	22	5	0
Methods (cumulative points) At farm = 10, farmers' market = 5, value-added retail or restaurant = 5	15	15	15	0	10	20	10	10
Staff: Will this require more labor than we currently have available? Yes = –10, no = 0	0	0	0	0	0	0	–10	–10
Price: Cost Ratio	2.5	2.9	5	1.1	1.3	1.7	1.5	4
Will I have to deal with legal stuff? Yes = –5, no = 0	0	0	0	0	0	–5	0	0
Total Points	44.5	54.9	52	6.1	43.3	48.7	16.5	14

Note: When using a marketing matrix to evaluate various enterprises, establish the "scoring" criteria based on what is important to you

5. ***Targeting.*** How will you reach customers? Can you meet the special needs of an ethnic group? We sold fresh ducks to the owner of a Chinese restaurant for his family's personal consumption at holidays. He could purchase frozen duck at wholesale, but he was willing to pay for fresh ducks as a special treat. Can you get a specialty item into a local restaurant? A restaurant-bar near us had too high a demand for us to sell them meat for their regular menu, but we were able to sell them Polish sausage, which they served as a special in the bar during *Monday Night Football*. If you're thinking of selling hunting privileges at your private pond, can you make arrangements with a chapter of Ducks' Unlimited? *Weight:* If you believe you can easily target your audience (maybe the state director of Ducks Unlimited is a friend), give yourself a 100. If targeting will be hard, make it a 0.
6. ***Advertising ideas.*** Where will you advertise? Is word of mouth how you plan to grow, or do you need a campaign to reach people? Can you contribute, or sell at cost, ground beef to a civic organization's food stand — say, the High School Booster Club, for burgers at football games — in exchange for the advertising? *Weight:* If your plan is to market breeding stock, and your adult daughter designs Web pages for a living, give yourself more points. If you live in a county in North Dakota with only 700 residents, and your advertising for on-farm hunting can't just be an ad in a weekly local paper, give yourself a lower score.
7. ***Competitors.*** Who is the competition? Other farmers who are direct-marketing in your area, grocery stores? What does the competition charge? How can you differentiate your product from theirs? *Weight:* If you're thinking about direct-marketing lamb, and you have a reasonably close, urban farmers' market that no one is selling lamb at, you might assign 10 points for competition. On the other hand, if you want to sell eggs at the same market, but there are already three farmers selling eggs there, you might assign 0 points.
8. ***Time.*** Do you have the time to market this, or to raise additional animals? *Weight:* If you have plenty of time for this endeavor, give it 10 points. If you can do it, but your time is already limited, give it 5 points. You just don't have the spare time you think this will take, give it 0 points.
9. ***Pricing.*** How much do you think you can charge? What is your cost of production? What is the ratio of price to cost (the ratio equals the price divided by the cost)? Chapter 13, Financial Planning, will help you estimate your cost of production. *Weight:* When assigning points for pricing, give the highest weight to the item with the highest price:cost ratio, and the lowest points to lowest price:cost ratio.
10. ***Limitations/concerns.*** Does this require special licenses, a new building, or special equipment? *Weight:* In your weight system, these items may have negative weights. The need for constructing a new building might be –25 points, the need to purchase a new freezer might be –10, and the need to apply for an annual license from the county may be –5 points.
11. ***Labor.*** Will marketing this require additional help? Are family members and friends able to supply the extra labor, or is it going to require hiring employees? *Weight:* If you need to hire employees, this will probably be weighted with a negative number.
12. ***Special bonus points.*** Are there reasons to give this option extra points? Perhaps the customer surveys you did at the farmers' market indicated that there was a void you would be able to fill. Or maybe you have always dreamed of opening a bed-and-breakfast, so this rates extra points for its dream value.

Now, to actually create the matrix, start with a large piece of paper (either a poster board or a roll of white freezer wrap will work well) and write each idea across the top, starting about 4 inches (10 cm)from the left side. Down the side, write all your criteria; then you'll work across filling in all the boxes for each idea. As you fill in the boxes, you'll add current and projected status, as well as comments and thoughts on the idea. In the lower right-hand corner of each box, write down the points. When you have filled in all the boxes, add the points to obtain a score for each idea. You now have a weighted evaluation of all your ideas.

(cont'd on p. 115)

FARMER PROFILE

Doc and Connie Hatfield, and Oregon Country Beef

The year was 1986, and with 10 years of operating their Brothers, Oregon, ranch under their belts, Doc and Connie Hatfield were disheartened: The ranch was going broke in a poor cattle market.

Connie and Doc began talking about markets and marketing. Then, on a trip to Bend, Connie got a wild impulse and went into a health club. The owner, "a twenty-something Jack LaLane by the name of Ace, came out and asked if he could help me," Connie recalls. "I explained that I was a cattle rancher and wanted to know what he thought about beef.

"My first reaction was pleasure, when he said, 'Oh, I recommend that all my clients eat beef at least three times a week,' but before the smile could break my face, he said, 'but I'll tell you, it's so hard to find Argentine beef anymore!'

"I was dumbfounded for a minute. 'Why Argentine beef?' I asked. Ace went on to explain the benefits of Argentine beef — lower in fat, no antibiotics, no hormones. Well, darn, I thought, we're going broke, and there are people looking for the very thing we produce, but we've just never marketed it."

Shortly before this, the Hatfields attended a holistic management meeting. "That meeting really helped us view ourselves as part of a bigger picture," Connie says. And that meeting introduced the Hatfields to some other producers who seemed to share our values.

"We hosted a meeting in our living room," Doc says. "Thirty-six folks (husbands and wives) were crammed in there, and we figured we had 10,000 mother cows in that room, so to speak." The thirty-six folks were part of twenty-three families, from fourteen family ranches that spread out over eastern Oregon. Some came from the mountain areas, some from the high desert, but they all had interest in grazing well to take care of their land, and they all had interest in developing a closer connection to consumers. After the first meeting, each family agreed to provide one pen of finished steers, and Connie was appointed to try and market them.

"The co-op started with a clear vision and a written goal, and with that vision and goal has been able to build strong relationships," says Doc. "We have a legal organization, but we have no contracts with anybody, and the organization doesn't own much of anything. The co-op itself has less than $5,000 in capital assets — mainly a computer — but we have manpower and brainpower within the group to make things happen."

And the group has made things happen. From a rather humble beginning of just five head for slaughter per week, the group has grown to butchering 120 head per week. And Oregon Country Beef, in its eleventh year, did over $4,000,000 worth of business, marketing their branded beef through grocery stores and restaurants on the West Coast and in Japan.

Traditionally, co-ops began by members purchasing a share. Then the co-op would buy the member's product, add value, market the product, and pay members based on profits. But as Doc explains, "Buying shares doesn't fit with our group's goal. It creates a disconnect, where members aren't directly responsible or accountable. Our members are committed to a long-term program and take responsibility for the quality of the product they put out, all the way to the dinner table."

Each member of the co-op, beside putting cattle in for sale, agrees to do the following:

1. Attend two major meetings each year, spring and fall.
2. Attend Annual Appreciation Day in the summer, where customers are invited (store managers and owners, restaurant managers and owners, and individual customers from those businesses) out to one of the ranches for a picnic and tour.
3. Spend a weekend yearly in Seattle, Portland, or San Francisco doing in-store meat demonstrations and visit.
4. Commit to the Graze-Well Principles (principles of livestock management that lead to healthy land, livestock, and people).

The co-op's success is due, at least in part, to Connie's natural talent as a marketer. When she was first delegated by the group to try to find markets, Connie went to the city to study her potential clientele. "I'd first go into a store, on a reconnaissance mission. I'd walk around the store for a few minutes, I'd buy something to drink, and then I'd go sit in the parking lot and observe. I was looking for stores with an affluent clientele that would possibly support our natural beef at a price that would pay us our cost of production, a return on our investment, and a reasonable profit," she explains.

When a store looked like it might fit her criteria, Connie would make an appointment with the store's management and owners. "Store managers and owners generally allow 5 minutes for 'vender meetings,' as they call them: It's the time they allow a three-piece suited salesman to pitch his product. I'd get in there, dressed in my denim skirt, and my 5 minutes would quickly turn into an hour. I wasn't just a salesperson — I was a business owner just like them. And I cared more about my product than any salesperson ever could."

Connie would always begin her contacts, not by trying to pitch what she had, but by inquiring what niche Oregon Country Beef could fill for the store. Store personnel gave her some guidance, and off she went to find a packer the group could work with. Another "5-minute meeting" with a packer turned into 45 minutes, and the group found the packer to handle their product. "We were off and running," Connie says.

"We network with our packer, distributors, and customers. They are honorary members of the co-op and receive copies of all our reports, detailing finances and planning. We don't have contracts with any of these folks, either — we don't feel we need them. We want to be partners with them in getting a high-quality product to consumers who want it," Doc adds.

"Our Japanese client, Kyotaru Co., taught us a lot about building relationships in our business. In Japan, there is a concept called *Shin-Rai.* It's hard to directly translate Japanese language and culture, but the idea behind *Shin-Rai* is that two businesses work together for the benefit of both. This is a very different approach from the American 'competition' paradigm."

Oregon Country Beef has grown slowly and steadily using its approach of partnering with other businesses to get a premium product to consumers, and by standing 100 percent behind quality at all steps in the process. But, as Doc tells me, it "hasn't been all sweet. Other groups can do what we've done, but if during development of your co-op somebody isn't cussing or crying, you probably aren't going to get anything done."

Several of those that scored highest may be worth pursuing in the here and now; those that scored lower may be worth pursuing in the future, but they aren't at this time! Take the ideas that have scored high from a marketing-plan approach, and work through the overall planning in part IV with them.

Pricing

Price is basically a factor of what the market will bear, or how much people are willing to spend for your product (Figure 9.5). Your earlier research should help you determine the price the market will bear, but here are some other points to think about when you set prices:

- ***State the price carefully.*** Sometimes the way you state a price is the key to success. We marketed home-grown garlic at a local flea market for a couple of summers. At first we posted our price at $5.00 per pound. Customers screamed highway robbery. Then we changed our pricing to $0.75 per bulb, and the same customers thought it a good deal for our nice fat bulbs. But guess what: $0.75 per bulb was $5.00 per pound.

Little * Wing * Farm

JANUARY 1995 PRICE LIST

HAM	$2.40
HAM STEAK	$2.60
BREAKFAST SAUSAGE	$2.90
BRATWURST	$2.50
POLISH	$2.50
ITALIAN	$2.50
BACON	$2.70
SHOULDER BACON	$2.85
PORK CHOPS	$2.00
PORK RIBS	$1.90
HAMBURGER SALE PRICE	$1.65
BEEF CUTS COMING SOON	
LEG OF LAMB	$2.95
LAMB CHOPS & STEAKS	$4.00
GROUND LAMB	$1.49

RAISED AND DISTRIBUTED BY
LITTLE WING FARM, VERNDALE, MN, 56481
218-445-5494
MEAT IS PROCESSED AND INSPECTED AT
MILTONA CUSTOM MEATS, MILTONA, MN

Figure 9.5. Set prices based on the market and the uniqueness of your products.

- ***Emphasize the uniqueness of your product.*** "Fresh, Free-Range Eggs" in large print, "$2 per dozen" in small print. Break open one of your eggs next to a store-bought model, then explain to customers that your beautiful golden yolk comes from the extra vitamin A, vitamin D, and beta-carotene that your chickens get running around in a sunny yard!
- ***Feel free to mark up.*** We priced our meat, which we sold to a strictly rural clientele, just slightly above the top end of grocery store prices. The "good" Polish sausage at Fred's Hometown Grocery sold for $2.39 per pound, ours sold for $2.50. If we had lived near an urban center, our markup would have been even greater.
- ***Sell higher than the top end of the butcher market.*** For animals we sold on the hoof (see chapter 10), our pricing was slightly higher than the top end of the butcher market, and the customer paid for the butchering. Butcher hogs, during a high market, sell for 65 cents per pound ($1.43 per kg) live weight. We sold half or whole hogs at 70 cents per pound ($1.54 per kg), and we delivered it to the butcher. This was our price, both when the market was high and when it was low (and butcher hogs were selling for 35 cents per pound [$0.77 per kg]).
- ***Prices must exceed cost of production.*** Ideally, cost of production includes not only fixed costs (loan and mortgage payments, taxes, insurance, and so on) and variable costs (wages for paid employees, feed, fuel and maintenance for machinery and buildings, et cetera), but also an allowance for family salaries. Chapter 13 will go into these calculations.
- ***Price consistently.*** Changes, when they are necessary, must be made slowly and in small increments.
- ***Run a sale.*** Don't be afraid to put an item on sale. Sales can be especially useful if you have too much of one item taking up freezer space.

CHAPTER 10

Legalities

We the People of these United States, in order to form a more perfect Union, establish justice, insure domestic tranquility, provide for the common defense, promote the general welfare, and secure the blessings of liberty to ourselves and our posterity, do ordain and establish this Constitution of the United States.

— Preamble to the Constitution of the United States

ONCE YOU DECIDE to move beyond conventional marketing, you enter a maze of laws and regulations. You'll have to interact with bureaucrats. And you'll court new levels of liability.

Our phone rang early one morning a few years back; the woman on the other end introduced herself as a pig farmer who lived in the next county. She explained that she had called the state Department of Agriculture and asked what she had to do in order to direct-market meat from her farm. The answer from the employee she spoke to was, "You can't!" Shortly after, while surfing the Internet, she came across the home page of the Minnesota Institute for Sustainable Agriculture (MISA). On a whim, she decided to give her quest one last try; she called the institute. A staff member, Debra Elias, said, "I know you can do it, but I'm not sure what you need to do. I know some farmers in your area who have done it, though, and they might just be able to help." Debra gave her our number.

For most farmers, including our neighboring-county pig farmer, working their way through red tape and bureaucracies is, at the very least, intimidating and frustrating; at the worst, it's a true nightmare. But don't give up; it can be done, and the payoff is worth the trouble.

Legal Process

Civics — ah yes, that class we are all compelled to take in high school, which we then quickly forget. Sure, we remember that the federal government is a three-pronged affair: executive, legislative, and judicial; we know that presidents are elected every 4 years; we know taxes are inevitable; and, we're pretty cynical about the integrity of our elected officials and bureaucrats. Beyond that, government is kind of a blur. But the best way to accomplish your goals and negotiate this legal maze is to have a good basic understanding of the processes that go on at federal, state, and local government levels. In other words, another civics class.

Before I go on, I want to make some completely personal observations about bureaucrats (before any of them who might be reading this say, "Hey, wait a minute, we're not all bad"). Although a few of them are complete jerks on a power trip, most are good folks who are just trying to do their jobs. And for the most part, they're trying to do it right. They usually believe that what they are telling you is correct, and in the best interests of everyone, but sometimes even the good ones are wrong. Don't be afraid to stand up for your rights: Be polite, businesslike, and firm when you have to.

Now back to the civics lesson. No matter what level of government entity you're dealing with, from a local town council up to the federal government, there are certain similarities in the way all of them operate. Each level operates under documents that spell out its rules for operation. In the case of towns and counties, either state statutes describe their function and authority, or a document called the Articles of Incorporation elaborates on their function. The states, like the federal government, follow a document called a Constitution. The Constitution is the highest law in the land, and although it can be amended, the process isn't easy. It tells the government, and the people, exactly how things will work.

Each entity has some type of executive branch: mayors, governors, and the president all represent the executive officer. The executives may recommend laws to the legislative branches; they may have veto power over laws enacted by the legislative bodies; they may be granted a vote to break a tie of the legislative body; but they don't actually create the laws. Their job is to see that the laws, once enacted by the legislative body, are implemented and enforced. Executives, with the approval of the legislative branch, hire top-level employees, or bureaucrats, to actually run day-to-day affairs.

The legislative body (the U.S. Senate and House of Representatives, state Senates and Houses of Representatives, boards of county commissioners or town councils) actually holds the power to enact laws. Laws may go by different names — *ordinances*, *statutes*, and *codes* are common examples — but their adoption must follow certain procedures designed to guarantee public input.

Although the executive branch can't actually adopt laws, there are procedures that allow its members to adopt regulations. These are designed to guide implementation and enforcement of the laws that the legislative branches adopt.

The judicial body is the courts. The courts act as arbiters between the executive and legislative branches; they determine if adopted laws are legal under the Constitution; they provide a vehicle for individual members of society to file grievances and collect compensations, both against each other and against government entities; and they provide the system to adjudge guilt or innocence under the criminal codes.

As citizens, we have rights and responsibilities. Laws and regulations are — in the best of worlds — supposed to specify, clarify, and enlighten us on our rights and our responsibilities.

Laws and Regulations

Understanding your rights and responsibilities is crucial to successful alternative marketing (and doesn't hurt one little bit in other aspects of life, either). The way to do it is to open the books — the law books, that is. (For the remainder of this chapter, the word *laws* should be taken to mean both laws and regulations.)

To find out what laws you might collide with, start with a search of the local library. Librarians can point you in the direction of the reference shelves, the place where books of statutes and ordinances are kept, and they're generally happy to help you figure out how to weed out the information you need if it's your first time reading these onerous documents. Reference materials either must stay in the library or are subject to a very short checkout time, like two days; bring some change for the copy machine. Also, all federal laws and some state laws are available via the Internet (see appendix E, Resources), though currently few local governments post their laws online. If you don't have access to the Internet at home, libraries, County Extension Service offices, and schools often offer public access.

Laws can be nested one within another. This means that local entities may adopt laws of other levels of government, or general codes, by reference in their ordinances; the referenced codes can have additional references within them. The Uniform Building Code, Uniform Fire Code, Uniform Electrical Code, and Uniform Plumbing Code are all typical examples of items that are often adopted by reference. Adopting by reference is done, in part, to save space: Each one of these codes runs to hundreds of pages (of really small print), and they can be mind boggling in their own right. For example, let's say you need a fire inspection for the type of operation you plan to run. The fire inspector will most likely judge your property based on the Uniform Fire Code, and the Uniform Fire Code adopts, by citation, the Uniform Electrical Code.

Right about now you're probably thinking, "Oh my word, I'm not a lawyer, I'm a farmer; I can't possibly cope with all these laws." Don't. It's not as bad as it sounds. And you don't have to be a lawyer to make sense of this stuff. Let's look at a couple of scenarios to give you an idea how laws work in practice.

Scenario 1: Starting an On-Farm Bed-and-Breakfast. You want to start an on-farm bed-and-breakfast. In this case, there are probably no federal laws to worry about. You aren't carrying on interstate commerce — since your product can't possibly cross state lines — so federal business laws don't apply. And you probably aren't engaging in any practices that could trip other federal laws; for example, you aren't processing foods or drugs, and you aren't spewing out air pollutants or water pollutants.

So in this scenario, the only laws you need to worry about are those adopted by local governments or your state. How do you find out about them? Start with the local municipal code for your town or county. Look in the table of contents or the index for headings such as BUSINESS, LICENSING, LODGING, GUEST HOUSES, and so on. Go to each reference you find, and read it carefully. If it sounds as though it may even remotely apply to your proposed operation, take notes or make a copy of it. If there are no local laws that apply to what you want to do, then check the state statutes.

It is common for this type of business to need some kind of license from the local government. To obtain the license, you'll need to file an application (along with a fee) with the clerk of the entity. The application form may require a sign-off from certain officials, such as those employed by the local fire department or the local health department; make sure you check on what's required by all local government entities that have any jurisdiction over your site, including fire districts, water and sewer districts, and town or county governments. Your property tax statement should indicate who out there in government land has some say over what goes on, on your piece of land.

An alternative to doing your preliminary research at the library is to call each government entity, and ask the staff what you have to do to start your bed-and-breakfast. But as they tell you what hoops you have to jump through, nicely ask them to show you in writing the law they are basing the requirement on, and ask for copies of the pertinent parts.

Scenario 2: Direct-Marketing Beef. You want to direct-market beef (or lamb, pork, or goat). Once you begin direct-marketing, you fall under the United States Code (a compendium of all federal laws). The part of the U.S. Code that deals with meat is Title 21, Food and Drugs. Chapter 12 of Title 21, "Meat Inspection," requires all carcasses intended for use as meat, or meat products, to be examined by a federal inspector. The examination is intended to "prevent the use in commerce of meat and meat food products which are unwholesome, adulterated, mislabeled. . . ." Interestingly, bison, deer, or elk don't fall under this law, by virtue of the fact that they weren't mentioned when the law was written; however, other federal laws apply to keeping these animals.

Beef can be sold for meat in one of two ways: on the hoof or packaged. Selling meat on the hoof involves less red tape than selling packaged meat, because there is a specific exemption (Title 21, U.S. Code, Section 623) that allows the owner of an animal to have it butchered for personal use without having it inspected. This means that you can have the animal butchered at a "custom slaughter plant," as opposed to a federally inspected plant.

Custom plants must occasionally be checked for cleanliness, by either a federal or a state inspector, but unlike federally inspected plants there are no inspectors checking the carcasses of each animal that is being butchered. Generally, many smaller packing plants located in rural communities fall into the custom category; with their small scale, they can't afford to pay for a regular federal inspector to be on site.

Selling on the hoof means you are selling the live animal to the customer, and he or she is responsible for having it butchered to his or her specifications. You may split an on-the-hoof animal between multiple customers — for example, two families could each be buying one-half of a hog from you — but you must charge them based on the live weight of the animal. And they must pick the meat up from the butcher themselves. When you meet these criteria (sold based on live weight, butchering per customer specifications, and picked up by the customer), the exemption from carcass inspection kicks in.

If you decide to sell packaged meat (cut and frozen, or processed into meat products such as jerky or sausage), each animal must be slaughtered at a plant that has a federal inspector available to view the carcass. Packages must be labeled. Labels have to say what the product is and who it is distributed by; the USDA logo must also be on the label — or the little circle shown in Figure 10.1.

If the package contains a product that has any added ingredients (even just a pinch of salt), the label must show all the ingredients, and it must be approved in a USDA office in Washington, D.C. To receive approval, a copy of the recipe must be sent in, along with a copy of the proposed label; ingredients must be listed in order based on the quantity used in the recipe — largest-quantity ingredient listed first, and smallest quantity ingredient listed last.

Labels may not include any misleading words, and all claims must be supported. Our Little Wing Farm labels said, "Raised and distributed by Little Wing Farm, Verndale, MN." We were notified that we could not use the words "Raised by" on the label until we filed a notarized affidavit stating that all meat used in our products would come from animals that were raised on our farm! The fellow who owned the packing plant we worked with told us about another farmer who used the words "Texas Red Hot" on a jerky label; the USDA disapproved it because it would be misleading on the product of a Minnesota farmer. So he resubmitted the label with the words changed to "Texas Style — Red Hot," and it was approved. When shopping around for a federally inspected packer, look for one who's willing to help you with label design and approval.

Federal law also requires that meat packages be given a safe food-handling label (Figure 10.2). This label, through words and pictures, tells the less-than-bright consumer that he or she should cook meat thoroughly, wash hands and utensils before handling, and keep the meat refrigerated. Personally, this is one regulation that I find ridiculous and insulting, but it is a requirement.

After the meat has been cut, processed, frozen, and labeled, there are laws that dictate how it must be stored and transported. Although a federal storage and transport law exists, the USDA has delegated the authority to implement and enforce this to most of the states, so you need to check what is required by your state's Department of Agriculture. Check the index of your state's statutes for such key words as MEAT, FOOD STORAGE, FOOD OR GROCERY, or call your state's Department of Agriculture. You may have to apply for a "Retail Food Handler's License" or the like.

Figure 10.1. A USDA-approved meat label must include the distributor's business name and address, the contents of the package (including any ingredients used in preparing the contents), and the USDA logo with the number of the plant where the meat was inspected. If the contents are a prepared product with ingredients, the label must be approved in Washington, D.C.; if it is strictly a raw-meat product, say sirloin steaks, the label can be approved by the on-site inspector at the packing plant where the carcass is inspected.

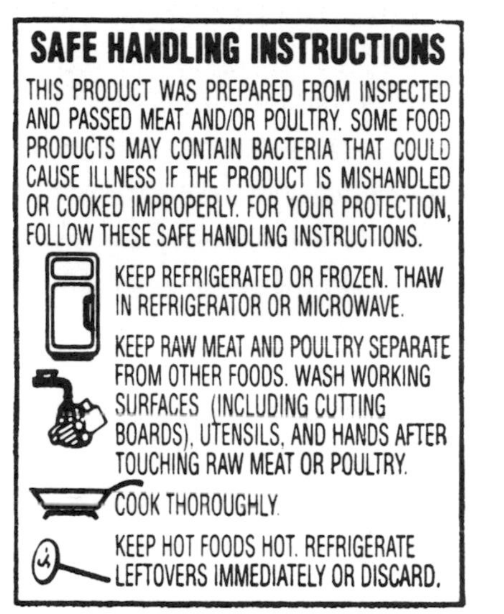

SAFE HANDLING INSTRUCTIONS
THIS PRODUCT WAS PREPARED FROM INSPECTED AND PASSED MEAT AND/OR POULTRY. SOME FOOD PRODUCTS MAY CONTAIN BACTERIA THAT COULD CAUSE ILLNESS IF THE PRODUCT IS MISHANDLED OR COOKED IMPROPERLY. FOR YOUR PROTECTION, FOLLOW THESE SAFE HANDLING INSTRUCTIONS.
KEEP REFRIGERATED OR FROZEN. THAW IN REFRIGERATOR OR MICROWAVE.
KEEP RAW MEAT AND POULTRY SEPARATE FROM OTHER FOODS. WASH WORKING SURFACES (INCLUDING CUTTING BOARDS), UTENSILS, AND HANDS AFTER TOUCHING RAW MEAT OR POULTRY.
COOK THOROUGHLY.
KEEP HOT FOODS HOT. REFRIGERATE LEFTOVERS IMMEDIATELY OR DISCARD.

Figure 10.2. Safe food-handling labels are required on all meat packages.

I'd done my research on the laws in Minnesota before I called the state to inquire about licensing, and it yielded an interesting little nugget: The state's Constitution (Article 13, Section 7) clearly states that a farmer does not need a license to sell "the products" of his or her "occupied" farm. But the employee I spoke with said I needed a Retail Food Handler's License. I then asked, "What about the Constitutional proviso that clearly states I don't need a license?" and read him Article 13, Section 7. There was dead silence on the other end until, after a few minutes, he said, "I'll call you back." The employee did call back a little while later; he hemmed and hawed and said, "Well, um, we think you need a license." I said thanks, hung up, and never got a license. During the ensuing years, anytime state employees said, "You need to get a license," we recited the Constitution to them and told them the state Department of Agriculture would have to take us to court if they wanted us to get a license. The subject was always dropped like a hot potato.

Now, even though we contended we didn't need a license, it didn't mean we were exempted from the rest of the regulations. The state still had the right to inspect our freezers to see that we were storing meat correctly — they just couldn't require us to obtain a license. Again, do your own homework, and when an employee tells you what you need to do, ask to see the laws in writing.

Finally, check to see if any local laws apply to what you're trying to do. If you want to open a retail meat store on your farm, you may have to go through the planning, zoning, or building department for approval.

Scenario 3: Selling Chickens and Eggs. You want to sell chickens or eggs. These products, like the beef in scenario 2, fall under Title 21, but there are generous exemptions for small-scale producers who wish to direct-market.

The federal exemption for poultry allows any producer who raises and slaughters less than 1,000 birds per year a complete exemption from any federal inspection. Producers who slaughter between 1,000 and 20,000 birds per year are exempted if they market the birds within the "local jurisdiction," which means the community you live in. If you live in one state but take your birds to a farmers' market in another, you are required to have them inspected if you sell more than 1,000 per year; if you sell all the birds locally, you can butcher up to 20,000 without inspection. This exemption only applies to poultry you have raised, and slaughtered, on your own farm, but it also extends to birds that you sell to commercial outlets — such as restaurants and hotels — as well as those birds you sell directly to consumers.

Federal law requires eggs to be candled and graded, but again, small producers are exempted. In the case of egg production, the federal exemption extends to producers with fewer than 3,000 laying hens.

Even though small-scale producers are generally exempted from inspection of poultry and eggs, you are still required to label your products. The label must clearly state your name and address. In the case of eggs, it's a good idea to date the carton.

Though most states generally accept the federal standards, not all do. Some states and local government entitites have adopted more stringent requirements than the federal government has for meat, chicken, and eggs, and it is your responsibility to know that! An old adage in the legal world says: "Ignorance of the law is no excuse." So play it safe and check your state and local laws — such as county health department laws — to be sure you are in compliance.

FARMER PROFILE

Tommy and Tommie-Lyn Cashion

Tommy Cashion began managing a ranch in Woodbury, Georgia, for an investor in 1981. Now, almost 20 years later, he continues to manage the ranch with the help of his daughter, Tommie-Lyn.

Originally, the ranch's primary enterprise was raising registered Brangus cattle (a breed that was developed from a cross between Brahman and Angus cattle), but the Cashions have diversified the operations substantially over the years. Today, the Flint Land and Cattle Company runs registered Brangus, commercial cattle, goats, and red deer (a European member of the elk family), as well as operating a packing plant and running a pub and retail sales outlet in Warm Springs, Georgia.

"A major operation for us for a number of years was raising commercial-grade beef heifers for replacement stock," Tommie-Lyn explains. "We'd purchase young heifers from cow-calf operators, raise them on managed pastures for about 18 months, breed them to our Brangus bulls, and then resell them to cow-calf operators. We hosted an annual auction for these replacement heifers, and the enterprise went pretty well until NAFTA [North American Free Trade Agreement] came along. NAFTA threw the beef market into chaos, and the replacement-heifer operation was obviously not going to be profitable anymore, so we began looking for an alternative enterprise."

For over a year, Tommy and Tommie-Lyn researched all kinds of alternatives. "We had to know each potential alternative really well, so we could make the best decision." They looked into beekeeping, alligators, catfish, elk, deer, hothouse tomatoes, blueberries, and more.

"During our research phase, we visited successful alternative operators and university personnel. We studied all the literature we could find on each potential enterprise. Each possibility was analyzed from a financial perspective — it had to be a moneymaker.

"We also wanted to incorporate a new component that would complement our existing operation; that had reasonable start-up costs [breeding-stock expenses, fencing, and so on]; and that had a variety of products to sell at the other end." The red deer seemed to fit the bill; aside from some additional higher fencing, the deer could commingle with the cattle and goats, could eat from the same pastures and use the same trace minerals. The Cashions could purchase three head of red deer breeding stock for what one elk would cost, and the animals yielded seven different products.

When the Cashions started the red deer operation, there were a few small breeders of exotic animals in Georgia, and the state didn't really bother them. But suddenly, when 1,000 acres were being fenced for deer and truckloads of animals were being dropped off, the state Department of Natural Resources came to life. Personnel came to the ranch and told the Cashions they couldn't raise red deer, because this was a wild species — and that even if they were allowed to raise deer, they sure couldn't sell them. "We'd done too much work on the project by that point in time to give up without a fight.

"It was an interesting experience. I didn't even know how to get to the state capitol, but suddenly I'm embroiled in a huge political brouhaha. We met with the state veterinarian and the state Department of Agriculture. We were able to get their support once we showed them that our operation was being run well and that we had adequate documentation for animal health purposes.

"Once the Department of Ag was on board, we contacted other producers of alternative species and agricultural groups to build a coalition. We spent over a year talking to senators and representatives, but we were able to get the 'Georgia Deer Law' passed in 1996, which vests control of seven species of exotic animals under the Department of Agriculture. There are regulations we have to follow that detail things we have to do, like fencing and health care, but with this law in place our right, and the right of other producers, to raise alternative species is secured."

Challenging Regulations

As you work your way through the system, you may come up against a government employee who says, "You can't do that!" or "You have to do this!", and what they're saying doesn't seem correct or reasonable. What are your options? Luckily, the way our laws are designed you do have some options. There is always the appeals process, which can go all the way to the highest courts, if you choose to take your grievance that far.

First, always ask that employees show you, in writing, the laws or regulations that they're basing their statement on. If you get the "It's our interpretation" line, remind them that their job isn't to interpret, it's to implement and enforce. Gray areas should be interpreted by the legislative body or the courts, not the staff, which is why we told state employees that they could take us to court over the Retail Food-Handler's License.

Second, ask what the formal appeals process is. At the local level, an appeal generally goes quickly to the elected board. At state and federal levels, appeals work through the chain of command, from lower employees to higher employees. The final step in the appeals process is the courts, but it's often quite expensive to go there.

If there is a dispute, document everything. Start when the mythical employee in the above paragraph says, "You can't" or "You have to." Write down immediately the date, time, name of the employee, and context of the conversation.

Review the law. During your review, read the "legislative intent." In federal laws, the legislative intent is called the "Congressional Statement of Findings." It defines why elected officials adopted the law. For example, the Congressional Statement of Findings regarding meat inspection says, "Meat and meat products are an important source of the Nation's total supply of food. They are consumed throughout the Nation and the major portion thereof moves in interstate commerce. It is essential in the public interest that the health and welfare of consumers be protected by assuring that meat and meat food products distributed to them are wholesome, not adulterated, and properly marked, labeled, and packaged. . . ."

Ask yourself whether the requirement the employee is making is reasonable, within the law itself and also within the legislative intent of the law. If you are still confident of your stand, then ask to speak to the next higher employee, working up the line. At each point, continue documenting what is said and done. Or if you're really confident, use the tack we did: "Take us to court if you want!"

If you have gone through all the employees, go to elected officials. Attend a town or county board meeting, or contact your senators' and representatives' offices. This step is where your previous documentation will help. Write out an outline to provide to your elected officials, detailing what has happened so far and why you think the requirement should be waived or changed. This step, particularly at a local level, often results in a change of the law, but these changes take time.

Elected officials respond to voters, particularly when there is some volume to the voice they're hearing — and by *volume*, I mean the number of constituents they are hearing from. When lots of potential voters (or organizations that represent voters) call up and discuss an issue, they get a much quicker response than one lone voice in the woods. If you opt to go to your elected officials, then, try to build support for your stand with other individuals or organizations. If the elected officials can't or won't do anything, your last stop for an appeal is the courts.

Although there isn't a law requiring you to use an attorney when you appeal through the courts, you'll probably need one. Once you venture into court, there are so many rules about how, when, and where to file forms that you'll need some guidance. If you can't afford to hire a private-practice attorney — who may charge anywhere between $60 and $300 per hour — check to see if there is a nonprofit law organization available to help. The Farmer's Legal Action Group in St. Paul, Minnesota, is one such organization. The best way to find these groups is to call the office of your state's Bar Association and ask if there are advocates available in your area.

(cont'd on p. 126)

FIGHTING A FACTORY FARM

This is the story of Podunk County. It provides an example of the legal process. Although Podunk County is fictional, the gist of the story is based on actual cases.

Podunk is a county of 12,000 people somewhere in rural America. The county has always been agricultural in character; wheat farming and beef ranching are the two major types of operations in the area. Two small feedlots, each with a maximum capacity of about 2,500 head of cattle, operate in the county. The largest employer is the railroad, followed by a feed mill. Most individuals in Podunk County make their living either directly from agriculture, or indirectly from the support industries operating in Podunk's three incorporated towns.

Like many small rural communities, Podunk loses many of its young people. They finish high school and go to college, join the military, or move to the nearest city in search of work; jobs are scarce.

A major corporation has bought up large landholdings in surrounding counties and installed large-scale confinement hog operations. In the surrounding counties, citizens initially greeted these operations warmly; they would create some jobs to keep young people in the area. Over time, though, sentiments changed. There weren't as many jobs as the citizens originally thought, and those that came in were low paying. The hog facilities had serious odor problems, reducing the quality of life for area residents, and at least one facility had a major leak in its lagoon, causing a fish kill in the creek.

Residents in Podunk feared that their county would be the next place the hog factories would move, so they got together a petition asking the county commissioners to outlaw factory farms in Podunk County.

As allowed by state law and Podunk's Articles of Incorporation, five county commissioners are elected by the citizens to act as the legislative body. The commissioners then elect a chair from among themselves. The chair serves the executive function, running the meetings, acting as the primary spokesman, and voting only to break a tie of the other four.

The commissioners couldn't simply adopt a law that said "no hog barns," because that would be arbitrary and capricious, and the courts would throw the law out. They decided the best method available to them to block any factory hog farms from coming into the county was to adopt a new zoning ordinance.

Zoning ordinances allow local and county governments to exert some control over new development within their communities. These ordinances provide public entities with a mechanism to review development plans, and issue — or deny — permits for the development based on specified criteria.

Prior to adopting the ordinance, the commissioners scheduled a public hearing to allow input on it. At the hearing, all interested members of the general public were allowed to speak — for or against the proposed law. Since this issue was controversial and most people had strong feelings on the subject, there was a large turnout at the hearing. Most citizens spoke out against the factory farms and in favor of the proposed ordinance; however, some citizens spoke out against the ordinance.

After the hearing, the commissioners debated the issue during one of their regularly scheduled meetings. During this process, which was also open to the public, they discussed the pros and cons of the proposal among themselves. After the debate, the commissioners voted to adopt the ordinance. The entire process, from the day the petition was presented until the final public notice was placed in the local newspaper announcing that the ordinance had been adopted, was documented by the county clerk.

The zoning ordinance identified certain areas of the county as Agricultural Districts. Others were identified as Residential Districts, and still others as Commercial Districts. The ordinance prohibited new agricultural enterprises from being started in any Residential or Commercial District; however, existing agricultural operations were grandfathered in, and they were allowed to continue operating in these districts. The ordinance required all new agricultural operations with more than 1,000 breeding animals or 3,000 finishing animals, regardless of species, to apply for a Special-Use Permit.

The new ordinance also included provisions allowing the county commissioners to hire a county zoning officer, and to appoint a citizen-based county zoning board. The first job for the zoning board, with the zoning officer's help, was to develop a regulation that clarified the conditions and criteria used in reviewing Special-Use Permits.

Like the ordinance, the proposed regulation had to go through a public hearing, public debate, and a public vote of the zoning board. Following zoning board approval, the regulation had to be approved by the county commissioners. In Podunk's case, approval became automatic unless the commissioners vetoed the zoning board's approval within 30 days. The commissioners didn't take action, so the regulation went into effect.

The regulation specified a number of different criteria and conditions an applicant would have to meet to obtain a Special-Use Permit for any type of large-scale animal operation. The process was designed to provide adequate review and public comment on each proposed Special-Use Permit. Criteria such as odor and noise thresholds, land area available for manure disposal, impacts on traffic and city services, and so on, were to be considered before the issuance of a Special-Use Permit.

The zoning officer's job was to apply all applicable ordinance and regulations on a day-to-day basis as the "enforcement officer." He had to review permit applications, act as staff to the zoning board, and ensure that terms of Special-Use Permits were complied with after the zoning board approved them.

Joe Greed, a local landowner who had hoped to sell his land to the big hog corporation, sued the county. Mr. Greed contended that the county had infringed upon his rights to sell his property, and had therefore reduced his property's worth — a "taking," in legal terms.

The district court found in favor of the county. The court's decision was based on the facts that the county had followed all the proper procedures in adopting the ordinance and the regulation; that the county was duly empowered to adopt such laws under overriding state statutes, the state's Constitution, and its own Articles of Incorporation; and that Mr. Greed's property still had value and could be used for agricultural purposes.

FARMER PROFILE

Ted and Molly Bartlett

Ted and Molly were both raised in the suburbs of Cleveland, Ohio, but they knew that they wanted to do something outdoors. Silver Creek Farm grew out of that desire.

In 1983, the Bartletts purchased their farm in central Ohio. It's a 75-acre diversified farm that incorporates managed grazing with a large-scale market garden.

"We got involved with managing grass pretty early after we bought this farm," Molly explains, "because Ted got interested in the fencing systems that were just starting to come out of New Zealand. He became an on-farm sales representative for Gallagher Fencing" [Hamilton, New Zealand].

Sheep have been their primary livestock species since the Bartletts began, but they also raise goats and poultry. "The first year, we marketed all of our lambs to friends and family. It was great: They all sold." But the next year, when the lamb crop was about ready and Molly began contacting people, many of them said they still had some lamb left from last year. "I began to understand that selling all our lambs directly was going to require a bigger direct-marketing program. We were going to have to be able to sell lamb by the package."

Molly began selling meat, but the label wasn't approved by the state or the federal government. "One day I got home from a meeting at the local County Extension Agent's office, where we had been trying to figure out just what we were going to have to do to be in compliance with the labeling laws. When I drove up to the farm, there was a guy sitting in his car outside. He got out, wearing a suit and a gun. We don't get many customers who wear a suit and a gun!" The gun-toting visitor arrested Molly and charged her with selling meat without a proper label. He confiscated meat and labels from the farm.

"Through all this, the worst thing was that we had such a hard time finding out just what was required. I wasn't allowed to design my labels, and I've never been able to keep any here at the farm; they have to stay at the butcher's facilities."

After 7 months, and many hassles, the Bartletts finally had their labels. "I guess that was about 10 years ago. We changed phone numbers a few years back, and I was really worried that that would send us into another round of trouble with the labels, but luckily it didn't."

The Bartletts continue to direct-market everything they raise. They have a CSA, and member families can purchase shares for lamb, chickens, and turkeys from the Bartletts, as well as their vegetables. The CSA members can also purchase beef shares that come from another farmer in the area.

"I'm determined to do as much direct-marketing as I can. I really think it's crucial in small-scale farming. But it's not easy. It takes time to deal with the regulatory people. To market meat, you need to be set up to accommodate your customers, and that means having freezers and storage facilities. We set up a small store in an old barn on the farm, and that works, but it takes a real commitment."

Advocates work on an ability-to-pay basis, and they are overworked and understaffed. The documentation you prepared earlier should again help them to help you. If you go through the courts and your appeal is denied, than you must do whatever the bureaucrat originally told you to do.

Liability

Unfortunately, we live in a time when there are almost as many lawyers as there are farmers! And lawyers need lawsuits, so liability is a much more serious issue than it was in the days when Ken's grandfather peddled his milk door to door from the back of a horse-drawn wagon.

Liability generally arises when someone fails to meet his or her responsibilities. The cattle break out and trample the neighbor's prize rose bushes — liability. The hired hand cuts off a finger in a piece of machinery that doesn't have proper belt guards on it — liability. A customer's child sustains a deep cut from a piece of broken glass while running around barefoot — liability. The ground beef you sold was contaminated with salmonella at the processing plant, and a customer's family became ill — liability. An employee verbally insults a customer — liability.

The best way to avoid personal liability is to pay attention to details. Make safety a top priority on your operation, studying each area of your farm and everything you do for possible hazards. Follow all laws; they may seem like a nuisance, but they offer you a degree of protection. If you have employees, develop written personnel and safety policies.

Safety policies can also extend to customers: "No shoes, no service" is an excellent policy on any farm. Place signs up for your customers to see. Areas that are off limits to customers should be clearly marked: DO NOT ENTER THE BARN — EMPLOYEES ONLY BEYOND THIS POINT. Also, don't be afraid to tell customers or their children to stop doing something that may be dangerous: "Please stop chasing the chickens!" If a customer is annoyed and leaves in a huff, count your blessings — you don't need people on your farm who aren't safety conscious, and who don't respect your rules.

Even if you are very safety conscious, accidents can still occur. And even if something isn't your fault, you may still be sued. Your property insurance company should be able to offer you extended personal liability coverage that will cover you for both employees and customers.

CHAPTER 11

Butchering & Processing

If you have enough do-it-yourself determination and a sufficiently mechanical mind to take things apart (in butchering you only take apart, you don't have to put back together), you can learn to butcher.

John M. Mettler Jr., DVM,
Basic Butchering of Livestock and Game

BUTCHERING INVOLVES the actual slaughtering and cutting up of animals for meat. *Processing* is taking raw meat and making some other type of product through the addition of other ingredients, as in a can of stew; or through specialized handling, such as smoking.

With the exception of poultry (which can be butchered on the farm for direct-marketing), I'm not going to go into the actual processes involved in butchering and processing. If you want to learn the techniques of butchering your own large animals, get a copy of Dr. Mettler's book (see the quote above), or check with your County Extension Agent to see if he or she can get you a booklet on butchering. These sources will do a far better job of describing the techniques than I could in this chapter.

What I *will* go over here are some things you'll need to know about butchering and processing for a direct-marketing operation. For instance, how much meat will a 1,000-pound (454-kg) steer actually yield? How do you respond when the butcher asks if you want the short ribs? And what do you say to first-time customers who are considering purchasing a lamb, but want to know how much meat they'll receive and how much freezer space it will take up?

Yields

When a butcher uses the term *yield*, he's generally referring to the hanging weight, which is what's left after the animal is bled, decapitated, gutted, and skinned. The term comes from the fact that butchers hang a carcass on a rail attached to the ceiling of the cutting room to make their work easier.

TYPICAL HANGING YIELDS

▸ Beef	60–65% of live weight
▸ Pork	70–75% of live weight
▸ Lamb	50–55% of live weight

Depending on the type of animal and its age, it can either be cut up immediately or hung in a cooler, for up to 2 weeks, to age the meat. Aging helps cure meat, improving flavor, texture, and tenderness.

Of course, when you are selling meat you aren't interested in the hanging yield, but in how much meat actually comes back. The meat yield depends on how closely fat is trimmed, and on how many bone-in cuts are prepared.

Let's look at a few examples: A 250-pound (113-kg) hog will hang on the rail at about 185 pounds (84 kg), and when you pick up the packages of meat, you will receive about 140 pounds (64 kg) of meat. A 1,000-pound (454-kg), grass-finished steer will hang around 650 pounds (295 kg) and yield 500 pounds (227 kg) of meat.

On average, cut and wrapped meat requires 1 cubic foot (0.03 m^3) of freezer space for each 35 pounds (15.9 kg) of meat. The meat from a whole steer will almost fill a 15-cubic-foot (0.45-m^3) chest freezer. The pig will require 4 cubic feet (0.12 m^3), which is about the size of the freezer space in a large combination refrigerator-freezer, but it is too much for a smaller combo unit. The lamb will fit into the freezer compartment of a small refrigerator-freezer.

Cutting Orders

When you drop an animal off at the butcher, he or she will want to know your "cutting orders." These are simply the directions that specify how you want the animal cut and wrapped (Figures 11.1, 11.2, 11.3).

Trimming

The first thing to tell the butcher is how you want the animal trimmed. *Trimming* is cutting away external fat. Trimmings can then be added back into ground meat, if your goal is to get more pounds back, but we always specify well trimmed, with no extra fat added to the ground beef. With these cutting instructions, our ground beef comes back in the 96 to 97 percent fat-free range. This is so lean that pan-frying requires us to put a spot of cooking oil in the bottom of the pan to keep the meat from sticking.

TYPICAL MEAT YIELDS

- Beef 45–50% of live weight
- Pork 55–60% of live weight
- Lamb 40–45% of live weight

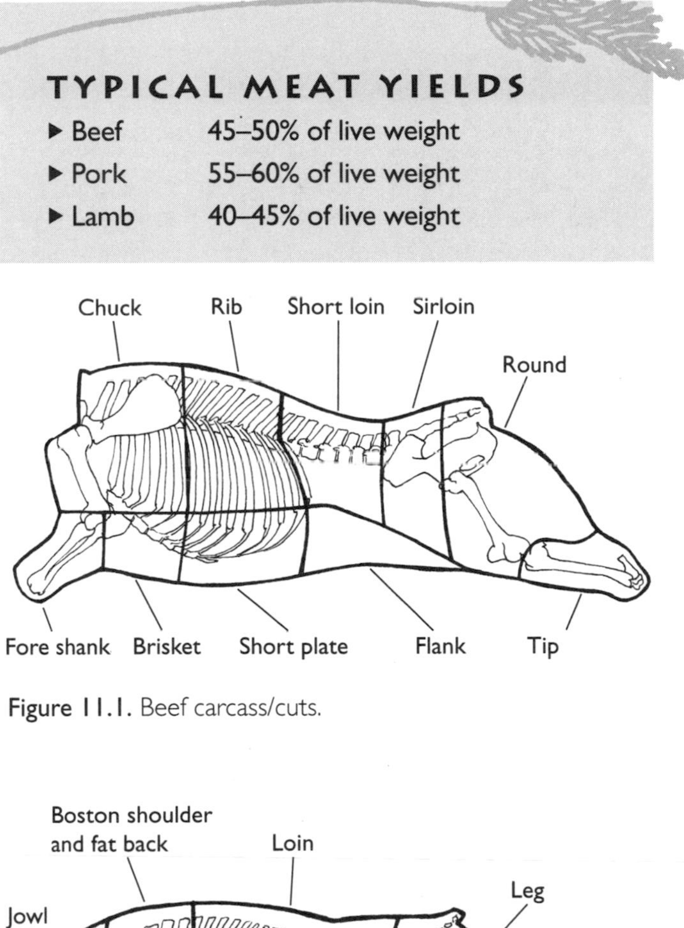

Figure 11.1. Beef carcass/cuts.

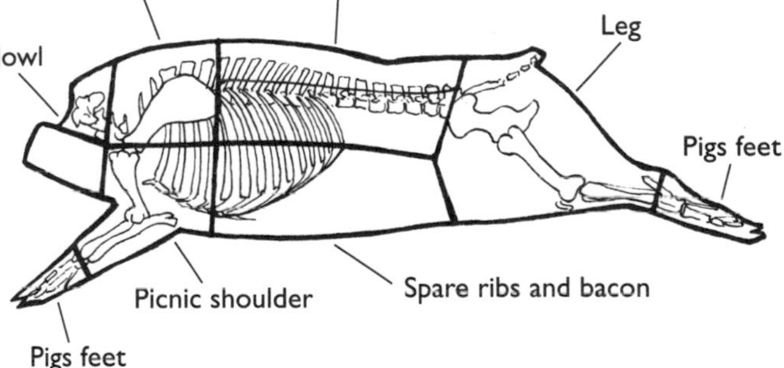

Figure 11.2. Pork carcass/cuts.

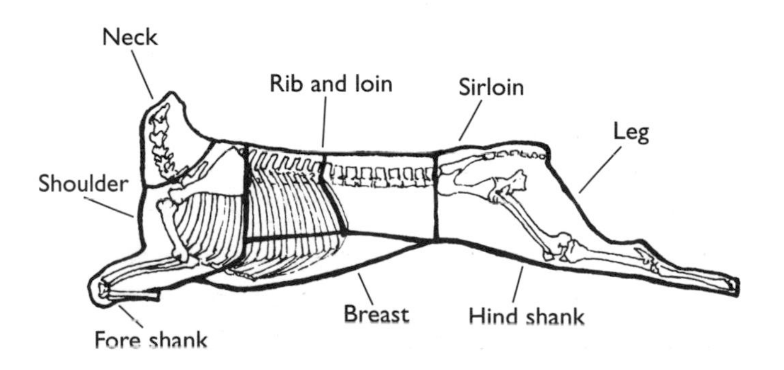

Figure 11.3. Lamb carcass/cuts.

Some processed meat products do benefit from the addition of some trimmings, though. We once had all-beef wieners prepared without additional fat. They were too dry! Luckily Karen, the woman who took cutting orders where we were getting meat butchered at the time, warned us about this, so we had only a few pounds made up to try them. Had we made up a large lot, our dogs and cats would have been happy, but we would have wasted our money.

Number, Thickness, and Packaging

The butcher also needs to know how many steaks, chops, and so on, you want in each package, and how thick to cut them. We found that for direct-marketing, packages with two steaks or chops each are a good size. Larger families can take out multiple packages. We also found that thicker cuts are more popular than thinner cuts. Three-quarters of an inch (2 cm) seems to be ideal for most steaks and chops.

Most of the meat from the round (the butt) and the chuck (the shoulder) is best ground. People purchasing directly from the farmer are generally the kind of people who can afford high-quality cuts — tenderloins, T-bones, sirloins, and rib eyes. They are not as interested in the low-quality steaks, like the rounds and the chucks. We did find that 2-inch (5-cm) thick "mini roasts" cut from the round, which fill the role of a traditional pot roast, sold moderately well, so we'd get a few of those off each steer.

Short Ribs

The first time we took a steer in to have it butchered, Karen asked if we wanted the short ribs. I stared kind of dumbly and said, "Well, I don't know. Why?" She went on to explain that most people didn't take the short ribs because they're tough and stringy, and there isn't much meat on them. On that outing, I opted not to take them.

The next time we took in a steer, when Karen asked about short ribs, I told her we'd keep them — I was thinking they'd serve as dog food if nothing else. I also told her we wanted all of the organ meat, even the tongue, as well as packages of soup bones and boxes of dog bones. Our policy had become: Get everything back, because something will eat it.

Short ribs are, true to their reputation, stringy and tough when prepared by most conventional cooking methods. But we discovered that they're actually pretty darn good if cooked in a pressure cooker. The technique: Put about 1 inch (2.5 cm) of water in the bottom of the pressure cooker, load in the short ribs, pour some barbecue sauce over the top, and pressure-cook them for about 45 minutes. After this treatment, they fall off the bone and taste great, though they are messy.

Pork Cutting Orders

Lamb, bison, deer, and elk all have cutting orders similar to beef, but pork is different. The butcher will want to know if you want hams or fresh roasts. Hams are just cured or smoked roasts. Most consumers want to purchase hams, but we did find some call for fresh roasts. Like the hams, bacon is cured side pork. We found that a lot of old-timers really enjoy side pork, so we always got a few packages, but again, bacon is far more popular. The front shoulder of a hog can be made into a picnic ham or sliced for pork shoulder bacon — nice lean meat much like Canadian bacon.

One point to keep in mind if you are marketing pork is this: Many people avoid cured meats because of concerns about additives such as salt and nitrates. If you run into some of these consumers, don't try to talk them out of their concerns; instead point out the availability of your fresh roasts, or research recipes that don't use additives.

With hogs, the butcher will also want to know if you want the pig's feet and knuckles back. We didn't have any market demand for these, but our dogs liked them.

There is a lot of trim on a hog (and it isn't all fat), so plan on getting lots of sausage made. All types of sausage are great direct-market items. Sausage sells at a fairly high premium, and the varieties are seemingly endless: Polish, bratwurst, Italian, breakfast, smoked; the list goes on. Consumer taste in sausage varieties varies depending on where you are located around the country, so discovering what's going to be most popular in your area may take some trial and error. Scope out area grocery stores to get some idea what is popular where you live.

Butchering Poultry

Butchering poultry isn't fun, but it does fall within the scope of on-farm, direct-market operations. If you're planning to butcher many birds, you need to develop an assembly-line approach to the job, or find a poultry packer. If you're starting out with a small number of birds for home use, or starting to assess your production and marketing strategies, the following directions work and will get you started.

Tools You Need

- *Three 5-gallon* (19-L) *buckets:* one to catch blood and guts, one for the hot water, and one for cold water.
- *Two sharp knives:* one paring-style knife about 3 to 4 inches (7.6 to 10.2 cm) long, and one thin boning knife about 4 to 5 inches (10.2 to 12.7 cm) long. If you don't know how to sharpen knives correctly, learn. (The best set of instructions I know of is in the *Joy of Cooking* cookbook.)
- Ax *(optional)*, for chopping the head off a bird, if that's your chosen method of dispatch.
- *Hot water* for scalding — the ideal temperature is between 130 and 140°F (55 and 60°C).
- A *heat source,* to continually heat the water for scalding.
- *An automated plucker (optional)* — available from NASCO (see appendix E, Resources).
- *Cold water,* for rinsing and cooling the bird — the colder, the better. Add some ice cubes to keep it very cold.
- A *trash can,* for feathers and offal. These can be buried in a compost pile after you're done, but they should be placed deep in the pile so they don't attract vermin, and so they heat up sufficiently to kill off bacteria.
- One *short, dull knife or tweezers.*
- A *small propane torch* to singe hairs.
- *Wax (optional)*, for removing pin feathers and hairs instead of the dull knife and propane torch. Wax is most often used when butchering ducks and geese.

Preparing the Bird

1. ***Starve birds for 24 hours before you butcher them.*** This helps pass out the contents of the digestive tract, so you have less to deal with. The birds should still have access to water.
2. ***Dispatch the bird.*** Though killing an animal is never a pleasant task, when done correctly each of these methods is quick and humane.
   — *Wring the bird's neck.* This is an old, somewhat lost skill. Pick the bird up by its head and swing it completely around in a 360-degree circle. If you use this method, hang the bird up after it's dead and cut its throat, as described below.
   — *Chop the head off with an ax* on a block. Also an old method, though somewhat messy.
   — *Hang the bird upside down* with its feet tied, or in a killing cone. Place a bucket beneath the bird. Using the sharp paring knife, slit the bird's throat by making a cut directly behind the lower jaw, but try to avoid cutting through the esophagus and windpipe. When you cut the bird's throat, you cut the jugular vein. Carefully placing the bucket underneath the bird will catch the blood and make cleanup easier. (*Note:* This is the most common method in use today.)

Killing cone.

3. ***Bleed the bird.*** After the bird is killed, allow it to bleed out for a couple of minutes. This step keeps the meat cleaner and less likely to spoil.
4. ***Scald the bird.*** This loosens the bird's feathers for easier plucking. Submerge the bird in the scald bucket for about 40 seconds by holding it by the feet and dunking it headfirst. (130°F [55°C] water works for a small bird, and 140°F [60°C] for a large bird). If your scald water gets cool, leave the bird in longer; if the water is too hot, the bird will be overscalded, which results in "cooked" meat. Swishing the bird around a little in the bucket helps move the heat around all the feathers.

When you're working with multiple birds, try to keep the scald water fairly clean; otherwise it can be a source of contamination. If the birds are dirty, hose them off prior to scalding.

Scald the bird.

5. ***Pick the bird.*** *Picking* means removing the bird's feathers, and it should be done immediately after scalding. When you're picking by hand, it's easiest if you hang the bird so you're working at a comfortable height. Pull the feathers down and away from the body. If the scalding temperature was right, most of the feathers will come off fairly easy. If you plan on butchering lots of birds, a rubber-fingered "plucker" makes the job easier.
6. ***Rinse.*** After most of the feathers have been picked, rinsing with cool water from a garden hose or under a faucet reveals any remaining feathers, as well as the pinfeathers.
7. ***Remove the pinfeathers.*** Pinfeathers can be either scraped off with a short, dull knife or picked out with tweezers. Any hairs can be singed with the propane torch (older birds have more hairs than young ones do). For waterfowl (which have lots of down and pinfeathers), dip each bird in hot wax (150°F [66°C]), then in cold water to set the wax, and peel the wax away. This should remove most pinfeathers. A second dip may be required.

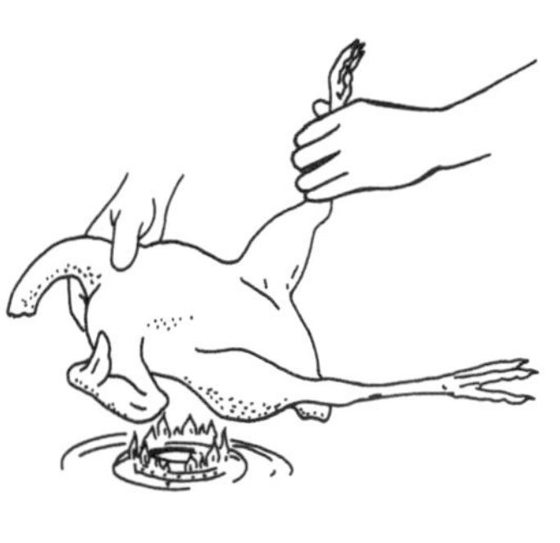
Torch pinfeathers.

Butchering the Bird

Note: You may find it easier to work on a cutting table than with the bird hanging. Try both approaches.

1. ***Remove the head.*** Cut just below the first vertebra in the neck and twist the head off.
2. ***Remove the neck:***
   — Insert the thin boning knife in the skin above the neck, at the shoulders, and cut forward to open the skin.
   — Pull the skin away from the neck.
   — Pull the crop, the trachea or windpipe, and the gullet away from the neck skin, and sever where they enter the body .
   — Cut the neck off where it enters the shoulders, by cutting the muscle around the bone and then twisting off. Wash the neck; then set aside in your bucket of ice water.

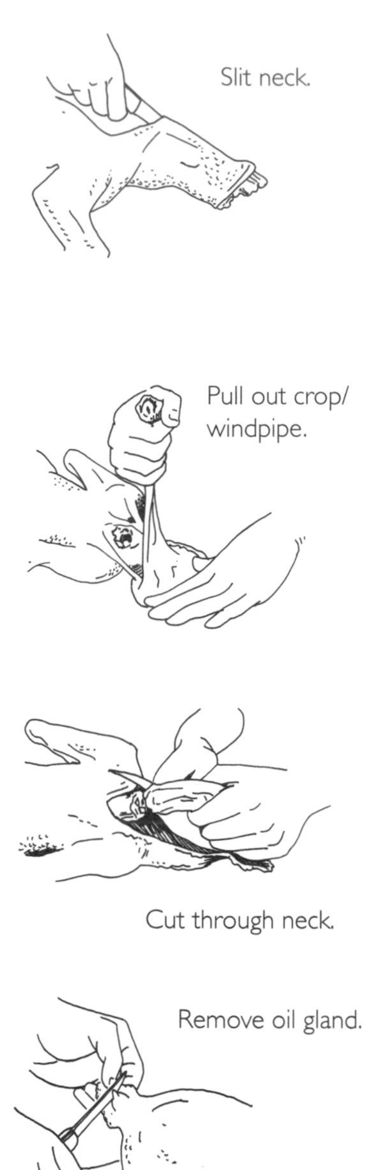
Slit neck.

Pull out crop/ windpipe.

Cut through neck.

Remove oil gland.

3. ***Remove the oil gland.*** About 1 inch (2.5 cm) in front of the oil "nipple" on the tail, make a clean cut all the way down to the vertebra in the tail. Then cut out the oil gland by cutting from front to back and scooping out the gland.
4. ***Remove the shank*** (the lower part of the leg and foot). Cut through the hock joint from the inside surface.

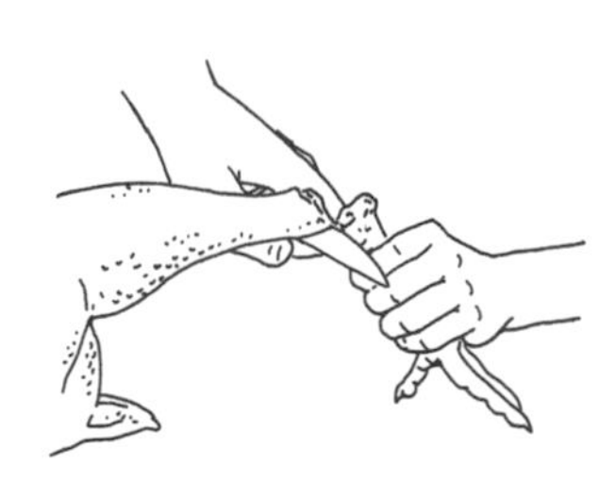
Remove shank.

Eviscerating the Bird

Evisceration — or removal of the animal's internal organs — takes practice, so take your time when you're learning to do it. If you accidentally cut through the intestines and release their contents onto the meat, you have contaminated it and should dispose of it. There are two methods for beginning the process. The first is appropriate for small birds that won't be trussed for roasting; the second should be used with large roasting birds such as turkeys.

1. ***For small birds:***
   — Starting just below the point of the breastbone, insert the thin boning knife just enough to penetrate the skin and muscle.
   — Cut down toward the vent (anus), and then cut all the way around it. When cutting around the vent, keep the knife pressed as close as possible to the back and tail.
   — Pull the vent, and a small section of the large intestine out of the way.
2. ***For large birds:***
   — Make a half-moon cut around the vent, pressing closely to the tail and backbone.
   — Insert your index finger into the cut and up and over the intestines.
   — Pulling the intestines down and out of the way, continue to cut around the vent until you have completed the circle.
   — Pull the vent and a short section of the intestines out of the way.
   — Pull the skin back toward the breastbone, then make a side-to-side cut about 3 inches (7.6 cm) wide and 2 inches (5 cm) below the breastbone.
   — Pull the bar of skin that remains backward and over the piece of vent and intestines.
3. ***Break all attachments and remove the organs.*** Insert your hand into the abdominal opening and gently work your way up and around, loosening the organs from the body cavity as you go. When all attachments are broken, scoop the insides out.
4. ***Remove the gizzard.*** Cut away the gizzard from stomach and intestines. Peel away the fat, split it open, and rinse well; add to the ice water.

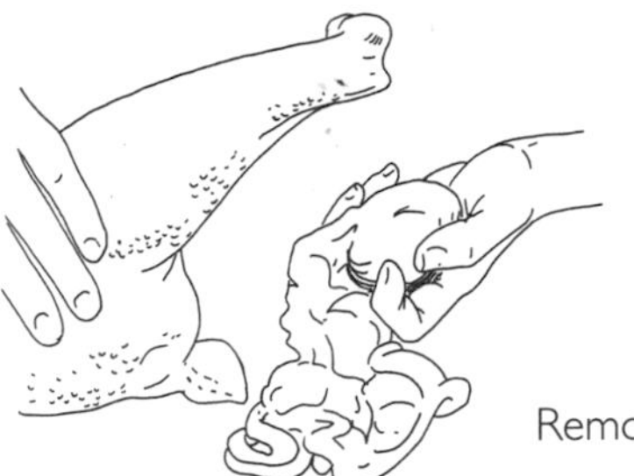

Remove internal organs.

5. ***Remove the heart.*** Trim off the heart sac and heavy vessels from around the heart, rinse, and add to the ice water.
6. ***Remove the liver.*** Pinch the gallbladder off the liver, rinse the liver, and add it to the ice water.
7. ***Remove the lining from the gizzard.*** Retrieve the gizzard from the ice water and peel its lining away by inserting a fingernail under the lining at the cut edge and pulling away. Return the gizzard to the ice water.
8. ***Check along the backbone for remaining organs and remove.*** Use your hands to check along the backbone for any remaining organs (the lungs and sexual organs are often still in the bird). Remove any that you find.
9. ***Rinse well inside and out, and chill.*** Rinse the bird well under running water, inside the body cavity and on the outside of the carcass. When the carcass is clean, place it in the ice-water bucket to chill. Chilling time varies depending on the size of the bird and the temperature of the water, but it may take several hours. The goal is to get the carcass temperature down to about 40°F (4° to 5°C), as quickly as possible.
10. ***Hang and drain the carcass.*** Once the carcass is adequately chilled, hang it to drain for about 10 minutes.
11. ***Wrap the organs.*** While the carcass is draining, wrap the heart, liver, and gizzard in plastic wrap. The neck and the wrapped organs can be inserted into the carcass.
12. ***Bag the bird and label.*** We find that extra-large, zip-seal freezer bags are convenient and work well. Label bags with the weight and date you butchered. Remember, if you plan to sell the birds, the label should also include your farm name and address.

Whether you plan to sell meat or just keep it for your own use, proper handling after butchering is important, both from a safety standpoint and for maintaining meat quality. Though some smoked meats can be kept for long periods without freezing, most of your meat will be frozen.

All meat should be chilled before freezing. Meat freezes faster and better if the freezer isn't packed too fully. The ideal long-term freezer temperature is 0°F (−18°C). To help prevent freezer burn, force out as much air as possible before sealing packages, and make sure packaging is secure. If you opt to use freezer paper to wrap meat, a double layer of wrapping greatly improves "keepability" in the freezer.

FARMER PROFILE

Cyd Osborne and Dave and Debbie Turunjian

In 1991 Cyd Osborne, who was raised on a farm, decided to try her hand at farming. She began raising free-roaming laying hens and marketing her eggs. Free-roaming birds, unlike free-range birds, are allowed to move around in a large area, but they are still inside a henhouse. By 1993, Cyd's brand — Nest Fresh Eggs! — was building a respectable market share in a few of the large grocery chains in the Front Range and mountain resort communities of Colorado.

Cyd's story is typical of what can happen to someone who develops a good market for a niche product: "The business quickly outgrew my ability to produce all the eggs and deal with the marketing. I decided I was good at the marketing, so what I needed to do was recruit some other farmers to help produce eggs. I knew if I just tried expanding by hiring help, they would never take as good care of my hens as I wanted, but by partnering with other producers I would get people who had a vested interest in the birds under their care."

Finding partners wasn't terribly hard. The poultry industry is one of the most vertically integrated segments of the food economy, and egg production is no exception; it had moved into the hands of one or two major producers, effectively closing the market to small and medium operations. Some of the smaller producers were still around, their barns now empty. Cyd contracted with three farmers. Her agreement stipulated conditions the birds had to be raised in — like no more cages and no more subtherapeutic antibiotics — and stipulted that the growers would "not compete." In return, Cyd agreed to purchase all their eggs and at a higher price than they would have received had they still been able to market their eggs through conventional channels.

Dave and Debbie Turunjian own and operate one of the partner farms. Dave and Debbie wanted to farm and bought out an existing egg farm. "Debbie is a true animal lover," Dave says, "and when she first walked into the old barn, with 30,000 hens in little cages, she cried." When they purchased their farm in Longmont, Colorado, they were selling to one of the middlemen farmers, but he went out of business, and they had nowhere to sell their eggs.

"We had already been talking to Cyd, because we wanted to change the operation, but it took some time. Our henhouse was empty for half a year as we remodeled. Then we put up a second house."

Each one of the Turunjian's henhouses is 14,000 square feet and accommodates about 13,000 hens. But the cost of remodeling and adding the second barn was steep. "We both have good off-farm jobs, or we couldn't have done it at this point in time. But we did want to farm, and we did want to do it differently, so we decided it was worth the expense."

PART IV

PLANNING

CHAPTER 12

Farm Planning

I have always rejected the idea that profit is the sole purpose of dairying or any other occupation. It is not the ultimate goal. Profit is rather something that, while giving satisfaction in itself, should be used to achieve a further aim. This further aim is good living.

Wilbur J. Fraser, *Profitable Farming and Life Management* (1948)

Anybody can buy a farm; but that is not enough. The farm to buy is the one that fits the already formulated general plan — no other! It must be positively favorable to the crop or animal to be raised. . . . To buy a place simply because it is "a farm" and then to attempt to find out what, if anything, it is good for, or to try to produce crops or animals experimentally until the right ones are discovered, is a costly way to gain experience. . . .

M. G. Kains, *Five Acres and Independence* (1935)

SMALL-SCALE FARMERS FALL more or less into two classes: those who were born and raised on the farm and have known little else, and those who come to the farm from "town" with little or no experience in agriculture. And the sad fact is that both classes fail regularly. Many born farmers are locked into a paradigm that is now dated — they do things the way their dads did, which worked well in the 1950s and 1960s when gas cost 25 cents per gallon, but those same approaches are breaking them now. And the townies are burdened with a false vision of country living that comes from the glossy pages of a magazine: the easy life of a gentleman farmer.

Farming is hard physical work, and it can be economically stressful. Although I've always hated the term agribusiness, because it implies corporations and large-scale farming operations, farming is a business nonetheless. Even on a small scale, it should be treated as a business if making a profit is one of your goals — and for Ken and me profit, not for itself but for our "good life," is definitely a goal. To be successful, all businesses — large and small are well advised to spend time on planning.

As I said earlier, there are various whole-farm planning models available for farmers and ranchers. These models are designed for various purposes; some, like Farm*A*Syst (a cooperative program of the U.S.

Department of Agriculture and the Environmental Protection Agency), are designed to help farmers cope with a wide variety of regulatory programs; others are single-dimension tools, such as the plans designed to help a farmer receive organic certification.

In our minds, the holistic management planning process is the most effective: It provides a process for making decisions based on what is important to us as individuals — our values and our vision for the future. It can bring other people into the process, and in fact encourages us to do so, but it is not predicated on an outsider's demands or views. The planning process that I describe in the remainder of this book is thus based largely on the holistic management model, though as I explained in part I, it is a slightly modified version for easier use.

Again, if you live in a brittle environment, manage many acres of land, have to cope with a high debt ratio, or have other significant management concerns, this process should be viewed as a starting point. As time allows, you should study and work with the full model and planning process as outlined in the *Holistic Management* textbook by Allan Savory (see appendix E, Resources), either through personal study or through attending holistic management classes. However, if you're just getting started, have a fairly small piece of land, live in a nonbrittle environment, or don't have an extremely tight budget, the remainder of this discussion may suffice.

Planning as a Process

Planning is a process. It involves gathering information, brainstorming, and finally formulating your information and ideas into some kind of usable form or plan. Then it requires monitoring, controlling or adjusting according to what you find in your monitoring, and replanning if things go far afield, as they sometimes do. Because it's a process, planning isn't something done in one quick session; it may take months to develop your first complete set of plans. Once it is developed, don't just set it in a drawer and forget about it; visit your plan often. Use it as an aid and it will help you succeed.

Planning, like goal setting, should be done with all the pertinent players involved: family, employees, and even knowledgeable and caring outsiders. You, as the ultimately responsible party, may have to make some final decisions, but your decisions will be better and receive more support from those around you if everyone is involved in them.

There are also various pieces of the planning process, though they all interrelate. In chapter 9, I discussed developing a marketing plan; this is a very specific type of plan that helps you identify possible products or services for enhancing income. Chapters 13 and 14 detail financial plans and biological plans (grazing plans and landscape plans in the full model). Chapter 15 provides monitoring information.

Financial planning allows you to allocate your resources — both money and labor for their best effect in implementing your goal. But remember, implementing your goal is what you are trying to do, so even if an operation like monocropping or confinement dairy looks good financially, if it doesn't move you toward your broad goal over the long term, then it isn't a good deal.

Biological planning allows you to harmonize your land and your animals so that you improve your bottom line while improving the environment. It helps you implement strategies that move you toward the landscape that you desire.

Holistic Guidelines

During the planning process and during day-to-day decision making, many of the holistic guidelines come into play, particularly the testing guidelines — sustainability, weak link, cause and effect, marginal reaction, gross profit analysis, energy/money source and use, and society and culture. The testing guidelines are implemented as a series of questions, though not each one of them will be necessary in evaluating every idea, enterprise, or tool.

The questions, if you have sufficient information, should rather quickly yield yes or no answers — Yes, this action will move me closer to my goal, or No, it won't. At times, the question may yield a "No, it won't" answer to one or more of the tests, yet you'll still have to take that action for some other reason (perhaps you have to do something to comply with a regulation, or you still need to use some borrowed money). The main point to remember is that the purpose of these guidelines is to make it easier for you to decide what to do, when to do it, where to spend your resources (time, money, and so on), and which tools to apply under a given set of circumstances. Let's look at each one.

Sustainability

If I take this action, will it lead toward or away from the future resource base described in my goal? This guideline helps you evaluate the long-term social and environmental aspects of a decision. For example, if one of your goals is to have a vibrant local community, then driving 200 miles (322 km) to shop in the city probably would fail this test. If one of your goals is to have an abundance of wildlife, using a chemical that can kill birds, fish, or beneficial insects will fail.

Weak Link

Does this action improve, or address, the weakest link in my operation? If you take a chain and stretch it until it breaks, the point at which it breaks is its weakest link. Repair the chain, stretch it again, and the new point at which it breaks is now its weakest link. There may be many weak links in a chain, but there is always just one weakest link; and the best way to use your limited resources to strengthen the chain is to fix the weakest link first. Once it is repaired, there is a new weakest link, and you can set out to repair that.

The weakest link in your operation may be social (the in-laws think you're crazy), biological (there are too many weeds, or not enough legumes), or financial (you aren't effectively converting solar energy into paper dollars). Using this guideline helps you aim your limited resources at your weakest link.

Determining the weakest link. When trying to figure out what the weakest link is in your operation, start by considering the three main areas of a livestock operation: energy conversion, product conversion, and cash conversion. *Energy conversion* is the change of solar energy into forage. Signs of a weak link in energy conversion include insufficient forage, low-energy forage, poor drainage, low plant species diversity, high supplemental feed costs, and excessive amounts of bare earth. *Product conversion* is the change from raw plant to a marketable product. Indications of a weak link in product conversion might include unused forage, low birth rates, poor gains, lots of illness, and high mortality. *Cash conversion* is the ability to take your product and market it for paper dollars. Weak links in this area are indicated by low prices, excessive middlemen between you and the consumer, poor understanding of marketing mechanisms, and regulatory roadblocks.

Whenever you possibly can, use your planning to address the weakest link first! This may take time. For example, if you decide that your weakest link is in the marketing — that you want to get out of the commodity system and into direct-marketing — it's going to take some time. You'll have a steep learning curve, so allow yourself the time needed to work through this curve.

Cause and Effect

Does this action address the root cause of the problem? Often an action — say, spraying weeds with an herbicide — will treat the symptoms of a problem without getting down to the real cause of the problem. Why are the weeds there? That's the problem.

It's human nature to like simple, quick fixes to problems, but most of the time a quick fix doesn't last long. When we moved to Minnesota, our farm had leafy spurge (a noxious weed) on it. The first spring that we were there, a crew from the township board showed up, spray cans in hand, to "control" our leafy spurge. We told them we didn't want them to spray it, and they said they wouldn't for now, but that we had to do something to control it. As we talked to them, we asked about the history of the infestation. They told us that it had been growing and spreading on our farm for 20 years, despite annual spraying. The first year we clipped it and dug it. After that, our sheep took care of it. The problem was that nothing that had lived on the farm in the past liked leafy spurge, so it was able to grow, and grow, and grow. Sheep do like it, though, so when we added them to our mixture, spurge no longer outcompeted the other plants in our pasture.

Marginal Reaction

Will this action provide the greatest return in terms of money or time? The marginal-reaction guideline provides a method for figuring out where you get the biggest bang for your buck. Application of the guideline basically requires you to ask yourself, with each dollar you spend (or with each unit of time you commit to a project), "Could this dollar (or hour) be spent better somewhere else?" This test is especially useful for comparing two separate actions. For example, "I can afford to spend $800 this year. Would it best be spent on fencing or on fertilizer?"

Gross Profit Analysis

Which enterprise, or action, will contribute the most money to covering overhead expenses? Like marginal reaction, the gross profit analysis is an effective tool for comparing separate actions (Figure 12.1).

In traditional cost accounting, fixed costs (see chapter 13 for an explanation of these) are apportioned to various enterprises: If half your land base is maintained as a hay crop, you would assign half your land expenses to hay production. Although this may be appropriate after the end of the year, when figuring cost of production, it doesn't provide an effective way for you to compare options before you begin production. By using gross profit analysis, you can figure out which of several options is likely to return the most surplus money, which can then be applied to reducing fixed costs.

Simply stated, a gross profit analysis takes all potential income from a given project, subtracts the variable expenses for that project, and results in a gross margin, or profit, between the two. Gross profit analysis can be used to compare the straight dollar reactions of various enterprises or ideas.

Gross profit analysis is a handy tool for making other comparisons as well. For example, if manpower is a limiting factor in your operation, you might choose to study the profit margin in dollars per man-hour for various enterprises.

Once you narrow your options down to a chosen enterprise, gross profit analysis can also be applied to that enterprise under varying conditions — say, to look at a worst-case, an expected-case, and a best-case scenario. This approach gives you a feel for what happens when cattle prices drop, or when high death loss hits your broiler operation.

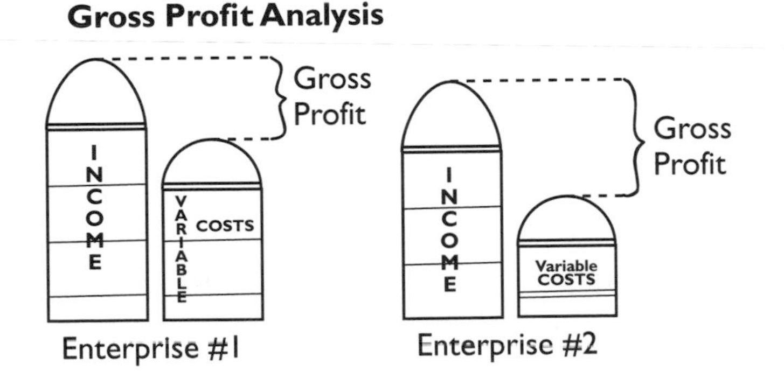

Figure 12.1. In comparing two different enterprises, the one that makes the most income doesn't necessarily make you the most money. Although Enterprise 2 brings in less income, it yields a greater gross profit because of lower variable costs.
(Modified from Sam Bingham, *Holistic Resource Management Workbook*. Covelo, CA: Island Press, 1990, p. 10.)

Energy/Money Source and Use

Is the energy or money used in conjunction with this action derived from the most appropriate source? Will the way in which the energy or money is used move me toward my goal? Energy from a renewable source (solar power, wind power, and the like) is more appropriate, we feel, than fossil fuel or nuclear energy. So our first choice is to use the most appropriate form when possible. Still, at times we all must use gasoline or piped-in electricity. When using these less appropriate forms of energy, ask yourself if they are being used the best way possible. Are you consolidating errands when you go to town, so you use less gas? Or if you are using a less appropriate source of energy, are you working toward a system that allows you to move to a renewable source of energy? When we purchased our place here in Colorado, the house was "off the grid." It came equipped with a gasoline generator, which is powering my computer as I write this, but we are working on installing a solar-energy system.

Money can be derived from the conversion of solar energy to grass, grass to animal, animal to marketable product, and marketable product to paper dollars. The internal dollars on a farm, or solar dollars, are those that come directly from this conversion process; these are the most appropriate dollars for a farm to use. Paper dollars are derived from off-farm sources. The more solar dollars you can use, the better, but at times paper dollars either become necessary or look like a good deal. For example, a government employee may tell you that you are eligible for cost share on some kind of on-farm project. It looks like "free money," but if it doesn't actually move you toward your goal and meet all the tests, it's best left to someone else. All free money has a string attached and what you find at the other end may be a very unpleasant surprise.

Society and Culture

Will this action lead to the quality of life I am seeking? Will it adversely affect the lives of others? This last testing guideline is fairly self-explanatory. Something that does not benefit society as a whole probably won't benefit you for very long either.

Nay-Sayers

Over the years I've met many a person who has said that planning in general, and holistic planning in particular, is unnecessary. Their argument usually goes something like this: "Heck, this stuff is just common sense. All this goal writing and planning, who needs it? We know our goals.

FARMER PROFILE

Kerry and Barb Buchmayer

After years of dairy farming in New York State, Kerry and Barb Buchmayer decided to sell. They were concerned with the high taxes and the lack of a future in farming for their children in New York.

Winter feeding also took a great deal of their energy, time, and money. "In New York, with its long, hard winters, we might get 6 months of decent grazing, but the remainder of the year we had to feed stored feeds. The work of milking was intense enough without the extra effort that winter feeding entailed," Kerry explains. "We began looking at areas of the country with longer grazing seasons, and settled on Missouri." The Buchmayers relocated in 1996.

When they first moved to Missouri, Kerry and Barb didn't plan on milking; their first thought was to raise replacement dairy heifers. But after a time, they decided milking might not be so bad in this climate, particularly if they did a few things differently this time around.

"We studied holistic management," Barb says, "and we realized our weak link was in the marketing.

"In New York, we sold milk that won all kinds of quality awards from our processor, but then it was just mixed with all the other milk. We were paid a slight premium, but not much of one. We knew if we were going to milk again, we had to improve the marketing of our product. And we also knew we were going to have to do it with a modern parlor."

In New York, the Buchmayers had milked in an old flat barn, and the milking chores took 6 hours per day. When they began reconsidering milking in Missouri, they visited several farms with "swing-over" parlors. The swing-over design allows the person doing the milking to work standing straight up; one person can milk seventy cows per hour.

The Buchmayers installed a swing-over 10-cow parlor constructed from used equipment on their farm, then began working toward the second part of their plan: direct-marketing. Kerry and Barb decided to go for organic certification, and to bottle milk on the farm.

"This was like something that was meant to be," Barb says. "It was the culmination of all our skills over the years and all our dreams."

They began the long and sometimes intense process of learning what it would take to bottle their own milk. "We had to find an organic certifying organization that we felt comfortable with; we had to find and purchase the equipment; we had to learn about the legalities of processing and selling our milk; we had to develop marketing plans, and obtain financing," Kerry says.

For the Buchmayers, the financing became a real stumbling block. Despite the fact that they had high

We work toward 'em. And this is just gonna take up valuable time I could be using to get the hay into the barn."

I've got to agree that a lot of this is just plain common sense. So is planning a trip — say, from Bangor, Maine, to Los Angeles, California, with a stop at Liberal, Kansas, to visit Cousin Bob and a side trip to the Grand Canyon along the way. Common sense says that all of these destinations are southwest of Bangor, so if you just head on out driving southwest, sooner or later you might complete the whole trip. But the odds of completing your trip, in a timely fashion, with minimal wrong turns and extra miles, are greatly increased if you purchase a map and plot out your course. Well, planning is like using the map: It is common sense and it enhances the trip, increases the likelihood that you'll arrive at the spot you set out for in the first place, decreases expenses, and just makes life on the road a sight simpler!

Tom Frantzen tells the story of how, after beginning formal planning, he figured out that every time he started a tractor, it cost him $17 per hour for fuel, maintenance (including his labor), tires, and so on. "Once I had that dollar figure in my mind, my tractor use went down considerably. When I did have to start the tractor, I'd try to have multiple jobs already lined out, so that it was used efficiently." The planning process may have been "just common sense," but by going through the exercise Tom was able to clarify in his own mind the effect of his actions on his ability to reach his goal.

equity, "the local banks just couldn't understand how we could hope to sell milk for $5 per gallon. They didn't understand that the markets for organic products are one of the fastest-growing segments in the food economy, and that in larger communities the demand is there," he explains.

And their preliminary contacts with grocers and consumers in Columbia, Missouri (a college town about two hours away), indicated that the demand definitely was there. "People really wanted our product to become available," Barb says. They were already purchasing organic milk, but it was being shipped in from Utah. There was interest in a local supplier.

Financing finally came from the local, rural electric co-op, which had a revolving loan program for helping small businesses in the area. "At first, the co-op staff said we had to get a cash-flow analysis from a commercial ag-economist at the university. The person we were supposed to deal with was very negative. He came from the 'get big or get out' paradigm, and he said that he thought organic foods were unnecessary. He came up with a cash-flow analysis that was lacking in basic economic principles and said we were going to lose our shirts, even though there were stores, and consumers, who wanted our product." Finally, Kerry and Barb told the staff at the electric co-op that they couldn't work with the economist, and wouldn't. "The staff passed the application on to the board without the cash-flow analysis. It turns out the board was much more interested in our equity position than what an economist at the university had to say, and they approved the financing."

Both Kerry and Barb indicate that they got more support, assistance, advice, and information from other producers and processors than from places like the university or extension service. "There were a few really helpful people in the sources where farmers traditionally turn for help and support, but for the most part the bureaucrats and specialists didn't want to be bothered with us. We had to develop our own network." They did that by contacting sustainable agriculture groups and attending grazing conferences. "Once we found a network of individuals who had done some of the things we were trying to do, we knew we were on the right track, and that we could succeed," Kerry says.

The Buchmayers offered a word of advice for anyone trying to start up an alternative enterprise: "Plan how long you think it will take you to develop your project, and then double your projection. We found that everything took longer than we thought it would."

Setting a goal is crucial to planning. The trip in the above analogy has a goal: Bangor to Los Angeles, with two stops in between. Sometimes heading out on a trip without a destination is an adventure, but it tends to cost more money and cause more stress than a trip with a set destination. Establishing a holistic goal gives you the destination. It will help you get to where it is you want to go.

Example Farms

Throughout the next three chapters I will discuss four example farms. Although these are fictional farms (I don't think anyone wants to share their personal financial information, in the level of detail I need here), they are fairly realistic examples. The box (at right) tells the story of our example farms.

Computers

Although not absolutely necessary, computers really come into their own for planning. A word processor will enable you to keep notes from brainstorming sessions, to keep a monitoring journal, and to complete other writing projects. Spreadsheets make financial planning much easier, allowing you to both record data and play with various scenarios, such as calculating gross profits under varying circumstances: worst case, expected case, and best case.

As I noted in part III, we also use our computer (with the help of a drawing program) to develop labels and advertisements. In this way we've been able to develop ads that are both eye catching and inexpensive.

Computers have become far cheaper in the past few years. One that can do everything you'll need can be purchased brand new for less than $1,500. And if that is too much for your current budget, shop around for a used model. The computer I use now is the first new one that I've ever owned, and it's my fourth. My fifth computer, a laptop I can take with me wherever I go, was also purchased used.

Computers get sold while still in good working order for next to nothing, because lots of "computer geeks" upgrade every year or two. Nothing is wrong

(cont'd on p. 144)

THE EXAMPLE FARMS

The Blacks. Miles and Gail Black are third-generation ranchers who have never known anything other than the mountain ranching community of central Colorado. The 2,800 deeded acres (114.8 ha) they now run, was homesteaded by Miles's great-grandfather at the turn of the century. Now in their mid-40s, they are trying to figure out how to accommodate their son, Mark, and his fiancée, Laura, who want to stay and ranch with Miles and Gail.

As well as running their deeded land, the Blacks also lease 20,000 acres (8,100 ha) of Bureau of Land Management land from the federal government. The lease entitles them to run 100 cow-calf pairs for 5 months of each year, if the grass holds out.

The Blacks have a relatively small line of equipment relative to the size of their spread — they keep one older tractor and some haying equipment, as well as a few older four-wheel-drive pickup trucks, two fifth-wheel stock trailers, and a stock truck.

The Joneses. Gil and Jenny Jones are family farmers who have run a 240-acre (98-ha) farm in central Wisconsin since their marriage in 1977. They both grew up on dairy farms in the area. They have two teenage boys and one teenage daughter: Gabe, age 17; Mike, age 14; and Cindy, age 15.

Until 1990, all the tillable ground (even some steep ground that was highly erosion prone) was planted to crops. They maintained a small beef cow-calf herd of about fifteen mother cows to graze in the low ground. Jenny had to work off the farm as a secretary in a local insurance office, and they were going deeper into debt each year. They were seriously thinking about selling the farm.

In the winter of 1990, they decided it was time to either sell or change their operation. They attended some workshops put on regionally by a sustainable farming organization. They began participating in a grazing group, they attended workshops on holistic management, and they began using planning and monitoring procedures like those outlined in the next three chapters. During the next decade, they were able to turn their operation around. In 1996, Jenny quit her job in town to work full time on the farm with Gil and the children.

Today, the Jones run a grass-based livestock farm, incorporating both cattle and sheep. The whole farm has been developed into pastures and hay fields. They use workhorses for cutting and raking hay, as well as in their garden. In 1997, they were able to begin working on direct-marketing some of their animals.

The Millers. Gary and Michelle Miller purchased their 60-acre (24-ha) farm in western Massachusetts in 1988. The Millers have no children. Neither Gary nor Michelle came from a farming background, but they saved their money for several years to be able to purchase their dream farm.

The land the Millers purchased was half open, half wooded with a mix of maples, oaks, and pines. The buildings were old and in need of repair, but Gary had the basic carpentry skills to do the job himself. The first 3 years that they owned the farm, the Millers kept working in Boston, commuting to the farm on weekends and for their vacations. They worked on the buildings, put in a large garden, and kept a pair of draft horses.

In 1991, the Millers made the jump, quitting their jobs and moving to the farm full time. They market garden produce, herbs, and flowers at a local farmers' market during the summer; sell firewood and maple syrup in winter; and raise chickens, cattle, and sheep to direct-market through the farmers' market and to a mailing list of customers. They currently rent an additional 40 acres (16 ha) of pasture from a neighbor. Their only equipment is one older tractor with a loader for moving manure and snow, and an older rototiller.

The Wilsons. Tom and Karen Wilson and their three sons — Rick, age 14; Jeff, age 12; and Byron, age 7 — live on a 200-acre (81-ha) farm in western Tennessee that has been in Karen's family for more than 200 years. The Wilsons returned to the farm from Nashville 7 years ago, when Karen's dad decided to retire. Tom was an engineer and Karen a high school English teacher, but they were both happy to give up their jobs for a life in the country, working the land. Getting their children out of city schools was also important to them.

Karen's parents wanted to buy a house in town and spend the winter traveling to southern Texas, so they had to sell the farm to Tom and Karen. The sale included all equipment and livestock. They also felt this sale would make their estate clearer when they died, so Karen wouldn't need to fight it out with her siblings. She and Tom would owe the estate payments on the loan, just as they now owed her parents.

During the years that Karen's dad farmed, he kept a few beef cows on the hilly pastures that couldn't be tilled, but primarily he raised crops on about 120 acres (49 ha) of the farm. He had a 30-acre (12-ha) tobacco allotment; the remainder was grown in a corn and bean rotation. Tom and Karen kept farming much the way her dad had done before them, but Karen still has to work as a teacher in the local school district, and each year they have to dip deeper into their savings just to live.

with their old computer; they just want a new one that is faster and has more bells and whistles. You don't have to have the newest, fastest model for farm use.

If you're intimidated by the thought of using a computer, you'll find that often the place you buy it from (or the person you buy from if it's used) will help you get started. If not, go to an adult education class, get the kids to help at first, or find a friend or neighbor who is familiar with computers. They really are quite user friendly once you get the basics down.

Calculations

There are lots of equations and calculations in the planning chapters. They all use basic math — nothing too fancy, but in case your math skills have gotten a little rusty, remember the following rules:

- Work that appears inside of (parentheses) should be done first.

  Example:

  $(2 + 3) \times 4 = 5 \times 4 = 20$

- When rounding a decimal to a whole number, the number 5 or greater is rounded up; anything below 5 is rounded down.

  Example:

  8.5 rounds to 9, when rounded to the nearest whole number

  Example:

  3.43 rounds to 3, when rounded to the nearest whole number

So that you can find the numbers easily when you want to work with them later, all the calculations used in part IV are listed in appendix F.

CHAPTER 13

Financial Planning

North American livestock producers are the most productive in the world. We are also the least profitable.

— David Pratt, "Farm Advisor" (*Livestock & Range Report No. 961*, University of California Cooperative Extension, Summer 1996)

Renting farm property has several outstanding advantages, among which perhaps most important are: 1. No capital need be invested in permanent features. 2. Such capital as the tenant does invest may all be for machines, tools, animals, portable houses and other "loose" property which he can take with him. . . .

— M. G. Kains, *Five Acres and Independence* (1935)

No unemployment insurance can be compared to an alliance between man and a plot of land.

— Henry Ford

HUMAN NATURE IS A STRANGE THING: No matter how much money most people (or businesses) make, their expenses climb to near or above that amount. This is part of why many high-salaried individuals file for bankruptcy despite big incomes. Financial planning should help you overcome this tendency, but keep in mind that traditional economic theories and traditional financial planning supported the very decisions that drove many farmers off their land over the past 50 years and left us with polluted water, polluted air, and erosion. This is because traditional financial planning looks at the bottom line as sacrosanct. Your financial plan, on the other hand, must look at finances in a broader sense, and should include not only paper money but also quality-of-life, social, and environmental aspects of your decisions. It isn't simply about record-keeping, accounting, or budgeting; it *is* about planning for income, planning for profit, and prioritizing expenses.

Balance Sheets

Assets are the economic resources that a business has available to it, and that are expected to provide some future benefit. In the case of a farm, land, buildings, equipment, livestock, and money in the bank are all types of assets. *Liabilities* are the debts of the business and may include mortgages, loans, and amounts of money owed to suppliers or employees. *Equity* is theoretically what the owner has already put into the business, or the difference between assets and liabilities.

Let's say you own a farm that is valued at $100,000 under current market conditions, and that you have a mortgage for $80,000 against it. This leaves you with $20,000 worth of equity. If land values go up, your equity increases, but if they drop so does your equity.

The balance sheet is an accounting form that lists all the assets on one side, the liabilities and equity on the other. The two sides must balance (Table 13.1). Now, if you study the balance sheets for the example farms, your first reaction will probably be that the Blacks are rich, but remember — assets, particularly land, are only of value when disposed of. This is what's known in agriculture as being land rich, cash poor. The Blacks have more than $1.5 million of equity, but the Joneses, the Millers, and the Wilsons all have more cash and other investments.

Each of our example farmers does have some owner's equity, which means that if they sold out, there would be some money left over after paying all the bills, but this isn't always the case in agriculture. Many, many farmers are stuck in the unenviable position of having negative equity: If they sell out, they still owe the bank money.

Table 13.1

Balance Sheets for Example Farms

The Blacks

Assets (in dollars)		Liabilities (in dollars)	
Land	1,400,000	Mortgage	127,000
House	18,000	Pickup loan	2,000
Facilities	120,000	Operating loan	8,000
Vehicles	28,000	Misc. bills	700
Tractors	4,000	BLM lease	3,190
Implements	2,000		
Breeding stock	90,000		
Cash in bank	1,200		
Total Assets	**1,663,200**	**Total Liabilities**	**140,890**
		Owner's Equity	**1,522,310**
		Liabilities & Equity	**1,663,200**

The Joneses

Assets (in dollars)		Liabilities (in dollars)	
Land	240,000	Mortgage	70,000
House	40,000	Misc. bills	6,000
Facilities	30,000	Vehicle loan	12,000
Vehicles	30,000		
Tractor	4,500		
Implements	2,500		
Livestock	45,000		
Cash	6,000		
Other investments	4,000		
Total Assets	**402,000**	**Total Liabilities**	**88,00**
		Owner's Equity	**314,000**
		Liabilities & Equity	**402,000**

The Millers

Assets (in dollars)		Liabilities (in dollars)	
Land	98,000	Mortgage	21,500
House	35,000	Misc. bills	1,300
Facilties	25,000	Annual lease pymt.	1,000
Vehicles	4,000		
Tractor	1,200		
Implements	800		
Livestock	3,750		
Cash	4,500		
Other investments	12,000		
Total Assets	**184,250**	**Total Liabilities**	**23,800**
		Owner's Equity	**160,450**
		Liabilities & Equity	**184,250**

The Wilsons

Assets (in dollars)		Liabilities (in dollars)	
Land	200,000	Mortgage	170,000
House	28,000	Car loan	8,000
Facilities	30,000	Farm loan	30,000
Vehicles	15,000	Coop bill	1,600
Tractors	40,000	Misc. bills for drying grain	800
Implements	10,000		
Livestock	12,000		
Cash in bank	1,500		
Stock portfolio	3,000		
Total Assets	**339,500**	**Total Liabilities**	**210,400**
		Owner's Equity	**129,100**
		Liabilities & Equity	**339,500**

Note: "Facilities" includes farm buildings (which includes housing for hired help), fencing, water systems, etc.

Land

Land is a unique piece of the agricultural pie. It is, of course, one of the main ingredients of a farm, yet it very rarely pays for itself in farm-generated cash flow. This paradox is the result of land's unusual role in the economic landscape.

Land Is a Nondepreciable Asset

Depreciation is the "costing out" of an asset over its "useful life," but unlike most other assets land's useful life is in perpetuity. Let's look at an example: You purchase a new pickup truck; it has a useful life of seven years. Each year that you own the truck, the Internal Revenue Service will allow you to deduct one-seventh of the truck's value from your income for tax purposes. This is depreciating the asset, and it is allowed because the IRS figures that at the end of seven years, your truck isn't going to be worth much of anything anymore. On a farm you can depreciate buildings, breeding stock, machinery, equipment, office machines, and vehicles. But land, in theory, retains its useful value indefinitely; so land can't be depreciated.

Land Appreciates Over Long Periods of Time

Land distinguishes itself from other assets not only because it retains its usefulness, but because it generally appreciates (or adds value) over long periods of time. During certain economic periods, the value of land can increase very dramatically (high inflation), though most often the increase is slow. Then there are some periods when land will lose some of its value (deflation). But if you purchase a piece of land and hold on to it for 30 years, the odds are it will be worth quite a bit more than what you originally paid for it. The appreciation of land value is generally caused by one of two things: anticipation of future profits or nonagricultural pressures.

Anticipation of Future Profits. The anticipation of future profits can drive prices up from year to year and, at times, cause spiraling inflationary and deflationary cycles with respect to land value, much like a self-fulfilling prophecy. This happened in the go-go days of the late 1970s and early 1980s. Prices began climbing, so farmers (and investors) thought, "If my land is worth X amount of dollars today, it will be worth X-plus tomorrow. If I buy more land, it too will increase in value." Unfortunately, what many farmers and investors found was that when the peak was reached, the prices fell just as dramatically as they had previously risen, and farmland prices in the late 1980s were at the bottom of the barrel. Many farmers, investors, and small banks lost everything as a result. Our Minnesota farm was a prime example: An investor purchased it in 1983 for $48,000; we bought it 6 years later for $30,000. The investor — who purchased numerous farms in the area during 1983 and 1984 — declared bankruptcy.

Nonagricultural Pressures. Nonagricultural pressures that may drive up prices include recreation, tourism, and development. Depending on where you live in the country, these pressures may drive prices completely off the scale. Nonagricultural pressures came to bear on central Minnesota farmland during the late 1990s. City people were purchasing farms in our area for hunting places or weekend retreats, almost doubling land values in the area over a year or two.

Impact on Farmers and Farms

Land's nondepreciable nature has a large impact on farmers and farms. In real economic terms, owned land is generally a drain on the wealth of a farmer; rented land is generally a boon. Land is a good long-term investment if you can afford to purchase it with after-tax income (profits), but if you don't have true after-tax income available, then renting is the better option. Rents are fully tax deductible from farm-generated income, and even if land prices are pressured by outside forces, rents remain based on agricultural operating income potential. Most often, the best approach is to blend a small deeded holding with rental lands. (Notice how many of the farmers and ranchers that are highlighted throughout this book rent at least part of the land they are using.)

The Balance Sheet. Develop your own balance sheet, but perform two different sets of calculations.

The first set should be based on current values of all assets; the second should be based on the lowest value of assets in the most recent economic cycle. For example, if land is currently selling near you for $1,200 per acre ($2,965 per ha), use that figure to calculate your current situation. Now, think back to what the worst prices were for similar land in recent history; if it sold for $700 per acre ($1,730 per ha) 10 years ago, calculate that situation — it gives you a worst-case scenario. If you do have to borrow money against land, never, never borrow more than the low value of the land. And when the cost of land is spiraling upward, *beware*. Hoard your money for the next time that the low price and the current price happen to be about the same — that's the time to buy. These are the times when the popular media, like local newspapers, are bemoaning the fact that "you can't give real estate away!"

Opportunity Cost

When you own an asset, there is always an opportunity cost on your capital. *Opportunity cost* is the amount of money you could earn with the money that is currently tied up in an asset if it were invested somewhere else. One of the safest long-term investments available in the world is U.S. 30-year Treasury Bonds, so these are a good benchmark for safe earnings, or a minimum return on investment.

The Wilsons have equity of $129,100. Subtracting the cash in the bank and stocks from their equity (because these should be earning a return) gives a farm-invested equity of $124,600. If this were invested in a T-Bond (currently paying about 5%), it would earn them about $6,200 in a year. This is their opportunity cost, and their operation should pay them at least this much above operating expenses each year. The Blacks' opportunity cost is $76,000 [(1,522,310 − 1,200) × 0.05]; the Millers' is $7,200 [(160,450 − 16,500) × 0.05]; and the Joneses' is $15,200 [(314,000 − 10,000) × 0.05].

In reality, when you invest in a business you have a greater risk than when investing in a T-Bond, so you should consequently aim for a higher return on your investment. Businesses typically seek to double the T-Bond rate or, in today's market, make a 10 percent return on their investment.

From the current farm equity that you calculated on your balance sheet, calculate the opportunity cost based on the T-Bond rate, then double this. If you look at this number and say, "Wow, there's no way I'm making this kind of return on my investment," don't feel bad. Few farmers do. However, you now have an amount to begin aiming for in your financial planning.

Zero-Based Budgeting

In the early 1980s, I was hired to manage a sanitation district in Frisco, Colorado. A sanitation district is a unit of government whose primary responsibility is operating wastewater treatment plants. I had worked up through various wastewater plants over the previous decade, so I was qualified as far as operations, maintenance, and laboratory work were concerned, but I had no real experience in the business-administration end of things — like preparing a budget.

Even though I was inexperienced in such matters, I knew that the district's previous budgeting process could use a little improvement. I called up the local certified public accountant who had done the district's audits (a legal requirement for all public entities in Colorado) and offered to buy him lunch if he'd let me pick his brain on how to go about improving the budget process. Pat was enthusiastic about my request: He'd been recommending better budgeting to the elected board of directors for several years.

With his guidance, I developed more detailed accounts; for example, instead of one line item for maintenance-related expenses, I broke out several different maintenance categories. In the case of a farm, your line items should be broken down so that you can track various enterprises or operations. Instead of one large category for "feed," break it down into feed categories for each class of animals being raised. Instead of one income line for "farm products," break it down into "cull cow sales," "cull sheep sales," "steer sales," "lamb sales," "beef sales (meat)," "lamb sales (meat)."

The next recommendation Pat made was to "budget income conservatively low, and expenses conservatively high" — and to do it with a technique known as zero-based budgeting. When budgeting conservatively low for income, you have to factor in such things as a higher-than-normal death loss, or lower prices. For budgeting conservatively high expenses, think about things like price increases. But the word *conservative* in both categories means stay reasonable — if feeder pigs are currently selling for $45, the 10-year high was $60, and the ten-year low was $20, then $30 is probably a conservative figure for budgeting income.

In zero-based budgeting, you begin each year as if you have no previous experience — or no known amount of income or expenses from previous years. You have to think about each item you will need, investigate prices, and develop the budget based on those numbers. Typically, most people budget — if they do it at all — by saying, "Last year I spent X number of dollars on something, so this year I'll probably spend X plus a little more." And, by God, that's what they end up spending — or even more than this. Zero-based budgeting takes more time and a good deal more thought, but it provides a superior product when you're done.

In a traditional budgeting paradigm, the Blacks might say, "Last year we spent $4,000 on fuel for trucks and autos, so this year we'll figure that we're going to need $4,500." This year, instead, the Blacks are going to sit down with pencil and paper and do some figuring (Table 13.2).

During our lunch discussion, Pat reminded me that this is a mental exercise in "best guessing" and, as he pointed out, even your best guesses can be wrong. His recommendation was that after you've developed your expense figures, add 5 percent for a cushion.

Table 13.2

Gasoline Worksheet for the Blacks

Gas for Miles's Pickup

- ☐ 2 town trips per week at 30 miles = 60 miles × 52 weeks = 3,120 miles
- ☐ Ranch miles = 50 per week × 52 weeks = 2,600
- ☐ 2 city trips per month at 220 miles = 440 x 12 months = 5,280 miles
- ☐ Total Miles's pickup = 11,000/12 mpg × $1.10/gal = $1,008.33

Gas for Mark's Pickup

- ☐ 1 town trip per week at 30 miles = 30 × 52 = 1,560 miles
- ☐ 1 city trip per month at 220 miles = 220 × 12 months = 2,640 miles
- ☐ Total for Mark's Pickup = 4,200/10 mpg × $1.10/gal = $462.00

Gas for Cars

- ☐ 2 town trips per week at 30 miles = 60 miles × 52 weeks = 3,120 miles
- ☐ vacations and long road trips = 5,000 miles
- ☐ Total for cars = 8,120/14 mpg × $1.10/gal = $638.00

Diesel for Stock Truck

- ☐ 20 trips to sale barn at 120 miles = 2,400 miles
- ☐ 20 trips from ranch to allotment at 30 miles = 600 miles
- ☐ Total = 3,000/12 mpg × $1.20/gal = $300.00

Miscellaneous Gas and Diesel

$500.00

Total: $1,008.33 + 462.00 + 638.00 + 300.00 + 500.00 = $2,908.33 × 1.05 = **$3,053.75**

The family members all agree they should be able to live within this budget for gas, which knocks almost $1,000 off what they spent the previous year. They see that this allows for plenty of town and city trips, but they will make a conscientious effort to keep their use within these numbers.

Cost of Production

There are two types of costs that all businesses incur in producing their product: fixed and variable. *Fixed costs* are those that must be paid regardless of what volume or type of production occurs and generally include things such as mortgages, long-term loans, insurance, full-time labor, and so on. No matter what enterprise you undertake (or if you choose to do nothing), these bills must still be paid. *Variable costs* are those costs that are only incurred as a result of a specific type or volume of production; these usually include feed, seeds, fertilizer, and veterinary expenses. These costs are affected by the actual production that is taking place (Figure 13.1).

The division between fixed and variable isn't always clear-cut: Some items may be fixed in certain

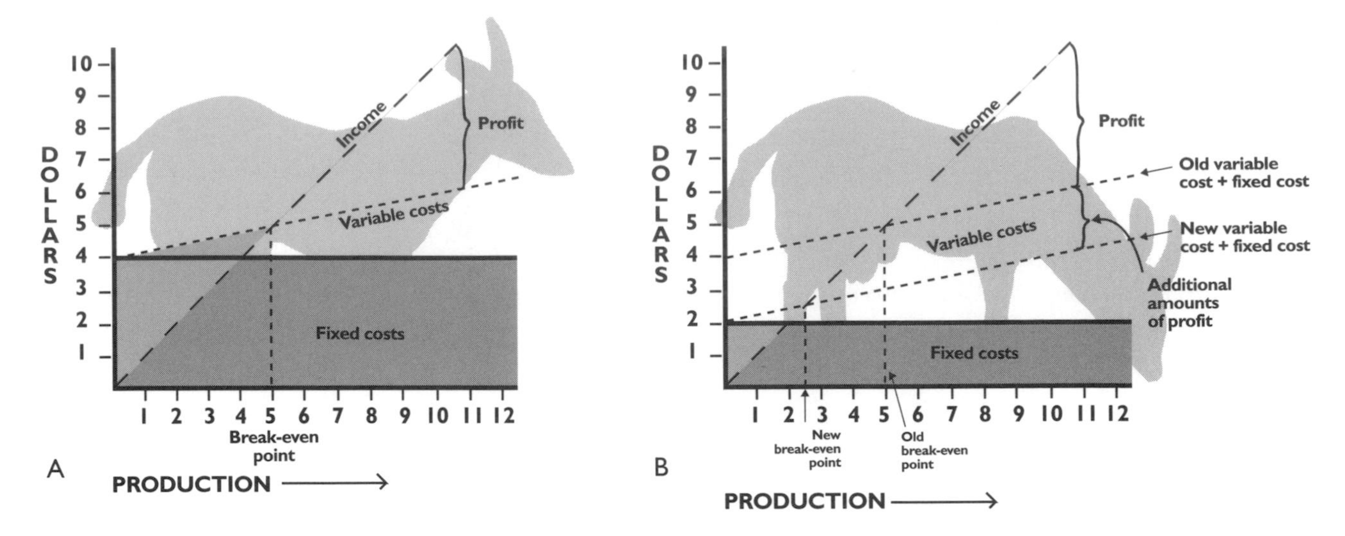

Figure 13.1. (A.) Income is generally a linear function that correlates directly to production. For example, if 1 unit of production creates $1 in income, 10 units will create $10, as is the case in figures A and B. Fixed costs are constant and are unaffected by production. **(B.)** Variable costs, such as income, are a linear function that correlate to production. For example, if it costs 20 cents in variable costs to produce 1 unit, then it will cost $2 to produce 10 units. In figure A, the farmer needs 5 units of production to break even, but by cutting fixed costs in half (figure B), he needs less than 3 units of production to break even.

(Modified from Sam Bingham, *Holistic Resource Management Workbook.* Covelo, CA: Island Press, 1990, p. 8.)

scenarios and variable in others. For example, dollars spent on the purchase of livestock are a fixed cost if you're purchasing breeding animals, but they are a variable cost when purchasing stockers.

Items that are normally considered variable may become fixed if purchased ahead of time in bulk. If a local feed dealer is going out of business and you borrow money from the bank to purchase his remaining minerals at a deep discount, the cost is now fixed; regardless of what type of livestock you choose to raise, the cost for the minerals has already been incurred.

The division between fixed and variable costs also depends on your individual values. Some people may consider expenses such as health insurance fixed costs — something they must have, and that must be paid for from their operations. Others may consider these types of family living expenses variable — something they'll pay out of their farm-generated income if there is enough left over; otherwise they'll pay from non-farm-generated income.

Finally, if you are considering a new enterprise that would require a large capital investment, the cost is variable until that money is actually spent. After the commitment has been made, it's a fixed cost. If you are considering building a hoop house in which to raise laying hens year-round, the cost is variable for planning purposes. Once you spend the money to build the hoop house, though, the cost is fixed.

Profit

Profit is what's left over (if anything) when you subtract expenses from income. You can increase profit by increasing income while holding expenses steady, holding income steady and decreasing expenses, or, best of all, by increasing income and decreasing expenses — this is when you really see your profits rise.

Plan for and work toward profit in a systematic fashion. Think about it in planning, before the year begins and before the expenses are budgeted.

Developing Your Own Plan

You now have enough background to begin the process of developing your own plan. This will take time and is best done in a series of sessions. Since this planning process isn't directly related to taxes (though by budgeting and monitoring, tax time may become less stressful), you can begin at any time of the year.

The point is to begin taking a stab at it — don't put it off until the first of the year. Use worksheets similar to those in Tables 13.2 through 13.8 to develop your financial plan.

Define appropriate line items in which to break down income and expense. Go into as much detail as you possibly can. Categorize line items for each enterprise, or possible enterprise.

1. ***Develop your own list of fixed costs that must be paid out of your farming operation*** (Table 13.3). As you develop the list, think about ways to reduce some of these costs.
   — Do you have more tractors than you need? Sell some and use the money to pay off loans.
   — Are you holding onto other depreciables that have little or no purpose? Sell them, too. "Heavy metal" (such as tractors, farm equipment, vehicles) must be kept to a minimum on a small farm. Try to avoid these things by renting them when you need them, or paying a contractor, or working with a neighbor.
   — Are you paying off debts on high-interest credit cards? Consolidate them with a lower-interest loan from the bank, and then tear up the cards.
   — Can you reduce mortgage payments by refinancing, or lower insurance expenses by shopping around? If so, do it.
   — Brainstorm ways to cut all fixed costs!
2. ***Estimate income for the next year on current enterprises*** (Table 13.4). Plan for your profit right now, by cutting your estimated income in half. One half is now allotted for profit, the other for expenses. If you have fixed costs in excess of the one-half mark, then initially you'll have to allot less for profit — say, one-quarter — but establish right now an amount for profit. A long-term financial objective should be to see your profit return a reasonable amount on your investment.
3. ***Begin budgeting for your expenses*** (Table 13.5). When you develop a budget for expenses, your objective is not to let your total costs (both fixed and variable) exceed the expense allotment you established in step 2. Using the zero-based approach, work out an estimate of expenses by month for the coming year. Again, brainstorm ways to cut all expenses. Study your variable expenses. Calculate what it costs you to use equipment, and then ask yourself if you can cut usage. When trying to allocate expense money, use the marginal-reaction test to evaluate where your money should be put for the best use.
4. ***Develop an account for family salaries.*** If early on you can't pay yourself as much as you budget, you should still strive for a system that pays you for your labor, plus a return on your investment.
5. ***Develop an account for depreciation.*** This is money you are setting aside to be able to replace or repair equipment and buildings when the time comes. Although in the early years, you may not be able to do so, it's a good idea to open an interest-bearing account into which you pay your depreciation expense at the end of each year. That way, in a year when you need to replace or repair a major capital item, you have money set aside for that purpose.
6. ***Budget for items used in bulk on an inventory worksheet.*** If there are items you purchase in bulk and use throughout the year, then budget for the use of these with an inventory worksheet.
7. ***Transfer information from worksheets to the main budget worksheet.*** Once you have developed worksheets for income and expenses, transfer these to the main budget worksheet. This is also the sheet you'll use to monitor finances.
8. ***Evaluate existing enterprises.*** As time permits, evaluate existing enterprises — or ideas for new ones — using the gross profit test. Those enterprises that score highest on gross profit and meet the other tests (holistic guidelines) in moving you toward your goal are the ones you should put the greatest emphasis on. But remember — make changes somewhat slowly and don't diversify too quickly. Jumping in and out of too many things too often is a sure way to drive yourself, and everyone around you, crazy.

Table 13.3

FIXED-COST WORKSHEET FOR THE MILLERS

Item	Rate	Cost
Farm payment	$278/mo	$3,336
Farm insurance	$670/yr	670
Lease	$1,000/yr	1,000
Health insurance	$570/quarter	2,280
Household expenses (electricity, heat, etc.)	approx. $165/mo	1,980
Total fixed costs		$9,266

Table 13.4

INCOME WORKSHEET FOR THE MILLERS

Item	Rate	Income
Garden Income		
Fresh produce sold at farmers' market & farm	25 weeks @ $60.00/wk	$1,500
Fresh herbs and flowers sold at farmers' market & farm	25 weeks @ $125.00/wk	3,125
Dried herb and flower wreaths and decorations sold at craft fairs	200 @ $35.00/each	7,000
Fall pumpkins sold at farm	300 @ $5.00/each	1,500
Total Income		$13,125
Woodlot		
Firewood picked up at farm	20 cords @ $80.00/each	$1,600
Firewood delivered	40 cords @ $120/each	4,800
Maple syrup sold at farmers' markets and craft fairs	100 qts @ $11.00/each	1,100
Total Income		$7,500
Livestock		
Chickens sold at farmers' market and picked up at farm	1,000 @ $7.50/each	$7,500
Lamb sold at farmers' market and picked up at farm	40 lambs @ $150.00/each	6,000
Beef sold at farmers' market and picked up at farm	6 steers @ $1,250.00/each	7,500
Total Income		$21,000
	Total Income	41,625.00
	Expense Allotment (Total inc. x 0.5)	20,812.50
	Variable Allotment (Expense allotment, fixed costs)	11,546.50

Note: The expense allotment does not include depreciation or family salary line items. Salaries for hired employees should always be incorporated into the variable expense budget. See Table 13.8 for depreciation, family salaries, and profit.

Table 13.5

Variable Expense Worksheet for the Millers

Item	Cost
Garden	
Seeds and bedding plants	$200
Garden supplies, including tools, hose, cloches, etc.	350
Packaging, including strawberry flats, labels, etc.	115
Production supplies for wreaths, including frames, ribbons, etc.	250
Gas for rototiller	120
Miscellaneous	500
	$1,535
Woodlot	
Chains and sharpening	$300
Gas/oil for chain saw	250
Canning jars and labels for syrup	60
	$610
Livestock	
Poultry	
☐ Chicks	$600
☐ Feed	300
☐ Butchering	800
☐ Maintenance on portable buildings	340
	$2,040
Cattle	
☐ Feed	$300
☐ Vet expenses	150
☐ Butchering	600
	$1,050
Sheep	
☐ Feed	$300
☐ Vet expenses	200
☐ Buchering	1,200
	$1,700
Horses	
☐ Feed	$400
☐ Vet expenses	100
	$500
Other	
☐ Fencing	$300
☐ Miscellaneous	400
	$700
Other	
Office supplies, including postage	$150
Farm-related phone	250
Advertising, including flyers, order forms, etc.	450
Repairs & maintenance on equipment & vehicles	600
Vehicle operating expenses, including gas & insurance	500
Education, including magazines, meetings, seminars, etc.	400
Miscellaneous expenses	1,000
	$3,350
Total Variable Expenses	**$11,485**

Monitoring Money

After you establish a financial plan, you must begin monitoring. Financial monitoring assures that you are following the plan; when things go wrong, it lets you make corrections early on so you can minimize the impact of problems.

Monitoring of finances requires that each month, as early in the month as possible, you put together the records of how much money came in the previous month, how much money went out, and what inventory items were used (Tables 13.6, 13.7). These figures should now be recorded on the main budget worksheet (Table 13.8).

Variations from the budget, either income shortages or expense excesses, require that some action be taken. For example, if the $30-per-pig income you budgeted for from the sale of weaner pigs suddenly falls to $20, you need to begin looking for areas within the expense list that you can cut back. Or from our personal experience: When the price of corn began skyrocketing, we should have sold our pigs immediately.

Troubleshooting

How do you know what to do in these dire situations? Begin by working through both the gross profit and marginal-reaction exercises. Take a best-, expected-, and worst-case set of numbers to run through the gross profit test. If the worse case loses money, it's probably a good time to get out; if the expected and best cases are going to lose money, it's definitely time to bail out of the enterprise! And when things are looking bad, or your plans aren't working out quite as you thought they would, go back to brainstorming with everyone. Sometimes a unique solution will present itself.

FARMER PROFILE

Bob Bowen and Anne Bossi

"I went bankrupt in 1989, on my first farm," Bob Bowen begins by telling me. "It was 100 acres, and I raised pigs commercially, and marketed them as a commodity.

"We [Bob and his first wife] bought the land when land values were really high. Then when the bottom fell out of land values around us, and the bank called the note, there was no way to pay it. I showed up at the bank one day with a truckload of hogs. I just wanted to give them back." He chuckles as he tells the story. "Of course, I'd called the local papers and the television station to tell them I was going to do it, so I had great coverage."

Unfortunately — or maybe fortunately, depending on how you view things — it didn't stop the bank from taking back the land. "After that first experience, I decided I was still going to farm, but I was going to do things real differently. For one thing, I decided I would avoid the government as much as possible — no loans, no subsidies. I also decided to look at what people like to eat and try to sell them that, instead of what I wanted to grow! And I decided that *I* was going to set *my* prices — not 'the market's.' My price would be based on how much money I needed to get out of a product to cover all my costs — even all the incidentals and fixed costs — plus pay a little profit. If I couldn't sell for that price, I'd find something else that I could. I charge what I have to, to make a profit, and I don't care what they charge at the grocery store down the road. If no one wants to buy at that price, I won't raise that item." For example, Bob and Anne charge $2 per dozen for eggs, and they sell lots of them.

Anne and Bob knew each other before he lost the farm, and his marriage, so shortly afterward, he moved onto her place. "Me and some of my pigs that were still left," he laughs. Over the ensuing years, the two have created a phenomenally successful farming venture with only 23 acres of land and a strong family partnership.

The first year out of his troubles, Bob raised chickens, but he had to get his customers to reserve (and pay for) them in spring because he couldn't afford to buy the chicks and feed otherwise. When I comment that this was almost an early variation on a CSA, he agrees, but says he's glad that they don't have to operate that way anymore. Still, the experience taught Bob and Anne that they can work with their customers when they need to raise capital.

When Bob and Anne wanted to raise additional capital to build a new barn, they sold "Barn-Bucks" to many of their regular customers. Each Barn-Buck had a $10 face value, and customers purchased them at face value. The customer was then entitled to redeem their Barn-Bucks, plus 10 percent value interest, on farm products 3 months later. "If someone purchased $100 worth of Barn-Bucks, 3 months later they could use them to receive $110 worth of chickens, eggs, or pork."

From Bob's first humble steps "out of the dark times," the business has "grown exponentially." Today, Bob and Anne market the pigs from 25 sows, 10,000 broilers, eggs (which are now done in a novel partnership with another farmer) from 1,200 layers, 750 turkeys, 400 ducks, and milk and cheese from goats. Anne runs a spring bedding plant business, but with plants from all over the world. And as well as marketing their own items, they also market beef, lamb, and rabbit from other area farmers. "These fellows grow a great product but aren't much into marketing," Bob quips.

Bob hits five farmers' markets per week during the good-weather months, and two indoor, year-round markets during the winter. He also has store and restaurant customers. "When I pull up, I can supply just about anything someone wants."

When I asked Bob if he and Anne — who also works as the Northeast field representative for Heifer Project International — ever feel the need to de-diversify, he tells me they already have. "We used to do the layers ourselves, and the rabbits ourselves. Now we've set other farmers up with those projects, but we still work on the marketing."

And Bob is one hell of a marketer. Before going into his first full-time farming operation, he sold used cars and insurance to farmers. "Selling insurance to farmers was sort of like an alcoholic selling beer. Each farm I went to, I fell in love with. I wanted to milk the farmer's cows instead of sell him insurance."

Bob observes, "Most farmers are just consumers. They sell commodities, like wheat, but then buy their bread. They sell at wholesale and buy at retail. It doesn't work. If you want to make money, start selling something directly, or work with someone who will.

"I do these marketing workshops around the country now. I tell people, selling is hard work, and the only way to really learn it is to do it. You have to stand out in the pouring rain. You have to try with every customer. You won't sell to all of them, but give it the old try. If a vegetarian tells me 'I only eat vegetables,' I tell 'em, 'Hey, that's all my chickens eat, too!' Sometimes they come back to eating meat when they learn how my animals are raised."

Note: Heifer Project International (Little Rock, Arkansas) works to fight world hunger, not by giving food but by giving the means of production. The group donates to those in need around the world, both livestock and training in practices that will sustain the animals and the people. In return, each recipient agrees to "pass on the gift" by returning a female offspring or by sharing knowledge in training others. The project was started by a midwestern farmer in the 1930s. During the Spanish Civil War, Dan West went to Spain as part of a missionary group. As he handed out cups of milk to war-ravaged children, he realized that what these people needed was not a cup of milk but a cow. He began by asking farmers back home to donate a heifer. Today, HPI operates in more than 100 countries around the world.

Table 13.6

Main Monitoring Worksheet for the Millers

Date	Payments to or from	For	Amount	Income		Expense	
				Farm	Personal	Farm	Personal
Jan. 2	Langston's Gift Shop	Wreaths	600.00	600.00			
Jan. 2	Dr. Jones	Vet service for sick cow	35.00			35.00	
Jan. 4	John Carpenter	Birthday present	25.00				25.00
Jan. 6	Feed store	Trace mineral blocks	7.29			7.29	
Jan. 28	Tom's Texaco	Repair alternator on truck	86.57			86.57	
Jan. 29	Hometown Grocery	Food shopping	128.67				128.67
Jan. 30	Sale barn	Steer	740.00	740.00			

Note: Each income and expense item is recorded twice, first on the Main Monitoring Worksheet and then on the Farm Monitoring or Personal Worksheet. At the end of each month, the categories are totaled and the information is transferred to the Annual Budget Monitoring Worksheet.

Table 13.7

Farm Monitoring Worksheet for the Millers

Date	Payment to or from	For	Amount	Income					Expense							
				Garden	Woodlot	Poultry	Sheep	Cattle	Garden	Woodlot	Poultry	Sheep	Cattle	Horses	Other livestock	Other farm
Jan. 2	Langston's Gift Shop	Wreaths	600.00	600.00												
Jan. 2	Dr. Jones	Vet for Buehla	35.00										35.00			
Jan. 6	Feedstore	Trace mineral blocks	7.29									2.43	2.43	2.43		
Jan. 28	Tom's Texaco	Alternator	86.57													86.57
Jan. 30	Sale barn	Steer	740.00					740.00								

Table 13.8

ANNUAL BUDGET MONITORING WORKSHEET FOR THE MILLERS

Category	Item	Budgeted	Actual			Actual			Actual		
			January	Year to Date	Percent to Date	February	Year to Date	Percent to Date	March	Year to Date	Percent to Date
Income	Fresh produce	1,500.00	0.00	0.00	0						
	Herbs/flowers	3,125.00	0.00	0.00	0						
	Dried wreaths	7,000.00	600.00	600.00	0.09						
	Pumpkins	1,500.00	0.00	0.00	0						
	Firewood @ farm	1,600.00	160.00	160.00	0.10						
	Firewood delivered	4,800.00	560.00	560.00	0.12						
	Maple syrup	1,100.00	220.00	220.00	0.20						
	Poultry	7,500.00	0.00	0.00	0						
	Lamb	6,000.00	0.00	0.00	0						
	Beef	7,500.00	740.00	740.00	0.10						
	Total	**41,625.00**	**2,280.00**	**2,280.00**	**0.06**						
Expense	Fixed costs	9,266.00	400.00	400.00	0.04						
	Depreciation	800.00	66.67	66.67	0.08						
	Family salaries	15,000.00	1,250.00	1,250.00	0.08						
	Garden	1,535.00	32.67	32.67	0.02						
	Woodlot	610.00	18.00	18.00	0.03						
	Poultry	2,040.00	0.00	0.00	0						
	Sheep	1,700.00	78.00	78.00	0.05						
	Cattle	1,050.00	120.00	120.00	0.11						
	Horses	500.00	45.00	45.00	0.09						
	Other livestock	700.00	67.98	67.98	0.10						
	Other farm expenses	3,350.00	237.47	237.47	0.07						
	Total	**36,551.00**	**2,315.79**	**2,315.79**	**0.06**						
Profit	**Income – Expense**	**5,074.00**									

Note: This monitoring sheet would be filled out each month of the year to help track what's actually happening on the farm.

CHAPTER 14

Biological Planning

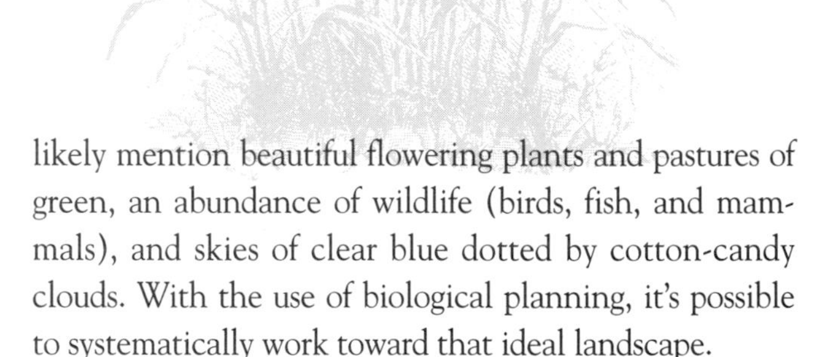

The untapped potential of maps in community discussions first came home to me as a teacher on the Navajo Indian Reservation. Once I put up both an aerial photo montage and a wall full of topographic maps to illustrate some tiresome point in a math book. Within a very short time finger smudges revealed a whole host of local issues I had not uncovered in years of daily contact with people. At first, people who knew nothing of contour lines, scale, symbols or English labels regarded my displays as incomprehensible bits of abstract art — but the minute they saw the connection to the land, they learned very fast. Soon grandparents who spoke only Navajo and had never darkened the door of a school pored over them, detailing range disputes, dried-up springs, forgotten cornfields, ancient ruins, the habitat of medicinal plants, and myriad other matters that without the graphic aid of a map they could not begin to communicate to a stranger.

— Sam Bingham, *Holistic Resource Management Workbook*

ASK ANY PEOPLE — urban or rural, rich or poor, from almost anywhere in the world — to tell you about their most desirable landscapes, and they'll describe lush grasslands, healthy waterways, and robust forests. They'll likely mention beautiful flowering plants and pastures of green, an abundance of wildlife (birds, fish, and mammals), and skies of clear blue dotted by cotton-candy clouds. With the use of biological planning, it's possible to systematically work toward that ideal landscape.

Biological planning helps you plan both your grazing and the landscape you want to see in the future. It can't be rushed. It may take you months to develop a complete plan, because throughout the process you'll be taking breaks to think about alternatives and scenarios. A good biological plan is like a good bottle of wine — it gets better with age.

The development and implementation of the plan can take years. The best approach is to do your development as both time and money permit; when done this way, development will usually pay its own way. You can begin implementing the grazing plan almost immediately, with the landscape plan following over time.

The Grazing Plan

A grazing plan helps you use the tools of rest, grazing, and animal impact to grow healthy land and healthy animals. Effective grazing plans balance forage with animals, which in turn improves the ecosystem processes and moves you toward your goal. And best of all, profits improve.

The grazing plan provides a method for forage budgeting and for timing moves of livestock through paddocks (either permanent or temporary). The grazing plan allows you to eliminate both overgrazing and overresting of the plants in your pastures by helping you answer the following questions:

1. How much forage do we have?
2. How much do our animals need to eat?
3. And how much are they actually eating?

Forage Quality versus Quantity

Before you can actually begin to answer these questions, you need to think about quality and quantity of forage, and how the two affect your animals. Both quality and quantity of forage change throughout the year as well as from year to year, but with good management, quality and quantity can be maintained at consistently high levels. The key is to have a diverse group of cool-season grasses, warm-season grasses, and legumes.

During the first flush of spring grass (which may come in February in areas of the Deep South and California, or mid-May in the far North), the grass puts out both its maximum quality and quantity for the entire year. As the growing year progresses, both quality and quantity decrease, though they may show a second spurt of growth sometime in fall.

Hay Equivalent. The quantity of forage can be measured as a hay equivalent. The *hay equivalent* (HE) can be used in two ways: first, to estimate the amount of hay, in pounds (kilograms), required to meet the dry-matter intake of animals; and second, to estimate the amount of forage production if a hay crop were being removed from a field. Even though the animals are on pasture and eating the forage directly, the hay equivalent is still convenient for calculation purposes. Average hay equivalent yields for pasture and hay ground can run from lows of less than 0.25 ton per acre (560 kg per ha) on rangeland or 1 ton per acre (2,242 kg per ha) on a poor pasture to a high of 10 tons per acre on very high-quality, well-managed, irrigated hay field.

Megacalories. Nutritionists measure the quality of pasture in Megacalories of energy (Mcal). The more energy that is available in the sward, the more Mcals the animal gets with each bite. As the grass matures, its protein and energy content decrease, and its fiber (cellulose and lignins) increases. This change reduces the nutrients and energy available with each bite. Pasture energy can range from lows of 0.1 Mcal per pound (0.22 Mcal per kg) of feed, up to highs of 0.8 Mcal per pound (1.76 Mcal per kg). The energy is always highest in plants that are young and vegetative, and that have not begun setting seed yet.

Cool-season grasses are higher in energy than warm-season grasses. This is because warm-season grasses have a thickened outer wall, or sheath, around the nutritious inner cells, which helps protect the plant from desiccation during very hot weather.

As energy is reduced, the animal must increase how much it eats to meet its needs — but at the same time the food, being more fibrous and less digestible, is passing through its system more slowly. This is the double whammy: The critter needs more food but can only eat less.

Impact on Animal Production. A study done by the Forage Systems Research Center at the University of Missouri illustrates the impact of forage quality on animal production. Between 1982 and 1985, researchers examined the performance of yearling steers and cow-calf pairs on two high-quality pastures (brome/red clover and orchard grass/red clover) and on lower-quality pastures of endophyte-infected fescue. (Fescue isn't necessarily a lower-quality feed, but when infected its quality drops significantly.) On both types of high-quality pastures, the animals performed about the same, but when they were placed on the low-quality pastures performance plummeted: Average daily weight gain dropped 22 percent for the calves and 51 percent for the steers on the low-octane pastures. (The calves didn't suffer as badly as the steers because their production was somewhat buffered by their mothers.) Conception rates (rebreeding) for the cows reflected the poorer-quality feed by dropping from 98 percent on the high-quality pastures to 74 percent on the low-quality.

The best way to obtain high-quality pastures is to grow a diverse polyculture of grasses and legumes. The grasses provide carbohydrates and the legumes provide crude protein. An ideal mix runs from 60 to 70 percent for grasses, and 30 to 40 percent for legumes. As your grazing-management skills improve, your pastures will tend to move in this direction naturally, because more even grazing of a sward leaves plenty of an open canopy, which allows the legumes to compete with the grasses.

Biological Cycles

Regardless of where you live in the country, there are predictable cycles for forage production. Grass growth begins in the spring of the year (the date of which depends on locale), and flushes in late spring or early summer. Growth usually tapers off during the peak heat of summer. Grass may have a second, smaller flush in late summer or early fall, and then becomes dormant for midwinter. (The exception are the southernmost areas of California.) Figure 14.1 shows a fairly typical growth curve.

The biocycle for mother animals peaks during early to midlactation. It tapers off during midlactation and is at its lowest point from after weaning until 2 to 4 weeks before parturition.

FARMER PROFILE

Karl and Jane North

"During the late 1970s, we spent 6 years learning about low-input farming in the Pyrenees Mountains of France. There isn't much farming left in that area of France, but what there is, is pre–industrial Revolution type of stuff. Farmers use donkeys and hand-tools," Karl explains.

Upon their return to the United States, the Norths purchased 60-acres of bare land in central New York State. "By purchasing bare land, we had the opportunity to build things just how we wanted them. We gave a lot of thought to how buildings should lie on the land to take advantage of the land's attributes. We built an energy-efficient passive-solar home and three-sided, open sheds partly from trees on our land."

The land is only about half open. The treed portion provides wind protection not only for their buildings, but also for the Norths' flock of sheep and their workhorses. "We have one tractor that we use for plowing snow in the winter, turning compost piles, and baling hay, but we use our horses for cutting hay, raking hay, and in the woodlot."

The Norths' primary enterprise is a seasonal sheep dairy. Forty-five ewes lamb on pasture in early May, and are milked in a six-stanchion parlor through the end of September. The milk is then made into cheese in an on-farm cheese plant. "Although we only milk for 4 or 5 months, we are able to carry an inventory of cheese year-round, because the types of cheese we produce require long aging."

Marketing is critical to the success of an operation like the Norths'. They market cheese, lamb, yarn, and tanned hides.

"About 60 percent of our cheese is marketed at the Ithaca, New York, farmers' market. The market has a cosmopolitan clientele of people who can afford $12 per pound for cheese." The rest of the cheese is marketed via mail order and an in-town outlet. The mail-order business grew out of write-ups in various magazines and books about cheese, and from taste-testings they did in New York City. "For us, mail order works. Our product is unique, and the customers are willing to pay not only for our cheese, but also for the shipping."

Ethnic markets also help Karl and Jane with marketing their lambs, which come in May and are sold or butchered when the grass runs out in December. "With this short window, most of our lambs weigh less than 80 pounds when we sell them. One-third of our lamb crop we direct-market, but the other two-thirds are sold to a packer. Luckily, we have an ethnic packer who likes lambs this size." Conventional packers want lambs up to 120 pounds.

"We really try to think in terms of a whole system. Holistic management has helped with that, but we were thinking that way even before we found it." The whole-system approach helped them look at all the products they could sell from their farm, including the yarn and the hides. "What most outsiders find unique about our operation is the harmony of it, how everything works together and how energy efficient it is."

From a week or two before parturition until 2 to 4 weeks after, the mother animal's energy requirements are soaring, yet her physical capacity to take in sufficient feed is limited. This is caused by the pressure the developing fetus is exerting on internal organs (most women who have had children can attest to this situation), and after birth it takes a while for the digestive organs to get back to where they can handle a maximum volume of feed. If the feed she is eating at this time is low in nutrients and energy to begin with, then the mother animal is running at a big loss; to make up the loss, she eats into her reserves of fat and muscle. Figure 14.2 displays a typical mother animal's cycle.

For breeding males, the demand peaks during the 6 to 10 weeks of the major breeding season. Growing animals have a relatively high demand year-round, if fast growth is a goal.

Energy in the forage supply must be capable of meeting the needs of stock during the various phases of life: growth, breeding, lactation, and so on. Matching the biological cycles of both animals and plants can minimize, or even eliminate, the need for supplemental feeds (Figure 14.3). This means that mother animals should be bred to calve/lamb/foal/kid, and so on, at the onset of the spring growth period for your area of the country. If you live in Mississippi, that may be March 1; in Minnesota, it's May 1. With the possible exception of a few pockets of California, having baby animals in January or February is off cycle for most places in North America.

Numerous university studies support the profitability of getting in sync with the natural cycles. Another study from the Forage Systems Research Center showed that weaning weights for calves born in spring (after March 15) weren't significantly lower than the weights of calves born in winter (518 versus 526 pounds [235 vs. 239 kg]). When factoring such issues as higher death loss for winter-born calves and higher expenses for feeding the cows of winter-born calves into the equation, the spring-born calves had a real profit advantage that greatly offset their slightly lower weaning weights.

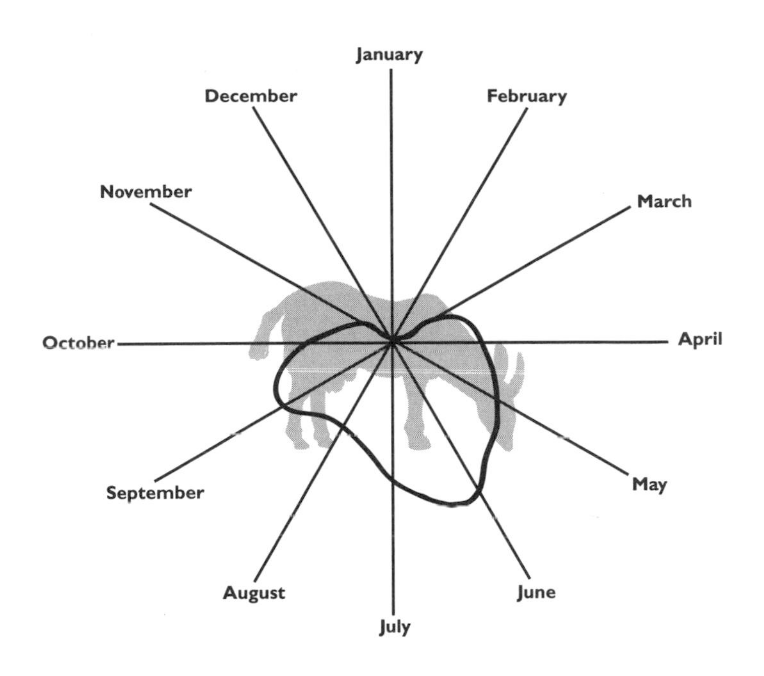

Figure 14.1. An approximate growth curve for an area of the country where grass begins to grow in early March and goes dormant by the first of November.

(Modified from David Pratt, "Cows, Grass, and Profitability," *Livestock and Range Report #961*, Summer 1996.)

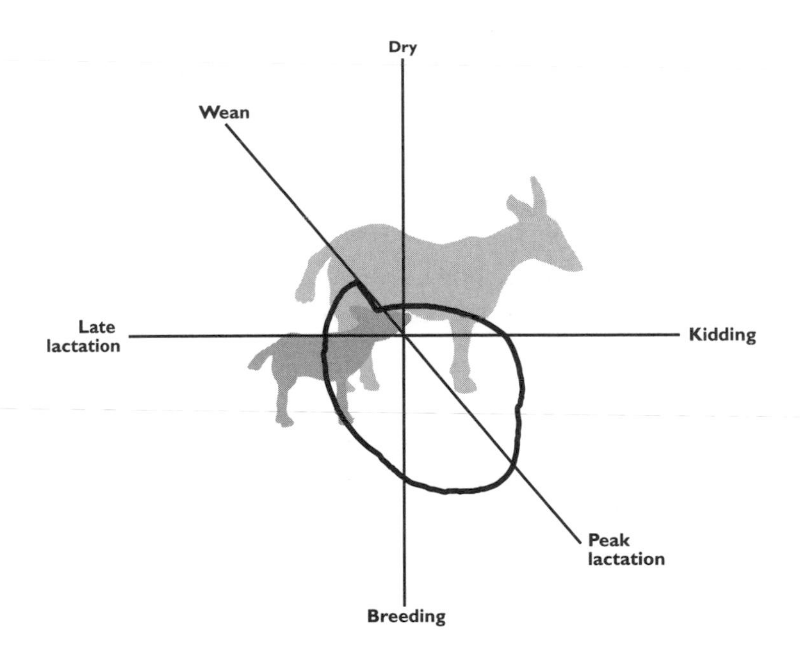

Figure 14.2. The nutritional needs of a mother animal during the course of a birth cycle. Production is optimized on a seasonal growth curve (see Figure 14.1) for a "best fit." If that were done with the growth curve provided in Figure 14.1, calving would begin about 10 days into March.

(Modified from David Pratt, "Cows, Grass, and Profitablility," *Livestock and Range Report #961*, Summer 1996.)

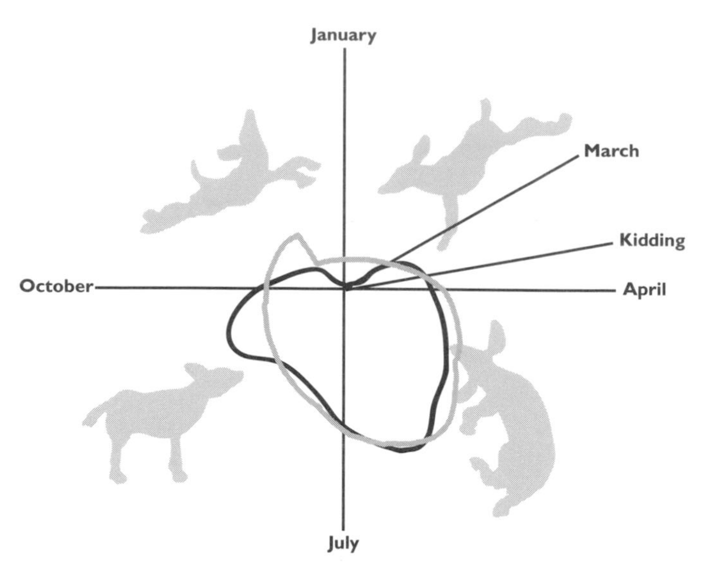

Figure 14.3. By superimposing Figure 14.2, mother's nutritional cycle, over Figure 14.1, seasonal growth curve, you can determine the ideal time for babies to be born in your area. In this example, kidding would begin around the third week of March. Remember, you're seeking the best fit between the two curves.

(Modified from David Pratt, "Cows, Grass, and Profitability," *Livestock and Range Report #961*, Summer 1996.)

Matching Animals to the System

There are a couple of fancy words, phenotype and genotype, that scientists use to describe living organisms. *Phenotype* means to show a trait, and *genotype* means that a trait is genetically passed on.

When matching animals to a grass-based system, you want animals whose phenotype and genotype are appropriate. In other words, you don't want an animal that has been "bred up" to require high inputs in a low-input system.

Holsteins are a perfect example: Many farmers switching to a grass-based dairy operation find that their Holsteins don't perform very well. Smaller cows, such as Jerseys, do far better in this type of system. Smaller cows produce milk profitably without the high inputs of purchased feed supplements and extensive veterinary services that the larger cows require. Again, university studies support "small is beautiful" animals in a low-input system. They don't produce as much but are profitable nonetheless, because they don't cost as much to maintain.

Finding good breeding stock for a grass-based system can be a challenge. If you can't find a farmer who has been practicing managed grazing for a while to purchase stock from, you may end up having to breed up your own line. Look for low-input animals — Hereford or Angus rather than Simmental or Charolais beef cattle, Jersey or Ayrshire rather than Holstein for dairy.

Winter Management

There are a number of approaches to winter management, depending somewhat on your climate. Many grass farmers grow an annual crop for animals to graze during the winter. In the North, where snow can become pretty deep, corn can make a good grazing crop. In more moderate climates, winter wheat is popular.

Stockpiled Forage. Some areas of the country lend themselves to grazing year-round. During the dormant season, the animals can graze stockpiled forage. If you live in a relatively low-snow-cover area of the West, this may work for you. The key to success is making sure you've reserved enough stockpiled forage during the growing season to hold you over. In nonbrittle environments like the eastern states, the nutrient value of stockpiled forage will drop from leaching. If you attempt to feed a stockpiled supply in these areas, you'll need to get a few samples run for nutrient value (see appendix E, Resources) and decide on an appropriate supplementation program.

Self-Feeding Silage Stacks. Another technique that's popular in the Midwest is self-feeding silage stacks made out in the field (Figure 14.4). One problem with this technique for small producers is that you have to harvest the crops to ensile — an undertaking that requires lots of specialized equipment. In some areas of the country, there are starting to be some contract operators available to do this for you.

"Wintering" Animals. Like people heading to Florida for the winter, some farmers are working out novel arrangements with farmers in other areas of the country to take their animals in winter. Herds are shipped south for winter and north for summer, and producers in both areas benefit.

Hay. For most small-scale producers, though, winter management involves feeding hay. Hay should be fed in various areas of the farm, to spread out fertility. Traditionally, farmers fed their hay in the same

Figure 14.4. In many areas, self-feeding silage stacks are an ideal way to feed animals in winter. Silage stacks are prepared at harvest, right in the fields. Animals are given access to the face of the stack by placing one strand of polywire across. As the stack recedes, the wire is moved back.

area, year after year. This results in excessive nutrient loads building up in those areas, higher leaching of the nutrients into groundwater, and higher runoff to surface waters.

To get the most benefit from your hay feeding, spread your feeding areas around the farm, particularly on fields you take hay off. Feeding on hay fields returns the fertility to the field it came from. Although some hay may be "wasted" when you feed on the ground over large areas, that "waste" will go right back into your soil for increased organic matter, and the fertilizer value of hay fed in the field can equate to $18 per ton ($0.02 per kg) of hay fed, when you factor in the value of the hay itself, the manure, and the urine that are returned to the field.

Starting the Season

At the very beginning of the grazing season, you don't need to subdivide — but this period is very short. We always open all the gates in late winter, and when the snow goes the animals begin moving around in a large circuit every day. They clean up leftover grass from the previous season and begin nibbling the first new shoots. If they are given a vast area during this time, which may only last 2 or 3 weeks, they don't damage any one part of the farm.

As soon as the grass gets going in earnest, begin subdividing, or closing the gates if the farm is developed into a full complement of permanent paddocks. In this early season, the goal is to move the animals quite quickly — every day is perfect if you can manage it — so they just clip the tops of the plants.

It's a good idea to start each grazing year with a different paddock. By alternating which paddocks you use to begin the year, you help break up parasite cycles, and you create an environment that favors plant diversity.

Sometimes, particularly in nonbrittle environments, this spring flush gets away from you. The grass comes on so quickly you can't possibly graze it off, so cut some for hay or haylage, or just to leave on the ground as green manure.

Estimating Forage Budgets

The first process in actually developing your grazing plan is to estimate a forage budget for your farm. This budget will help you get a feel for the amount of forage your livestock will require each month of the year, and it will help you get a feel for an appropriate stocking rate for your land. Forage budgets can be based either on actual numbers or on averages. (In chapter 15, you'll learn techniques that will allow you to fine-tune the numbers for your operation, but start out with the averages if you don't already know actual numbers.) Typically, forage budgets are prepared during the winter for the coming year. Then through monitoring, you make appropriate adjustments during the growing season.

Creating a forage budget begins with a chart, similar to the one in Table 14.1. This forage budget is based on the Joneses' farming operation. The forage production chart uses the following information:

- ***The field.*** Give some description of each field or area on your farm. Try to group areas that have similar production.
- ***The crop.*** Describe what is growing on the field.
- ***Number of acres (or hectares)*** in this area. This number can be a rough estimate.
- ***Number of permanent paddocks.*** If this area is already broken down into permanent paddocks, enter that number here.
- ***Acres (or hectares) per permanent paddock.*** This is simply an average, so divide the number of acres in this field by the number of permanent paddocks in the field.
- ***Average yield.*** Unless you know an actual value for hay equivalent (HE) in tons per acre (or in kg per ha) from your monitoring program, use the values provided in Table 14.2. In this example, we assume that the Joneses have a firm feel for production in each field, as they have been monitoring for several years.
- ***Forage production.*** This value represents the total HE yield for this field in tons (kg). This figure is calculated by multiplying the number of acres in the field by the average yield. (This value is used later in other calculations and is denoted by the symbol FP_{HE}.)

Table 14.1

Jones Farm Forage Budget — Planned Forage Production (FP)

Field	Crop	Number of Acres (no. ha)	Number of Permanent Paddocks	Acres Per Permanent Paddock (ha/paddock)	Average Yield HE in Tons/Acre (HE in kg/ha)	FP Total HE Yield in Tons (in kg)	FP Per Paddock
Homestead	Lawn	4.0 (2)	1	4.0 (2)	2.7 (6,053)	11 (9,979)	11 (9,979)
High ground — low fertility	Permanent pasture — primarily grass	62 (25)	3	20.7 (8)	1.7 (3,811)	105 (95,275)	35 (31,745)
Old pasture — moderate fertility, moderate moisture	Permanent pasture — good mix of grass and legumes	27 (11)	2	13.5 (6)	2.0 (4,484)	54 (49,324)	27 (24,489)
Creek bottom, subirrigated — good fertility, excellent moisture	Permanent pasture — good mix of grass and legumes	43 (17)	3	14.3 (6)	2.9 (6,502)	125 (113,375)	42 (38,094)
Subirrigated pasture — good fertility	Hay/pasture — mixed grass and legumes	20 (8)	1	20 (8)	2.4 (5,381)	48 (43,048)	48 (43,048)
Crop ground — flat, good fertility	Hay/pasture — planted this spring to mix 65% grass, 35% legumes (this will yield higher in subsequent meetings)	50 (20)	2	25 (10)	2.3 (5,154)	115 (104,328)	58 (52,153)
Wooded area — moderate fertility	Trees — some grass under canopy, and scrub that needs knocking down by grazing animals	34 (14)	1	34 (14)	0.7 (1,568)	23 (20,866)	23 (20,866)
Average		34.3 (13.9)	1.9	21.4 (9)	2.4 (5,348)	68.7 (62,313.8)	34.9 (31,482)
Total		240 (97)	13	150 (61)	16.7 (37,437)	481 (436,195)	244 (220,374)

Table 14.2

Hay Equivalents

Type of Pasture	Hay Equivalents/Acre
Rangeland, arid climate, poor stand	0.25
Rangeland, arid climate, moderate stand	0.50
Rangeland, arid climate, good stand	0.75
Subirrigated field, permanent pasture, arid climate, good stand	5.00
Irrigated field, planted and fertilized, arid climate, excellent stand	10.00
Permanent pasture, humid climate, poor stand	1.00
Permanent pasture, humid climate, moderate stand	1.75
Permanent pasture, humid climate, good stand	2.50
Permanent pasture, humid climate, excellent stand	3.50
Planted field, fertilized, humid climate, good stand	4.00
Planted field, fertilized, humid climate, excellent stand	6.00
Planted field, fertilized and irrigated, excellent stand	7.50

Notice in this example that there are about 163 tons (147,871 kg) worth of production from fields that are defined as hay/pasture. Checking the feed requirements in November–March, the time of year that hay is required in central Wisconsin, we see that it will take about 148 tons (134,266 kg) of hay for winter feed — so Gil and Jenny's annual budget looks good for both the growing season and the winter.

The second part of forage budgeting looks at forage requirements, and requires a chart similar to Table 14.3. The forage requirement chart uses the following information:

- ***The number of animals in each class.*** In practice, the classes are broken down in more detail than I've done in this example: A class would be established for breeding males, as well as for breeding females; weanlings and yearlings would be broken into separate groups; and if you are running a dairy, high-producing cows would be separated from low-producing cows.
- ***The average weight*** is your best guesstimate of the average weight of all the animals in a given class. Table 3.1 (p. 24) can be used to help develop this number if you don't have a good sense of your animals' approximate weights.
- ***The total weight.*** Multiply the average weight by the number of animals within the class of livestock.
- ***The intake factor*** is a multiplier based on the percent of body weight that each class requires in dry matter each day. This can be taken from Table 14.4.
- ***The hay equivalent*** (HE). Multiply the intake factor by the total weight of animals within a class to come up with a hay equivalent. Remember though, that the HE is simply a measure of quantity. For animals to perform well, the feed needs to be of high quality, and when it is the animals can eat even more than the percent factors given. In fact, fast-growing and hard-working animals can sometimes eat as much as 6 percent of their body weight per day when feed quality is at its best, which translates into high production (growth, milk, fiber, breeding).

Table 14.3

Jones Farm Forage Budget — Planned Forage Requirements (FR)

	January	February	March	April	May	June	July	August	September	October	November	December
Seasonal; activity Biocycles	Breed sheep →			Calving →	Lambing →		Breed cattle →			Wean lambs → Wean calves →		Breed sheep →
				Horses working moderately →								
Cows, number	40	40	40	40	40	40	40	40	40	45	45	45
Average weight (lb)	900	900	900	950	960	975	1000	1025	1050	900	900	900
Total weight (lb)	36,000	36,000	36,000	38,000	38,400	39,000	40,000	41,000	42,000	40,500	40,500	40,500
Percent factor	0.0250	0.0250	0.0250	0.0275	0.0300	0.0300	0.0300	0.0300	0.0275	0.0250	0.0250	0.0250
Cow HE	900.0	900.0	900.0	1,045.0	1,152.0	1,170.0	1,200.0	1,230.0	1,155.0	1,012.5	1,012.5	1,012.5
Growing cattle, number	26	25	24	23	22	21	20	19	18	48	47	46
Average weight (lb)	500	500	515	550	560	600	650	690	730	400	430	460
Total weight (lb)	13,000	12,500	12,360	12,650	12,320	12,600	13,000	13,110	13,140	19,200	20,210	21,160
Intake factor	0.035	0.035	0.035	0.035	0.035	0.035	0.035	0.035	0.035	0.035	0.035	0.035
Growing cattle HE	455	438	433	443	431	441	455	459	460	672	707	741
Horses, number	4	4	4	4	4	4	4	4	4	4	4	4
Average weight (lb)	1,200	1,200	1,200	1,200	1,200	1,200	1,200	1,200	1,200	1,200	1,200	1,200
Total weight (lb)	4,800	4,800	4,800	4,800	4,800	4,800	4,800	4,800	4,800	4,800	4,800	4,800
Intake factor	0.025	0.025	0.025	0.030	0.030	0.030	0.030	0.030	0.030	0.030	0.025	0.025
Horse HE	120	120	120	144	144	144	144	144	144	144	120	120
Sheep, number	50	50	50	50	50	50	50	50	50	50	50	50
Average weight (lb)	140	140	150	150	160	170	180	190	200	140	140	140
Total weight (lb)	7,000	7,000	7,500	7,500	8,000	8,500	9,000	9,500	10,000	7,000	7,000	7,000
Intake factor	0.0300	0.0300	0.0300	0.0300	0.0325	0.0350	0.0350	0.0350	0.0325	0.0300	0.0300	0.0300
Sheep HE	210.0	210.0	225.0	225.0	260.0	298.0	315.0	333.0	325.0	210.0	210.0	210.0
Growing sheep, number	30	30	20	5	5	5	5	5	5	90	50	50
Average weight (lb)	85	100	120	120	125	130	135	140	140	50	65	75
Total weight (lb)	2,550	3,000	2,400	600	625	650	675	700	700	4,500	3,250	3,750
Intake factor	0.04	0.04	0.04	0.04	0.03	0.03	0.03	0.03	0.03	0.04	0.04	0.04
Growing sheep HE	102	120	96	24	19	20	20	21	21	180	130	150
AUs (sum of all AU)*	63	63	63	64	64	66	67	69	71	76	76	77
Total HE/day (sum)	1,787	1,788	1,774	1,881	2,006	2,072	2,134	2,186	2,105	2,219	2,180	2,233
Total HE/month**	55,397	50,050	54,982	56,423	62,184	62,160	66,162	67,777	63,147	68,774	65,396	69,226
Tons HE/month	28	25	27	28	31	31	33	34	32	34	33	35

Average animal units 68

Estimate of annual forage required 371

HE = hay equivalent; AU = animal unit.

→ sheep/lambs → cattle → horses

* Arrive at total AU by adding total weight per animal class for all classes, then dividing by 1,000.

** (total HE/day x no. days)

Note: The complexity of this chart prohibits metric conversions.

Table 14.4

Intake Factors for Calculating Forage Requirements

Class of Livestock	Percentage of Body Weight	Multiplier for Body Weight
Large species (cattle, horses, bison) at rest	2.5%	0.025
Large species at work (lactating or bulls in service)	3.0%	0.030
Large species, growing	3.5%	0.035
Small species (sheep, goats, llamas, deer, elk) at rest	3.5%	0.030
Small species at work	3.0%	0.035
Small species, growing	4.0%	0.400

Calculating Carrying Capacity

The *carrying capacity* is an approximation of how many animal units your farm can carry. There are two different ways to calculate carrying capacity: the first is more accurate and is based on your forage budget figures; the second is a more "quick and dirty" number, to be used in cases when you don't have a forage budget. Either way, the carrying capacity represents a theoretical stocking rate. We'll look at examples of each type.

The More Precise Method. With the estimates for forage production (FP_{HE}) and forage requirements (FR_{HE}) that you calculated in Tables 14.1 and 14.3, estimate carrying capacity with the following formula:

$$CC = \frac{(AU \times FP_{HE})}{(FR_{HE} \times 1.25)}$$

where CC = the carrying capacity in animal units;
AU = the animal units you calculated in Table 14.3;
FP_{HE} = the forage production in hay equivalent units (tons [kg] from Table 14.1); and
FR_{HE} = the forage requirements in hay equivalent units (tons [kg] from Table 14.3).

For example, using the Joneses' figures from the previous charts, carrying capacity would equal 70 AUs. The Joneses have 68 AUs, so they are slightly understocked. The calculation is:

$$CC = \frac{(68\ AU \times 480\ FP_{HE})}{(371\ FR_{HE} \times 1.25)}$$

$$= \frac{30{,}558\ AU}{436.25}$$

$$= 70\ AU$$

Note: In this equation the hay equivalent values cancel out, and only animal units remain.

Reserving extra carrying capacity isn't a half-bad idea, especially when you're starting out. This calculation is designed to provide a 25 percent reserve, which allows feed for wildlife and some cushion for those times when the animals consume higher percentages of feed. (If you want to increase the reserve percentage, increase the 1.25 variable accordingly: 30 percent would be 1.3. If you don't want to reserve as much, lower the variable: 15 percent would be 1.15.) Gil and Jenny shoot for a 25 percent reserve but have even a little more than that, as they still have feed for two additional animal units. If they have a bumper crop that exceeds their estimate, they take off extra hay. If they have a dry summer, they know that a surplus can act as a buffer.

The Quick and Dirty Method. The second version of the carrying capacity formula does not provide the level of accuracy that the first version does, but there are cases where it comes in handy. The formula for the rougher estimate is:

$$CC = \frac{FP_{HE}}{7.5^*}$$

where CC = the estimated animal units, and
FP_{HE} = an estimated forage production

*Or divided by 6,866 if you are working in metric figures.

Let's look at a few examples.

Example 1. Tom and Karen Wilson are thinking about converting all their tillable ground to pasture and purchasing additional livestock. They know their tillable ground is of high quality, and they have a long growing season, so Tom estimates they could yield 4.5 tons/acre (10,089 kg/ha) from these fields. First he multiplies 4.5 tons/acre by 120 acres (49 ha) to come up with 540 tons (494,361 kg) of FP_{HE}. Next, he divides this figure by 7.5 (6,866) to come up with an estimated carrying capacity of 72 AU.

Example 2. A doctor from Boston just purchased the farm across the road from Gary and Michelle Miller. He's made a proposition: They can use his land for grazing for free, if they will keep an eye on two horses he wants to buy for his wife and daughter. The doctor will benefit by having the farm "farmed," because if no agricultural pursuit is taking place there, the county assessor will raise his taxes. Gary and Michelle tell the doctor that they'll get back to him in a week or two.

The farm has about 45 acres (18 ha) of reasonably good permanent pasture, is well fenced on the perimeter, and has water available in the pasture. Gary and Michelle don't need any additional pasture for their existing herd, so if they take the doctor up on his offer, they'll have to purchase additional animals. After some discussion, they decide that if they do run animals on the doctor's farm, they will run stockers, as additional breeding stock would increase their winter chores more than they could cope with.

Gary and Michelle estimate that the average yield on the doctor's farm would be about 3.5 tons per acre (7,847 kg/ha), so the carrying capacity would be 21 AU (3.5 x 45 / 7.5) or (7,847 x 18 / 6,866). But that carrying capacity would be for the full year, and by using stocker animals, Gary and Michelle will only be responsible for the animals for about half the year. This means that they'll be able to double the carrying capacity during the fraction of the year they'll run animals on the doctor's farm, so they could carry about 42 stocker animals for 6 months — quite a boon! (If you are thinking about running stockers for a fraction of a year, multiply the carrying capacity by 12, and then divide the number of months you plan to keep the stockers.)

Before calling the doctor to accept his offer, Gary and Michelle use the planning guidelines in chapter 12 to see if this enterprise will move them closer to their goal. In this case, it appears to be a winner, so they call the doctor to let him know they'll work with him.

READING TOPOGRAPHIC MAPS

If you've never studied a topographic map, looking at one for the first time may be a little intimidating. Once you get the hang of it, though, topos are easy to use and they supply a remarkable amount of information.

Topographic maps are often referred to as quadrangles. Each quadrangle is designated by a name, generally the name of a prominent town or geological feature within that quadrangle (Figure 14.5).

Quadrangle simply refers to an area that is bounded by lines of longitude running north and south, and lines of latitude running east and west. Every single point on the earth can be described by the intersection of lines of longitude and lines of latitude.

Lines of longitude and latitude can be expressed in terms of degrees, minutes, and seconds; 360 degrees makes a complete circle. There are 60 minutes in each degree, and 60 seconds in each minute. Graphically, a minute is noted by the ' mark, and a second is noted by " mark. So a 7.5-minute map shows an area s of a degree tall by s of a degree wide, and 7.5 minutes could also be written as 7'30". The four corners of the map give the degrees, minutes, and seconds north of the equator, or west of the prime meridian (a longitudinal line that goes through London, England). As I sit typing, south of Hartsel, Colorado, I am at 38°57'10" north of the equator, and 108°47'05" west of the prime meridian.

In the late 1700s, a Public Land Survey was started, and as a result topos for most U.S. states show section, township, and range lines, though some maps of eastern states are not configured this way. The *sections* are 1-square-mile blocks, and are numbered from 1 to 36. Each *township* contains 36 square miles. Through much of the country, legal descriptions of all larger tracts of land are based on the section, township, and range description.

A graphic display on the lower portion of each quadrangle map shows the direction of both true north and magnetic north. True north runs in the same direction as the lines of longitude. Magnetic north is different from true north because compasses aren't attracted to true north — they are attracted to the magnetic pole, which is located to the west of Hudson Bay in Canada. The difference between true north and magnetic north is called the *magnetic declination,* and changes as you move around the country.

Maps all have a scale, which establishes the ratio between the distance shown on the map and the distance in real-life terms. Typical scales for topographic maps are 1:24,000 (7.5 minute), 1:62,500 and 1:250,000. When the scale is 1:24,000, 1 inch (2.5 cm) on the map equals 24,000 (60,000 cm) inches on the ground, which is equivalent to 2,000 feet (600 m). A graphic scale is printed on the bottom of maps that shows feet, miles, and kilometers.

Other information displayed at the bottom of most topographic maps includes a location map, description of symbols used on the map, and information about when and how the map was prepared. The names of surrounding quadrangles are written around in the edges of the map in parentheses.

One of the most useful things about a topographic map is its visual representation of contours, or elevations. Contour lines are light brown, squiggly lines that connect all the points located at a given elevation. Contour lines form a V when they cross streams; in fact, even if a water body isn't shown, if you see is a V in a contour line you'll know that water runs down at that point on the ground. A closed circle represents the top of a hill. Hatch marks along the edge of a contour line represent a depression.

A number known as the contour interval is given on the bottom of the map. If the given interval is 40 feet (12.2 m), for example, then each contour line represents land whose elevation is 40 feet (12.2 m) higher, or 40 feet (12.2 m) lower, than that of the line next to it. Contour lines may be heavy (these give the contour's elevation) or lighter (these don't). If a point lies on a contour line, you can read its elevation directly from the line. If it is between two contour lines, you can estimate its elevation as some amount between the two lines shown.

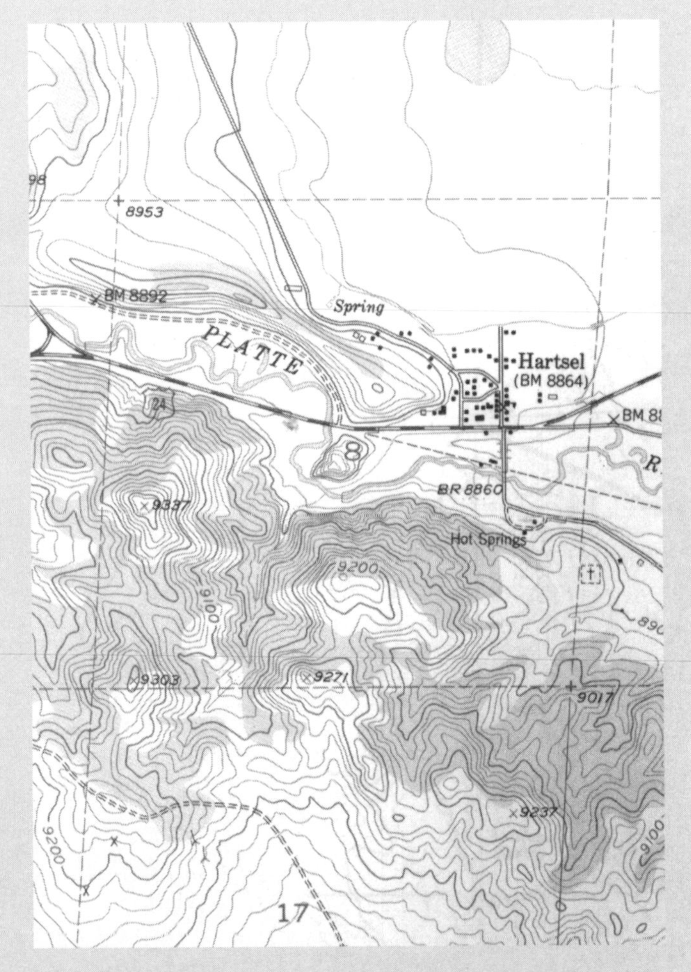

Figure 14.5. Topographic maps indicate road classifications, bodies of water, and land elevation.

(Reprinted from Hartsel Quadrangle c.1956, U.S. Geological Survey, Denver, CO.)

Developing Your Grazing Plan

To develop your own grazing plan, first go through the forage-budgeting process. Depending on circumstances, you may make a conscious choice to purchase in your winter forage by buying hay (or round bales of cornstalks, or silage, and so on). This decision makes good sense for many small operations; it allows you to maximize your carrying capacity during the growing season. If you choose to follow this route, you can make adjustments to your grazing-season carrying capacity by multiplying the carrying capacity by 12 and then dividing by the number of months you'll graze, like the Miller's did for their stocker operation.

Once you have assessed your forage requirements, you can establish a carrying capacity for your land. Establish a reserve quantity that you feel comfortable with. Depending on your circumstances, anywhere between 15 and 30 percent reserve is a good place to shoot for.

Set up a planning schedule like the one in Table 14.5. It helps you plan where you'll feed in winter, approximately when you expect to begin grazing, when you expect to end, and which paddock you'll begin with.

The Landscape Plan

The process of landscape planning begins with mapping the farm. Mapping is a practice that allows you to design a well-thought-out, ideal landscape. It is an integral step in long-term development.

The mapping process is easiest if you first acquire all the existing maps and aerial photos you can find for your farm — though if your farm is very small (maybe less than 20 acres, or 8 ha), you can probably do the map without this step. The U.S. Geological Survey (USGS) has both contour maps and aerial photos for most of the United States. Contour maps show elevations, watercourses, and general vegetation. For the highest level of detail in a USGS map, request a 7.5-minute series topographic map, which has a scale of 1:24,000 (1 inch to 24,000 inches or approximately 2,000 feet). In the western states, the Bureau of Land Management and the Forest Service also have good maps.

Gather as many maps that show your farm as possible. When you set out on this task, you may be surprised at just how much information has already been generated about your piece of land; for example, the USDA has soils maps for most of the country that describe the characteristics of the soil in detail. Consider your search an adventure.

Once you've put together your map collection, study it. Tape copies on walls, darken in your property lines, and try to absorb as much of the information as possible. Try to picture your farm within the larger area of the map: how it sits in the watershed, its proximity to existing natural features, industries, communities, roads, and so on. For a visualization exercise, look at the maps and ask yourself, "What would this look like if I were a bird, flying over the whole thing?"

After you've allowed time to reflect on your landscape — a week, a month, maybe more — it's time to begin mapping what you want to see in the future. Mapping can easily be done on a computer with a drawing program, and this approach allows you to print out working copies for everyone to use. When using a pencil and paper, you'll need to make copy-machine copies for some steps. It is easiest to begin mapping with paper that has a grid superimposed on it; ask for engineering-ruled pads at an office-supply store. If working with pencil and paper, also purchase plastic sheets, like those used for overhead projection slides, and grease pencils or dry-erase markers. These can be used for superimposing various ideas on one map.

Mapping Your Farm

On your first map, show property lines. Show existing natural features, such as ponds, wetlands, hillsides, or wooded areas, and any areas where serious disturbance or erosion has taken place, which will need to be repaired. The only human-made improvements that should be shown on this map are those that absolutely can't be moved (such as your house) or that you have absolutely no control over (for example, a major county road that cuts the property in half). Existing

Table 14.5

Planning Schedule

Paddock	January	February	March	April	May	June	July	August	September	October	November	December
Forage cycle	NA	NA	Snow off around 3/15 — allow entire farm to be grazed	Slow growth most of month — begin moving through paddocks 4/1	Flush begins about 5/1	Flush ends about 6/20	Slow growth	Flow growth	Growth spurt	Growth slows	Snow cover about 11/20	NA
Average recovery period (RP) (days)	NA	NA	45	30	15	18	30	30	28	45	75	NA
Desired grazing period (GP) (days)	NA	NA	NA	2	1	1	2	2	2	3	4	NA
Total paddocks (TP)*	NA	NA	NA	16	16	19	16	16	15	16	20	NA
Homestead	Not in use	Not in use	Not in use	Not in use	Graze	Graze	Not in use	Graze	Not in use	Not in use	Graze	Not in use
1a	Not in use	Feed hay	Open graze	Graze	Graze	Graze	Graze	Graze	Graze	Graze	Graze	Feed hay
1b	Not in use	Not in use	Open graze	Graze	Graze	Graze	Graze	Graze	Graze	Graze	Graze	Not in use
1c	Not in use	Not in use	Open graze with this pasture	Begin grazing	Graze	Graze	Graze	Graze	Graze	Graze	Graze	Feed hay
2a	Feed hay	Not in use	Open graze	Graze	Graze	Graze	Graze	Graze	Graze	Graze	Graze	Not in use
2b	Not in use	Feed hay	Open graze	Graze	Graze	Graze	Graze	Graze	Graze	Graze	Graze	Not in use
3a	Not in use	Not in use	Open graze	Graze	Graze	Graze	Graze	Graze	Graze	Graze	Graze	Not in use
3b	Not in use	Not in use	Open graze	Graze	Graze	Graze	Graze	Graze	Graze	Graze	Graze	Not in use
3c	Not in use	Not in use	Open graze	Graze	Graze	Graze	Graze	Graze	Graze	Graze	Graze	Not in use
4	Not in use	Not in use	Open graze	Hold for hay	Hold for hay	Hold for hay	Hold for hay	Hold for hay	Hold for hay	Graze	Graze	Not in use
5a	Not in use	Not in use	Open graze	Hold for hay	Hold for hay	Hold for hay	Hold for hay	Hold for hay	Hold for hay	Graze	Graze	Not in use
5b	Not in use	Not in use	Open graze	Hold for hay	Hold for hay	Hold for hay	Hold for hay	Hold for hay	Hold for hay	Graze	Graze	Feed hay
6	Feed hay	Not in use	Open graze	Not in use	Graze	Graze	Graze	Graze	Graze	Graze	Graze	Not in use

NA = not applicable.

$*TP = \frac{RP}{GP} + 1$

fences, watering points, field roads, and the myriad other "improvements" that are found on most farms should be left off, so they don't fog your vision of what you want the land to look like in the future (Figure 14.6). If you are working on a computer, print out enough copies of this first map for everyone in the family, hired hands, or other folks who are helping you with planning; when working with pencil and paper, take this version to town and make copies on a copy machine.

Now each person should take their map and begin drawing what they would like the landscape to look like if they could come back in a 100 years: a series of ponds along an existing stream, healthy trees growing on the hills, areas of open grasslands. Once everyone has gone through this exercise, compare notes. Chances are the visions won't be too far apart; if there are major discrepancies, put everyone's versions on a wall and study them for a week or two. Then join together again and try to decide if one version, or a combination of several, best meets the goal you've already set as a group. Redraw a permanent copy of your future look, or vision map, to keep as a reference.

Next, take one of the extra copies of your first map and use the data you've developed from studying the topographic map of your farm to approximate drainage lines, or the places that water will run downhill. On all but the flattest pieces of ground, there are multiple drainages; some may be home to a permanent stream, others may simply be runoff drainage paths during rain and snowmelt. Use a heavy line for big drainages and creeks, and a dashed line for minor drainages. Walk your land, after a rain or during snowmelt, with a copy of the drainage map to confirm that you show the drainages fairly accurately.

Take time to study your drainage-area map and your vision map. Once you feel comfortable with how the drainage and your vision wed to each other, lay out major areas of the farm: woodlots, crop areas if you plan to do some cropping (which should be rotated into pasture for at least 2 out of every 5 years), permanent pastures, field roads. This layout should include existing features as well as features you want to develop in the future. For example, maybe you want to set aside an area for woody agriculture (fruit and nut trees), or a spot to build a rental guest cabin.

SOURCES FOR MAPS AND AERIAL PHOTOS

- County governments often have maps that show roads, section lines, and property lines; check first with the assessor's office
- Local soil and water conservation districts
- USDA offices — Natural Resources Conservation Service (NRCS) and Farm Services Agency (FSA) offices are located throughout most of the country
- State foresters
- State mineral offices
- U.S. Coast Guard, if your property is on a navigable water body

Paddock Design

Now it's time to begin designing fencing systems and water systems. The first step in this process is to determine how many *permanent* paddocks you ultimately want to place on your site. As you learned in chapter 4, the more paddocks you have, the better, especially up to thirty. But large paddocks can be subdivided into smaller paddocks using portable temporary wire.

A good way to determine an appropriate number of permanent paddocks is to use the relationship that exists between average recovery periods (RP), grazing periods (GP), and total paddock numbers (TP), which can be expressed mathematically by the equation:

$$\frac{RP}{GP} + 1 = TP$$

The *recovery period* is the time, in days, that it takes for plants to recover their energy after grazing, and the *grazing period* is the time, in days, that stock normally remain on a paddock. When using the equation to estimate the number of permanent paddocks you want to eventually design for, use the average sea-

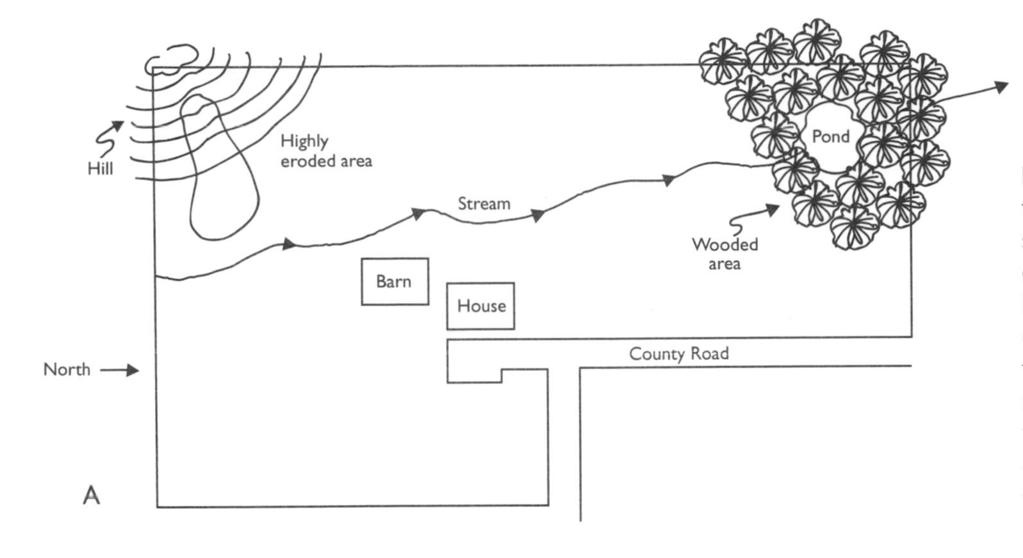

Figure 14.6. Mapping the Wilson's farm. **(A.)** This is an initial map showing major features of the existing landscape as well as those human-made features that really can't be relocated, like the house, the barn, and a county road. This map doesn't show human-made features that can be relocated, nor does it show current fields and fencing.

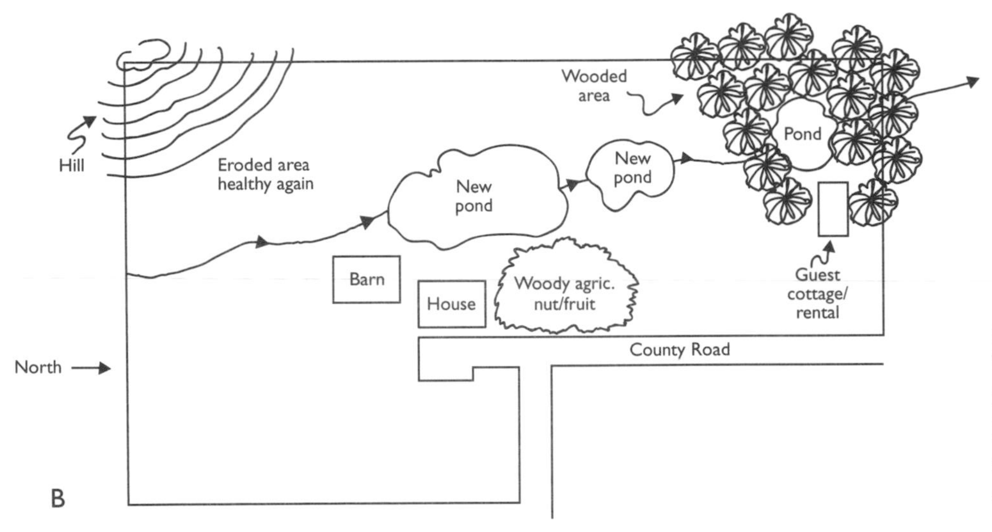

(B.) This is the landscape the Wilson's want to develop in the future; it is their "vision" map. Notice details like ponds, a guest cottage for agritourism, and an area for woody agriculture.

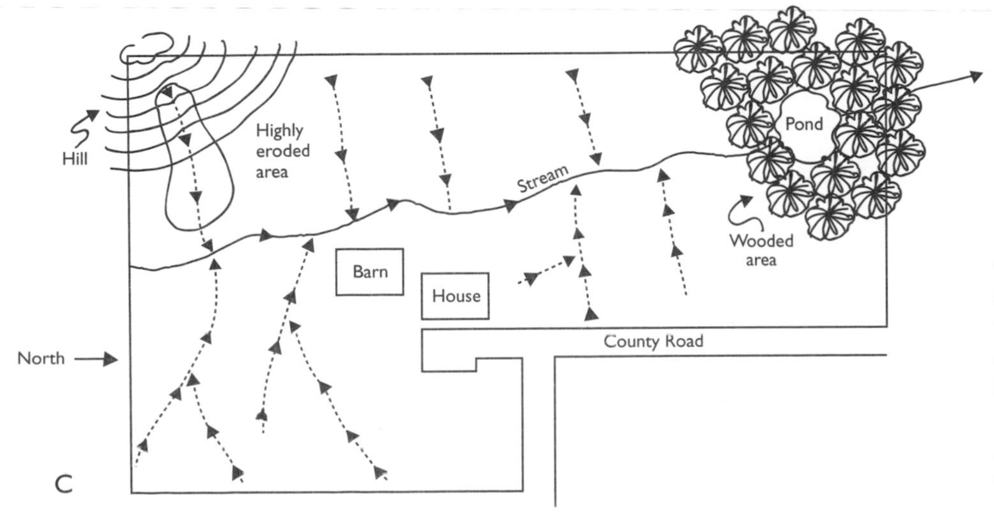

(C.) This map shows the drainage areas on the Wilson's farm. Once the vision map and the drainages are determined, the Wilsons can begin designing permanent fencing, new roads or trails, water points, and other long-term design features.

sonal recovery period, and an average grazing period that will work for you and your animals.

Following are a few examples.

The Millers. The Millers estimate that their average recovery period for the year is about 28 days, and they normally try to move their cattle, sheep, and horses at least every 2 days. Plug these numbers into the equation and you come up with

$$\frac{28}{2} + 1 = 15$$

If they develop fifteen permanent paddocks, for most of the growing season they won't need to use temporary subdivisions — they can simply open and close gates. They plan to develop eight paddocks immediately; they will develop the other eight as time and money permit.

The Blacks are in a far more brittle environment, and estimate that their average recovery period is probably closer to 60 days during the growing season. They feel that moving their herd over the larger areas that they are dealing with will work out best with a 3-day grazing period during the growing season. They come up with a figure of twenty-one paddocks for the growing season:

$$\frac{60}{3} + 1 = 21$$

The dormant season is long in the Blacks' brittle area, but they can graze for most of it, and the quality of the graze remains adequate even after the forage has gone dormant. They estimate that the average recovery period during the dormant season is 180 days, and they feel a good grazing period during this time will be 2 days. They calculate that they'll need ninety-one paddocks for their dormant season.

$$\frac{180}{2} + 1 = 91$$

For the Blacks, running temporary fencing on the scale of land they are dealing with would place unreasonable demands on their labor during the growing season, so they plan to develop the twenty-one permanent paddocks as quickly as possible. During the dormant season, they'll use temporary wire to subdivide the permanent paddocks, as needed. On their leased public land, which is used only during the growing season, they plan to use herding to control the animals' access to the grass.

The Wilsons. The Wilsons live in an area that has an average recovery period of 32 days for the entire growing season, which is quite long. They'd like to be able to move their animals every day — both for the higher production they feel it will give them, and to give their children a meaningful part in the work of the farm. Running the calculation, they come up with thirty-three paddocks. Since this seems like a high number of permanent subdivisions, they decide that they will subdivide into sixteen paddocks, and plan to use temporary subdivisions on a regular basis throughout the year.

Once you determine the number of permanent paddocks you eventually want to develop, look at your map for ways they might blend in with your land. In nature, there are almost no straight lines, so quit thinking in perfectly straight lines that run north to south, or east to west. Look at fencing designs that take into account the lay of the land. This doesn't mean that the fences have to be squiggly, it just means that you don't have to design paddocks that lie in a series of perfect squares. At the same time, square — or almost square — paddocks have some design advantages: The cost of fencing is lower on a paddock that has sides that are approximately the same in length as they are in width, and animals graze better when the proportions of the sides are fairly close.

Come up with two or three possible designs for your farm that appear to work with the landscape. Then evaluate each to see which one will actually be the most cost effective. This requires making some estimates about what fence and water systems will cost to run, and applying your results to the different possible designs.

Although it may not be exact, a good rule-of-thumb figure to use for estimating fencing costs for single-strand electric, including gates and corners, is 15 to 20 cents per linear foot ($0.50 to $0.67 per meter). Double-wire fencing will run from 20 to 25 cents per foot ($0.67 to $0.83 per meter), and triple wire will run from 25 to 30 cents ($0.83 to $1.00 per meter). Multiply the number of feet (or meters) of fence in each design by the appropriate figure to estimate your fencing costs.

Water systems may be harder to evaluate, especially if new wells will have to be drilled. For estimating the cost of buried water lines, use $1.50 per foot ($5.00 per meter).

Once you have worked through the calculations, you should be ready to choose the design that best meets your needs, and that does it at the lowest cost (Figure 14.7). After you've chosen your landscape plan, you can begin implementing it — but don't worry if you can't do it all at once. Your plan tells you how you want things to look in the future, not how they have to look today. (*Note:* These costs may vary significantly in countries other than the United States.)

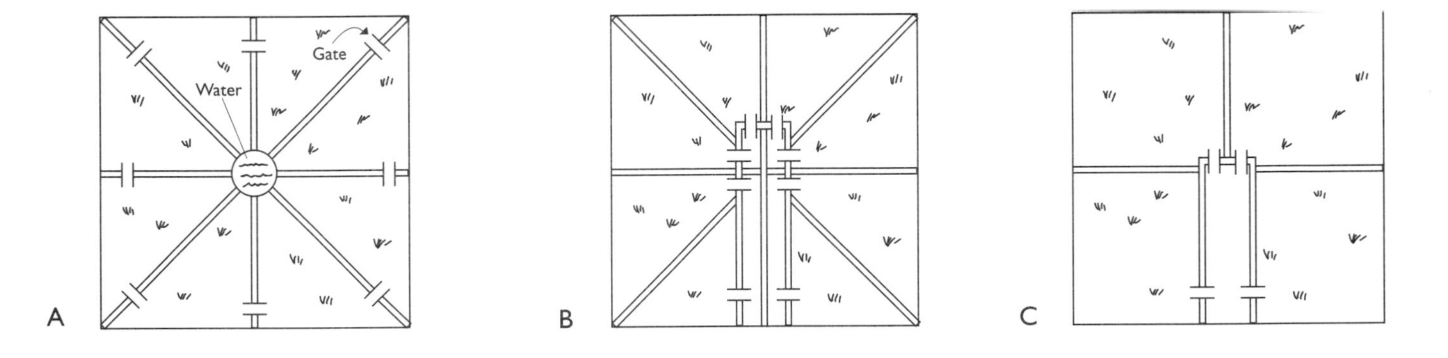

Figure 14.7. There are many ways to subdivide paddocks. Here are three designs that work well for subdividing a square field. **(A.)** Radial with central water. Note that the gates are away from center. By keeping gates away from the center, damage around the water point is reduced. **(B.)** Radial with a double central alley. This design is excellent for breaking up a large pasture into eight paddocks. The double alleyway minimizes damage and provides for flexibility during very wet periods. **(C.)** Square with a single central alley. The single alley works well when subdividing smaller pastures that won't receive as much pressure as a large pasture.

(Modified from Sam Bingham, *Holistic Resource Management Workbook.* Covelo, CA: Island Press, 1990, pp. 148, 149.)

CHAPTER 15

Monitoring

Interrelationships among ranch resources, such as people, finances, land, vegetation, climate, animals, and time, as well as activities and external influences must be understood and taken into account by the decision maker. The impact of each decision must be evaluated in advance and the outcome monitored. The decision maker must also be able to anticipate and implement timely changes to optimize outcomes.

— R. K. Heitschmidt and J. W. Stuth (ed.),
Grazing Management: An Ecological Perspective

ALTHOUGH THIS CHAPTER is entitled "Monitoring," it's really about monitoring, controlling, and replanning when necessary. If the world were a perfect place, then monitoring wouldn't be necessary — plans would work out perfectly — but unfortunately, the world is less than perfect. "The best-laid plans of mice and men" often do go astray. The three steps (monitoring, controlling, and replanning) are essential to success. As with planning, monitoring is performed for finances, grazing, and the landscape. Monitoring empowers you to fine-tune your operation.

Daily Records

A journal should be used for keeping daily records of importance to your operation. Record the paddocks that the animals are in, the stock densities in the paddocks, where animals are moving to and from, the length of grass as animals move in and out of the paddocks. Use it to record pertinent weather information. Use it to record individual animal information, including birth records, breeding records, culling decisions, vaccination records, and so on. (If you use the five-subject divided notebooks, the information can be easily organized into different sections.)

We also record our thoughts and discussions in our journal, things we come across in reading that we want to think more about, and information we learn at seminars or pasture walks.

Our journal is like a diary of farm life. We mention things like wildlife spottings, when various flowers begin opening, or when the various species of birds return and leave each year.

As you begin keeping a farm journal, you'll be amazed by the insights it will yield. It will help you understand what is taking place on your land, and when you discuss your progress with others it will give you documentation to back up the claims you make.

The three-ring-binder can be used for keeping all of your year's records in, including photos, computer- or hand-generated worksheets, maps, and, at the end of the year, the journal itself. This gives you one organized source for information.

MONITORING ESSENTIALS

You'll need a few items to monitor effectively. We use these:

- A journal. We simply use lined and divided five-subject notebooks — the kind you buy for the kids to start the school year. Nothing fancy, nothing expensive. We find it easiest to keep one per year.
- A three-ring-binder, preferably a fat one.
- Kitchen scale, preferably one that includes grams — most new kitchen scales do.
- Microwave oven.
- A hula hoop. Yup, that's right, a hula hoop. Preferably bright colored.
- A piece of light rope and tape. Use these to divide the hula hoop in half.
- A yardstick or meter stick.
- Some brown paper lunch sacks or large plastic food storage bags (the zip-seal type is convenient).
- A camera. A cheap 35 mm will do, with color film.

Monitoring Forage Quantity

Everything in grass farming starts with forage production, and optimizing your timing for keeping the grass growing at its best is the objective. The first step in the process is to determine how much forage is available. Available forage (AF) can be monitored in several ways, ranging from high tech to estimating a ballpark number.

The high-tech method of analyzing forage availability includes the use of a rising plate meter, but this is an expensive way to find out how much grass is in a field. These meters run several hundred dollars.

A method that works well for rough estimates of AF is this trick: Each inch (cm) of growth in a fairly well-sodded field will contain about 300 pounds (136 kg) of available forage per acre (hectare). If the pasture has 8 inches (20 cm) of standing plant matter, that's 2,400 pounds (8 x 300) (2,720 kg; 20 x 136) of available forage. If the field is not quite as well sodded, estimate a percentage of cover and multiply by that amount. For example, if the sod covers about 75 percent of the ground (25 percent bare spaces between plants), multiply your 2,400 (2,720) by 0.75. In this case, the field has about 1,800 pounds (2,040 kg) of available forage. Using the estimating approach is good for day-to-day monitoring, but at least a couple of times a year it's a good idea to run your own sample.

Evaluating a Sample

Running your own sample isn't hard to do, and it isn't expensive. You can run samples in the kitchen, using a kitchen scale and microwave oven.

By doing these samples, you'll be able to confirm that the estimates you're making are pretty close; if they're far off, your samples will allow you to develop an adjustment factor to bring them back in line. These samples should be run once during the high growing season, once in the low growing season, and once in the dormant season.

Start by weighing the bag you plan to use for collecting a sample on a kitchen scale; write this number down. Then head out to the field with your hula hoop, your yardstick, and your sample bag. Select a paddock that you are about to graze. Estimate how much forage you think you have in the paddock using the 300 pounds per inch per acre (136 kg/cm/ha) method.

Now toss your hula hoop out into the field. Wherever it lands is where you'll actually collect the sample. Using your yardstick (or meter stick), hand-clip a sample from a 1-foot-square (900-cm^2 or 30-cm x 30-cm box) area inside your hula hoop. Repeat this step two more times — so your bag will contain the clipped grass from 3 square feet (2,700 cm^2) worth of sample pasture.

Bring your sample back to the kitchen and weigh the bag and its contents on the kitchen scale. From this number, subtract the weight of the bag, and multiply the answer by 14,520 (37,037). (This factor is derived by dividing 43,560 square feet per acre by 3 square feet of sample or by dividing 100,000,000 square centimeters per hectare by 2,700 square centimeters of sample.) This amount equals the total pounds (kg) of forage available in the field on a per-acre or per-hectare basis, including water in the forage.

Finally, to determine the available forage dry matter, stir up the contents of your bag. Weigh out a 3.5-ounce (100-g) sample onto a paper or microwave-safe plate. Clip this sample so that all pieces of grass are between ¾ inch and 1¼ inches (2–3 cm) long, and spread them out evenly on the plate.

If the sample is a hay sample, which has already dried partway, initially dry for 3 minutes on the microwave's highest setting. If the sample is fresh forage, the initial drying should be done for 5 minutes.

After the initial drying, reweigh the sample, and then dry of for a minute more on high. Reweigh again. If the weight has changed, repeat the 1-minute drying and reweighing process until two successive weights are equal.

As Jim Gerrish says, "If you smell smoke during this exercise — open the microwave door and put out the fire!"

When the sample is fully dry, the weight in grams will be equal to the percent of dry matter. Multiply the percent of dry matter by the pounds of forage per acre (or kg/ha) you calculated when you first weighed the sample in the bag, to arrive at the dry matter pounds per acre (kg/ha).

Grain Samples. Grain samples can be tested for dryness using the same basic technique. Spread a 3.5-ounce (100-g) sample thinly on a plate, and follow the instructions above, but adjust initial drying time to 1 minute and subsequent drying times to 30 seconds.

Correction Factor

If the sample you have run yields a significantly different number than your estimate, you can develop a correction factor. The correction factor is derived by dividing the quantity of forage in your sample by the quantity you estimated. The Wilsons provide an example. Tom and Karen estimated that their pasture has 90 percent coverage, so they multiply 300 by 0.9 to get 270 pounds per inch (or 136 kg x 0.9 to get 122 kg/cm) of forage. Next, they measure sward length at 10 inches (25 cm), so they come up with available forage dry matter at 2,700 pounds (270 x 10 = 2,700) or 3,050 kilograms (122 x 25 = 3,050). After taking a sample, they discovered that their pasture actually contained about 2,450 pounds per acre (2,709 kg/ha). Their correction factor is 0.91 (2,450/2,700; or 0.89 [2,709/3,050] for metric units). Now each time they estimate forage, they should multiply the estimate they come up with by the correction factor of 0.91 (or 0.89 for metric units).

Monitoring Forage Quality

Unfortunately, there's no easy way to test forage quality at home. But laboratories are available to run samples through an extensive battery of tests that will tell you about energy, major nutrients, and micronutrients available in your pastures. Check appendix E, Resources, for an organization that can give you the names of testing labs in your area.

Monitoring Forage Intake

In chapter 14, I gave a chart of estimated dry-matter intakes for various classes of livestock, but as I said there, critters can eat a whole lot more than those average figures, especially when feed quality is high. When feed quality is very low, they can eat far less (Figure 15.1). One way to fine-tune your forage budgets in future years is to monitor actual intake a couple of times during the year. Again, time your monitoring to match the major cycles — high growth, low growth, and dormant season.

Intake can be measured by several different units: pounds (kg) of dry matter intake per head per day, pounds (kg) of dry matter intake per animal unit per day, or as a percent of body weight. Since the forage budget uses an intake factor based on percent of body weight, that's the measure we'll seek. There are five steps used to

Figure 15.1. The amount of feed an animal needs to take in and the amount it is capable of taking in are both functions of feed quality. With high-quality feed, the animal needs to eat less **(A)** than if the quality is poor **(B).** The problem is that when the quality is poor, the animal is physically unable to digest as much feed in the course of a day **(C),** so it is not meeting its nutrient requirements.

determine approximate intake factors for your herd: Determine paddock size, determine pounds (kg) of dry matter, estimate the total weight of the animals in the paddock, measure sward length as animals enter the paddock and leave the paddock to calculate the amount of dry matter consumed and calculate the intake factor.

1. To estimate the size of the paddock, pace off the number of steps in the length of paddock before you allow the animals to enter it. Try to take natural-size steps. Then pace the number of steps in the width of the paddock. Most adults have a natural step length of about 2 feet (61 cm). To determine your own normal step length, pace off a known distance of at least 20 feet (6.1 m) — say, along the side of a wall — and divide the distance by the number of steps. If you do this several times, the average value is your step length in feet (m).

Multiply the number of steps (S) along the length of the paddock by your step length (SL); then multiply the number of steps along the width by your step length. You now have the approximate length (L) of the field in feet (or m), and the width (W) of the field in feet (or m). Multiply these two numbers together (L x W) to estimate the area of the paddock in square feet (or m^2). Divide the square-foot value by 43,560 square feet per acre to come up with the acreage in the paddock (or the m^2 value by 10,000 to arrive at ha).

Let's look at an example. Gary Miller checks his step length along the side of his chicken shed. The wall is 22 feet (6.7 m) long. Gary paces the side of the shed in about ten steps. His step length (SL) is 2.2 feet (67 cm) per step (S).

$$SL = \frac{22\text{ feet }(6.7\text{ m})}{10\text{ S}}$$

Gary paces the length of his paddock, and comes up with 143 steps, so the length equals 315 feet (95.8 m).

$$L = \frac{143\text{ S} \times 2.2\text{ ft }(67\text{ cm})}{\text{S}} = 315\text{ ft }(9581\text{ cm } \textit{or } 95.8\text{ m})$$

He paces the width of this field at 125 steps, which equals 275 feet (83.8 m). With calculator in hand, Gary multiplies 315 feet (95.8 m) by 275 feet (83.8 m) to find 86,625 square feet (8,028 m^2). Finally he divides 86,625 by 43,560 square feet per acre to arrive at about 2 acres. (To find hectares, he divides 8,028 by 10,000 to get 0.8 ha.)

$$315 \times 275 = 86{,}625\text{ sq. ft.} \quad \textit{or} \quad 95.8 \times 83.8 = 8{,}028\text{ m}^2$$

$$\frac{86{,}625}{43{,}560} = 1.98\text{ acres} \quad \textit{or} \quad \frac{8{,}028}{10{,}000} = 0.8\text{ ha}$$

2. After you have the paddock size established, you need to determine how many pounds of forage are available in this paddock. Use the 300 pounds per inch, per acre (or 136 kg/cm/ha) estimate, or monitor intake on the same days of the year that you monitor the quantity of available forage to get an even more accurate number. You determine the total available forage dry matter in the paddock by multiplying dry matter per acre (or ha) times the number of acres (or ha).

In Gary's case, he has taken samples as outlined in monitoring quantity, and determined that there are 2,565 pounds of available forage dry matter per acre (2,837 kg/ha) in this paddock. Since he has 2 acres (0.8 ha) in this paddock, its total available forage dry matter equals 5,130 pounds (2,300 kg).

3. The next step is to let your animals in to graze as normal, but before allowing the animals in, measure the grass length (GL) of the sward. Then estimate the total pounds of livestock in the paddock — use the weights in Table 3.1 on page 24. This is as far as you can go until the animals leave the paddock.

Gary measures his grass length at 8 inches (20 cm). He estimates the weight of animals in the paddock at 23,200 pounds (10,524 kg) (2 draft horses at 1,600 pounds [725.7 kg] each; 20 ewes at 120 pounds [54.5 kg] each; 35 lambs at 40 pounds [18.1 kg] each; 12 cows at 900 pounds [408.2 kg] each; 12 calves at 250 pounds [113.4 kg] each; and 4 steers at 600 pounds [272.2 kg] each).

4. As the animals leave the paddock, measure the grass length again. To calculate how much of the forage the animals consumed, divide the grass length as the animals leave by the grass length when the animals came in, and multiply this number by the total available forage you calculated in step 2. This calculation gives you the dry-matter disappearance (DMD).

When Gary's animals leave the field, there is 4 inches (10 cm) of grass left. He calculates that his herd consumed 2,565 pounds (1,150 kg) of forage while they were in the paddock.

$$\frac{4\text{ in}}{8\text{ in}} \times 5{,}130\text{ lb of dry matter available} = 2{,}565\text{ lb DMD}$$

or

$$\frac{10\text{ cm}}{20\text{ cm}} \times 2{,}300\text{ kg } 1{,}150\text{ kg DMD}$$

5. Now, to find the intake factor (IF) for a herd, first multiply the total pounds of dry matter disappearance (DMD) consumed by 0.75, and then multiply the total weight of the herd (HW) by the grazing period (GP) in days that the herd was in the paddock. Now divide the first number by the second.

$$IF = \frac{(DMD \times 0.75)}{(HW \times GP)}$$

The 0.75 that you multiply DMD by in this equation is another correction factor. It takes into account the fact that in a sward of grass, there is more weight of grass closer to the ground than at the top of the plant. For example, if you have 8 inches (20 cm) in the sward that averages 300 pounds per inch (132 kg/cm), the top 4 inches (10 cm) of grass may contain only 225 pounds per inch (99 kg/cm) and the bottom 4 inches (10 cm) may weigh 375 pounds per inch (165 kg/cm).

Gary's intake equals 0.041, or 4 percent, of his animal's body weight during his two-day grazing period. He calculated this by:

$$IF = \frac{(2{,}565 \times 0.75)}{(23{,}200 \times 2)} = 0.041 \text{ or}$$

$$IF = \frac{(1{,}150 \times 0.75)}{(10{,}524 \times 2)} = 0.041$$

Weather Trouble

Monitoring forage quantity regularly during the growing season allows you to make adjustments when Mother Nature doesn't cooperate with your original plans. And make no mistake about it — no matter where you live, there are occasions when she is a curmudgeonly old bird and doesn't cooperate. Every place has its droughts, and every place has its washout periods. Both are at the minimum a nuisance, and at their worst a disaster.

Your first protection against the ravages of the weather is healthy land. When the ecosystem processes are operating well, the land has resilience; it's less impacted by weather events.

Heavy Rain

During heavy rains, healthy soils with good vegetative cover absorb water like a sponge. And during droughts, healthy fields remain shaded and cool, retaining what moisture they have for much longer periods than do those that have been abused.

The second thing that helps you cope with the vagaries of nature is having a polyculture of plants growing. When your fields are populated by a wide variety of different grasses and legumes, you'll find that something will grow in most conditions — wet or dry. An abundance of legumes, which tend to be more deeply-rooted than the grasses, will help move what water is available from deeper in the soil matrix up to the surface.

When bad weather does strike, and your forage budget (which looked so good when you prepared it in February) is suddenly just ink on paper, you have to adjust. The sooner you do, the less strain it will cause emotionally and financially.

During very wet periods, the grass can get ahead of you quickly, but getting into the field to cut hay becomes almost impossible. Keep moving animals as quickly as possible through the paddocks. Consider adding a few extra stocker animals. If you can't afford to purchase more animals, look for an area farmer who keeps growing animals in confinement — you may be able to work out a mutually beneficial relationship. If the feed gets just too far ahead and you can't add animals to take up the slack, drop one or more paddocks completely from your movement. It's better to have a few paddocks that become overly rank than to have the whole pasture go too far. When things dry out, the reserved paddocks can be cut for hay or green manure, or chopped for haylage.

Drought

Droughts are harder, because you have a forage shortage. There are two ways of handling this situation: Purchase additional feed, or destock. The first option isn't so bad if the shortage is very minor, but if you are falling far short of production it's best to destock early. In fact, the earlier you begin destocking, the fewer animals you'll have to remove altogether during an extended drought.

An example, using numbers from the Joneses' forage budget, may help you see why this is so: Under normal circumstances in most of North America, almost 60 percent of annual forage production grows during the April-to-June period. But in this example it's July 1, and April, May, and June were extremely dry in central Wisconsin; there's no sign of relief.

Gil has already grazed the grass a little closer than ideal, and the first cutting of hay, which he expected to yield about 85 tons (77,110.7 kg), came to only 70 tons (63,502.9 kg) — and part of this was from a subirrigated (naturally wetter) field. Common wisdom says that Gil can make up the hay shortage by selling off more young stock at weaning than he planned for in his budget. But Gil knows that this approach doesn't help the current shortage of grass; and if he continues grazing as hard as he has been, the yield will drop even farther (because of insufficient plant energy and recovery periods). Monitoring indicates that by slowing down the grazing a little, and giving extra time for recovery, the end-of-year yield will be about 340 tons (308,443 kg) of forage. Grazing can be slowed either by supplementing with purchased feed, or by selling some animals now.

Gil and Jenny sit down at the computer and begin experimenting with various scenarios. Table 15.1 shows their budget for forage requirements based on the immediate sale of five cows and their calves, two steers, and five ewes and their lambs. These sales result in the forage required being about 340 tons (308,443 kg). Gil hauls the animals to the sale barn.

What would have happened if Gil and Jenny had decided to wait it out, hoping against hope for late-summer and fall rains to make up the shortfall? If those rains didn't materialize — and the odds are that they wouldn't make up a shortage even if they did come — the end-of-growing-season yield would probably be down around 300 tons (272,155 kg), because they were grazing too hard. Table 15.2 shows how heavily they would have to cull in fall to make up for holding on in July. By waiting, they have to sell 100 percent of their calf and lamb crops, and more breeding stock — from both the cow and sheep categories. Table 15.3 summarizes the end-of-year impacts of July sales versus October sales, compared to their original plan.

The moral of the story: If you have to destock, *do it early!* One other point about early destocking that doesn't appear in this example is that you often get better prices. As a drought progresses and more people come to terms with their shortage of feed, they begin dumping animals en masse, thereby driving the price down. The time to sell is before everyone else has figured out that there is a problem, or while they are still holding out for divine intervention.

Table 15.1

Jones Farm Forage Budget Requirement Based on July Sales

	June	July	August	September	October	November	December
Seasonal; activity Biocycles	Lambing →; Horses working moderately →	Breed cattle →	→		Wean lambs →; Wean calves →; →		Breed sheep →
Cows, number	40	35	35	35	35	35	35
Average weight (lb)	975	1,000	1,025	1,050	900	900	900
Total weight (lb)	39,000	35,000	35,875	36,750	31,500	31,500	31,500
Percent factor	0.0300	0.0300	0.0300	0.0275	0.0250	0.0250	0.0250
Cow HE	1,170	1,050	1,076	1,011	788	788	788
Growing cattle, number	21	18	16	14	40	35	35
Average weight (lb)	600	650	690	730	400	430	460
Total weight (lb)	12,600	11,700	11,040	10,220	16,000	15,050	16,100
Percent factor	0.035	0.035	0.035	0.035	0.035	0.035	0.035
Growing cattle HE	441	410	386	358	560	527	564
Horses, number	4	4	4	4	4	4	4
Average weight (lb)	1,200	1,200	1,200	1,200	1,200	1,200	1,200
Total weight (lb)	4,800	4,800	4,800	4,800	4,800	4,800	4,800
Percent factor	0.030	0.030	0.030	0.030	0.030	0.025	0.025
Horse HE	144	144	144	144	144	120	120
Sheep, number	50	45	45	45	45	45	45
Average weight (lb)	170	180	190	200	140	140	140
Total weight (lb)	8,500	8,100	8,550	9,000	6,300	6,300	6,300
Percent factor	0.0350	0.0350	0.0350	0.0325	0.0300	0.0300	0.0300
Sheep HE	298	284	299	293	189	189	189
Growing sheep, number	5	5	5	5	50	50	50
Average weight (lb)	130	135	140	140	50	65	75
Total weight (lb)	650	675	700	700	2,500	3,250	3,750
Percent factor	0.03	0.03	0.03	0.03	0.04	0.04	0.04
Growing sheep HE	20	20	21	21	100	130	150
AUs	66	60	61	61	61	61	62
Total HE/day	2,072	1,907	1,927	1,826	1,781	1,753	1,810
Total HE/month	62,160	59,125	59,734	54,775	55,196	52,598	56,110
Tons HE/month	31	30	30	27	28	26	28

HE = hay equivalent; AU = animal unit.

→ sheep/lambs → cattle → horses

Average 63
Estimate of Annual Forage Required 339

Table 15.2

Requirements Based on October Sales

	September	October	November	December
Seasonal; activity Biocycles	Horses working moderately →	Wean lambs →; Wean calves →; →		Breed sheep →
Cows, number	40	20	20	20
Average weight (lb)	1,050	900	900	900
Total weight (lb)	42,000	18,000	18,000	18,000
Percent factor	0.0275	0.0250	0.0250	0.0250
Cow HE	1,155	450	450	450
Growing cattle, number	18	0	0	0
Average weight (lb)	730	400	430	460
Total weight (lb)	13,140	0	0	0
Percent factor	0.035	0.035	0.035	0.035
Growing cattle HE	460	0	0	0
Horses, number	4	4	4	4
Average weight (lb)	1,200	1,200	1,200	1,200
Total weight (lb)	4,800	4,800	4,800	4,800
Percent factor	0.030	0.030	0.025	0.025
Horse HE	144	144	120	120
Sheep, number	50	20	20	20
Average weight (lb)	200	140	140	140
Total weight (lb)	10,000	2,800	2,800	2,800
Percent factor	0.0325	0.0300	0.0300	0.0300
Sheep HE	325	84	84	84
Growing sheep, number	5	0	0	0
Average weight (lb)	140	50	65	75
Total weight (lb)	700	0	0	0
Percent factor	0.03	0.04	0.04	0.04
Growing sheep HE	21	0	0	0
AUs	71	26	26	26
Total HE/day	2,105	678	654	654
Total HE/month	63,147	21,018	19,620	20,274
Tons HE/month	32	11	10	10

Average AU 56
Estimate of annual forage rquired 300

Table 15.3

SUMMARY OF A DROUGHT'S IMPACT ON END-OF-YEAR ANIMAL UNITS DEPENDING ON SALE IN JULY OR OCTOBER

	Planned Number	If Sold in July	If Sold in October
Cows	45	35	20
Growing cattle	46	35	0
Horses	4	4	4
Sheep	50	45	20
Growing sheep	50	50	0
AUs at end of year	77	62	26

Stock Density

In chapter 3, I said that stock density is a measure of AU per acre (or per ha) on a given paddock at any one time. It can also be reported as pounds of livestock per acre (kg/ha) at a given time.

Management improves as stock density increases. In pounds of livestock per acre, desirable stock densities can run anywhere from 10,000 to 80,000 pounds per acre (11,208 to 89,666 kg/ha), depending on time of year, pasture condition, and grazing period.

You can calculate an appropriate stock density by using the following equation: Stock density (SD) equals the result of the available forage (AF) multiplied by the utilization rate (UR), then divided by the result of daily intake (DI) times the grazing period (GP).

$$SD = \frac{(AF \times UR)}{(DI \times GP)}$$

The utilization rate is the amount of plant material you want to take off. Somewhere between 50 and 60 percent utilization is about right.

The DI is the amount of feed (as hay equivalents) per day that the average animal consumes. It's derived by dividing the required feed in HE/day by the AUs on your forage budget.

The grazing period is one you determine.

As an example, let's say Gil and Jenny are calculating a stock density for one of the two paddocks in field 2 in May. Since it's early in the season, the paddock has about 6 inches (15 cm) of forage growth, and they plan to take 50 percent of it in this grazing. They are aiming for a GP of 1 day. Use the HE/day and AU figure from May on the forage budget to calculate daily intake. Plug in the numbers:

$$AF = \left(300 \frac{\text{lb/in}}{\text{acre}} \times 6\text{in}\right) = 1{,}800 \text{ lb/acre}$$

$$DI = \frac{1{,}902 \text{ lb/day}}{61 \text{ AU}} = 31 \frac{\text{lb/day}}{\text{AU}}$$

$$SD = \left(1{,}800 \frac{\text{lb}}{\text{acre}} \times 0.5\right) \div \left(31 \frac{\text{lb/day}}{\text{AU}} \times 1 \text{ day}\right)$$

$$= \frac{900 \text{ lb/acre}}{31 \text{ lb/AU}}$$

$$= 29 \text{ acre/AU}$$

Or in metric:

$$AF = \left(136 \frac{\text{kg/cm}}{\text{ha}} \times 15 \text{ cm}\right) = 2{,}040 \text{ kg/ha}$$

$$DI = \frac{863 \text{ kg/day}}{61 \text{ AU}} = 14 \frac{\text{kg/day}}{\text{AU}}$$

$$SD = \left(2{,}040 \frac{\text{kg}}{\text{ha}} \times 0.5\right) \div \left(14 \frac{\text{kg/day}}{\text{AU}} \times 1 \text{ day}\right)$$

$$= \frac{1{,}020 \text{ kg/ha}}{14 \text{ kg/AU}}$$

$$= 72 \text{ ha/AU}$$

You come up with SD = 29 AU per acre, or approximately 29,000 pounds per acre (72 AU/ha or 32,659 kg/ha).

FARMER PROFILE

Alan and Sharon Hubbard

Alan and Sharon Hubbard don't own any land, but they ranch on more than 6,000 acres in northeastern Kansas. Their 375-acre home place is part of a trust from Sharon's family; the remainder is land they lease from seventeen different landlords. "Buying land for ranching and farming doesn't really pay," Alan says. "Beyond owning a small piece for a headquarters, it's better to lease."

I first heard about Alan from Jerry Jost at the Kansas Rural Center. Jerry told me, "Alan is one of the most sophisticated managers in Kansas. He does extensive financial planning and monitoring, and he is finishing up a study in partnership with Kansas State University that looks at time management and production on set-stocked ground versus intensively managed grazing ground."

Alan elaborates on the study. "We're just about done.... It looked at how long it takes to care for the animals in the two systems, and at pounds per acre of beef productivity off both approaches. We've found — and it surprised us — that it takes less time to care for animals that are grazing intensively than it does to take care of those on the set-stocked land, and that we produce almost twice as many pounds per acre of beef where we are managing the grazing."

Alan and Sharon began developing their financial planning program almost a decade ago. "We do detailed enterprise analysis regularly, both on current enterprises and on possible other enterprises. Over the years, the time we've spent on financial planning has paid off well. When we first started, it did take a lot of time; now that we know what we're doing, the time commitment isn't so great."

As a result of their planning, the Hubbards have done many things over the years that raised a few eyebrows. "I guess it's actually getting better, because this year, when we brought in 160 Spanish meat goats from Texas, the neighbors only looked slightly baffled. When we began subdividing fields and running 100-plus head on only a 20-acre piece of land, it was the talk of the town! They didn't understand that those 100 head wouldn't be on that piece of ground the next day. They were sure we would ruin the land and ourselves.

"Then we sold all our equipment, except one loader tractor for moving any purchased round bales, pushing snow, and things like that. Selling the equipment set off another wave of shock." Today, the Hubbards' equipment lineup consists of one loader tractor, one four-wheel-drive pickup, one four-wheeler, and one stock trailer.

"We use horses quite a bit for working on the ranch. But originally, the horses were like farm equipment: they cost us money. Since we wanted to keep them, we did some enterprise analysis on horses.

"When you can 'stack' enterprises within existing enterprises — like figuring out a way to make some money with the horses we were keeping anyway — that's when you can really increase profitability. We now keep a breeding stud, and are breeding and selling colts. He is also used for some stud service for other horse owners. And Sharon has begun developing a trail-ride business using the horses. By using our enterprise analysis approach, we've been able to put the horses in the black. Someday they'll go from paying their own way to being more profitable."

During the winter, the Hubbards spend lots of time planning and monitoring. They run their cow herd in groups of 80 to 100 animals. Groups are established according to management needs: mature cows (4 years old or more), young cows and heifers (2- and 3-year-olds), and young heifers. They keep a detailed ledger for each group: how much they're eating, what they cost to maintain.

The Hubbards run their own herd and custom stockers. Most stockers come from area ranchers and farmers, who retain ownership. The Hubbards are paid a per-pound fee based on the gain the animals make.

One year, Alan needed some more grazing animals to match his forage. He ran several scenarios through his enterprise analysis (including purchasing stocker steers, stocker heifers, bred cows, and open cows) but bought 100 open cows and then bred them for fall calving. "We strictly do spring calving for our own cows, but these cows sold well to farmers that wanted fall-calving animals."

Alan concludes by saying, "I can't stress too much the importance of enterprise analysis for profitability. We need to look at all we do and ask ourselves if it meets our life goals, and if it's profitable. Should I keep bulls? Should I raise these heifers up or buy replacements? Should I keep this hay equipment and put up hay, or buy my hay in? When you study these questions in detail, you can make good decisions."

Paddock Sizing

From the stock density, you can now determine the ideal paddock size (PS) to meet the criteria you've established. The paddock size equals the actual total number of animal units in the herd divided by the desired stock density.

$$PS = \frac{AU}{SD}$$

Again Gil and Jenny plug the numbers in and come up with a figure of approximately 2.1 acres per paddock.

AU in May (taken from Table 14.3) = 64
SD (as just calculated) = 29

$$PS = \frac{64\ AU}{29\ AU/acre} = 2.2\ acres$$

Or in metric:

AU = 64
SD = 72

$$PS = \frac{64\ AU}{72\ AU/ha} = 0.9\ ha$$

On a 240-acre (97-ha) farm like the Joneses', to develop 2.2-acre (0.9-ha) paddocks would require over 100 subdivisions. The best way to achieve these high levels of subdivision is through a combination of permanent paddocks and temporary fencing: The most flexibility is provided by establishing 8 to 21 permanent subdivisions, and then using temporary fencing to further subdivide these during the year. But initially use temporary subdivisions for the entire farm, until you've developed the landscape plan. Your landscape plan will help you define the best long-term placement for permanent fences and other improvements.

Long-Term Monitoring

When we purchased our farm in Minnesota, the land was hurting. By the time we left, the ecosystem processes were operating at a much higher level, yet we really didn't have good evidence of this. We could tell people that there was a much greater variety of plants living in the riparian area than when we first got there, but we couldn't provide specifics. We could say we saw more wildlife, but we didn't have documentation. Now we are beginning our monitoring immediately, so we'll be able to define the improvements as they take place.

Long-term monitoring requires that you select some permanent points for performing the monitoring. You can choose these to represent different soils, or different niches — for example, sandy areas versus loamy areas, or river bottomland as one niche and hillsides as another. How many points you choose will depend on your own motivation, but at the minimum try to establish 3 points, at the maximum, 12 points. Monitoring will require about 2 hours per point each time you monitor it.

Your long-term monitoring needs to be done at approximately the same time every year. During the spring flush is an ideal time for your main information-gathering session, but if you're a little more ambitious, do it once during spring flush and once in late summer. The more data you gather, the better you'll understand what's happening on your land.

Once you select the monitoring points, mark each one permanently with a T-post, pile of rocks, or some other permanent fixture. One of the sites we've established now is where a power pole comes onto the land. The point simply needs to be some kind of well-defined spot that you can find easily.

At each of your marked points, begin by taking pictures of the landscape in all four directions — north, south, east, and west. If there are established features on the landscape — say, a big tree or a building — try to include them in a photo. They will give a sense of depth to the picture and will make it clear to anyone that views your pictures in the future that all the photos are of the same spot. The view should be established so you are really recording what's going on with the land; don't include too much sky.

Next, take a few photos that are pointing down at the ground a few feet away from your marker. Placing a ruler or a pen in these photos will give an idea of scale and perspective.

As you take photos, mark in your journal what they are. For example, "Picture 1 is at the permanent marker in the south pasture, and is facing south; picture 2 is facing north." If your camera doesn't have a date function on it, be sure to date and label your photos when they are returned.

Your photo records don't have to be limited to your permanent monitoring points. On monitoring days, we take several rolls of film from all over the place. Just make sure you keep a record of what they are, so they can be labeled when they are returned.

Next, while standing at a permanent marker, take your hula hoop and wing it out somewhere into the pasture. Where it lands is where you will record information on plant numbers, type, spacing, soil cover, capping, and any living organisms you see in your journal. Make notes that describe the condition within the circle of your hula hoop as best you can, recording the information on Table 15.4. The following list will help you develop the information:

- ***Ground cover.*** Make an eyeball estimate of how much bare soil is within the hula hoop. Estimate, too, how much litter (sticks, dead vegetation, manure) there is, and what type. For example, "At data point 1, in the south pasture, the cover is about 85 percent and there is heavy litter of dead vegetation and manure. At data point 1 on the hillside behind the house there is about 60 percent ground cover, and just a little litter of dead vegetation and no manure showing. Data point 1 on the bottomland shows almost 100 percent coverage, with little litter of any kind."
- ***Soil capping or pugging.*** Where the soil is bare, are lichens, mosses, or algae growing? These keep new plant seeds from sprouting. Or is the bare soil so compacted that water can't penetrate? In areas that tend toward wetness, is there pugging? (*Pugging* is a condition caused by animal's hooves — it's like little hummocks or mounds that develop where the animals step.)
- ***Plant numbers and types.*** Count the plants along the piece of rope that divides your hula hoop in half. For example, "Seven clumps of bluegrass, five clumps of orchard grass, three grass plants of unknown type, five white clover plants, two bird's-foot trefoil plants, three forbs (taprooted weeds), one baby pine tree. . . ." To the best of your ability, identify the plants. If your hula hoop landed up in the air, suspended on a bush, assume the bush to be the middle of the circle and work out in a line on either side of it in the direction of the rope. Make notes about the age and health of plants — if the perennials are old and showing decay, for example, or if the brush is young.
- ***Living organisms.*** Record any living organisms you see either within the circle itself or as you study the area around you. Look for insects, worms, snakes, frogs, birds, and mammals. Define what you see as specifically as possible if it's directly related to your site. For example, a flock of geese flying overhead doesn't relate to your site, but six geese nesting at the pond in the south pasture do.
- ***Erosion condition.*** Record any evidence that the soil is washing in rain, or appears to be remaining in place; any evidence that there are flow patterns of litter and soil on the land; and so on. For this, use a scale of 1 to 5, with 5 being highly erosive and 1 being resistant to erosion.

When you've completed recording the information, throw your hula hoop out again and record the information for that spot. Repeat this so that you have at least three data sets for this monitoring point. Then perform the same procedures for each of your permanent monitoring points.

After you've gathered all the information for your monitoring period, try to summarize it in your journal. For example, "The hillside behind the house is in poor condition, with lots of evidence of erosion, little diversity, and little litter." That field would be a good contender for feeding winter hay!

Your long-term monitoring record will clearly reveal what is happening on your land. Use it to show others the impacts of your decisions and management, or use it for the best reason of all — your own satisfaction.

Table 15.4

Long-Term Monitoring Record

Date ____________

Monitoring Points	*South pasture* *Point 1* *Data set 1*	*South pasture* *Point 1* *Data set 2*	*South pasture* *Point 1* *Data set 3*	*Riparian area* *Point 1* *Data set 1*	*Riparian area* *Pcint 1* *Data set 2*	*Riparian area* *Point 1* *Data set 3*
Ground Cover	*85% heavy litter (dead vegetation and manure)*					
Capping/Pugging	*some lichen and moss; no soil cracking*					
Plants	*7 blue, 5 orchard, 3 unknown, 5 white clover, 2 birds'-foot, 3 forb, 1 pine*					
Living Organisms	*6 geese nesting at pond; many butterflies*					
Erosion	*little erosion*					
Comments	*field is looking very good*					

Animals

The following pages provide some basic information on the most common livestock species: cattle, chickens, goats, horses, pigs, and sheep.

For each species there is a sampling of some major and minor breeds, but these are not intended to be complete lists of all breeds available for that species. For instance, there are more than fifty recognized breeds of cattle raised in the United States.

The feed chart for each species provides basic nutritional information, but these charts are based on "average" figures. Individual animals can deviate plus or minus 20 percent from these figures, depending on their metabolic rate, the weather, their general health, and the phase of the moon. But by using these figures — in conjunction with data in appendix B, Composition of Common Feedstuffs — to calculate a feed program, you'll be well on your way to maintaining healthy animals. (See page 210 for basic metric equivalents.)

Cattle

Scientific name: *Bos taurus* (European breeds) or *Bos indicus* (Asian breeds — especially Brahman and Zebu types)

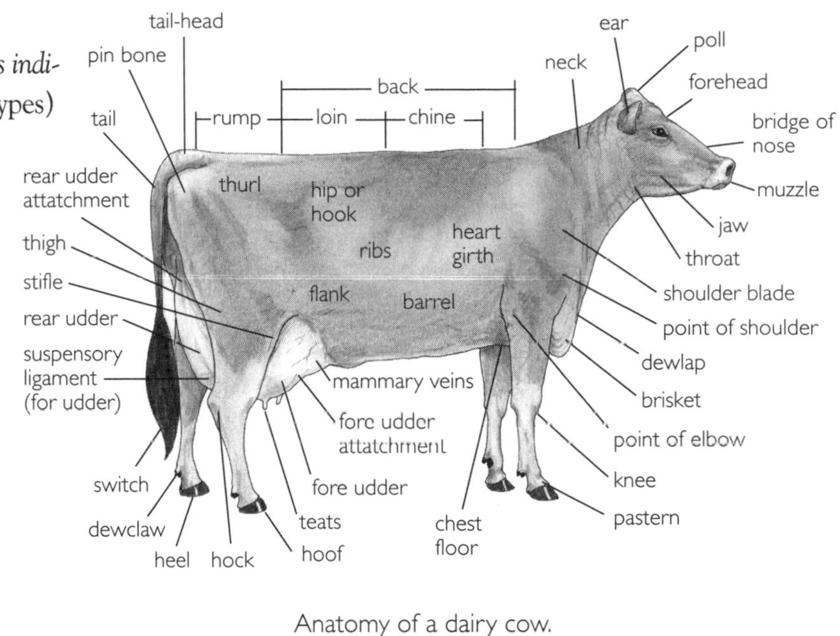

Anatomy of a dairy cow.

Terms:

Bull = Immature or mature male
Calf = Newborn
Calving = Act of giving birth
Cow = Mature female
Free martin = Heifer born as a twin, with the other twin being a bull (these heifers cannot breed)
Heifer = Immature female
Polled = Animal that naturally has no horns
Springer = Cow or heifer nearing calving
Steer = Castrated male

Reproduction:

Gestation = 305 days
Normal birth = 1 calf

Tons of manure produced annually per 1,000 pounds of stock: 15

Kilograms of manure produced annually per 1,000 kilograms of stock: 30,000

BREEDS

Major
Angus (beef)
Hereford (beef)
Holstein (dairy)
Jersey (dairy)
Limousin (beef)
Simmental (beef)

Common
Ayrshire (dairy)
Brahman (beef)
Brown Swiss (dairy)
Guernsey (dairy)
Highland (beef)
Milking Shorthorn (dual)
Texas Longhorn (beef)

Minor
Dexter (dual)
Dutch Belted (dual)
Galloway (meat)
Milking Devon (dual)
Normande (dual)
Pinzgauer (meat)
Red Poll (meat)

Top senses = Sight, smell

Fencing: Dairy cows can easily be kept with a single wire once trained. For stockers or cow-calf, plan on two wires for trained animals and four wires for untrained animals.

Housing: Beef cattle are extremely hardy and require no housing unless calving will take place in winter. Provide a windbreak. Dairy cattle that will be milked through the winter will need some type of housing in the extreme North; however, alternative housing — like a hoop barn — should work. Through the central part of the country, a windbreak should be sufficient.

Other Thoughts: I love cows! We had cows that were as friendly as pet dogs. A good alternative for a homestead operation that just wants to produce its own milk and meat is one of the dual-purpose breeds, or a Jersey cow. Let the cow raise her own calf; milk her once per day for your own milk. During the early part of the calf's development, you'll get enough milk for making butter, cheese, and yogurt; when the calf is near weaning, you may only get enough for your coffee, or a couple of glasses of milk per day.

Table A.1

CATTLE FEED*

For Growing Beef Cattle:

Live weight, in pounds	300	400	500	600	700	800	900	1,000
Dry matter lb	8.1	10.3	11.9	13.6	15.4	17.1	18.5	19.8
Crude protein %	14.0	13.5	13.0	12.5	12.0	11.5	11.0	10.5
Crude protein lb	1.1	1.4	1.5	1.7	1.9	2.0	2.0	2.1
TDN %	78.0	77.0	76.0	75.0	74.0	73.0	73.0	73.0
TDN lb	6.3	7.9	9.0	10.2	11.4	12.5	13.5	14.5
Energy Mcal	5.1	6.1	6.9	7.7	8.2	8.8	9.3	9.8
Calcium %	0.5	0.5	0.4	0.3	0.3	0.2	0.2	0.2
Phosphorus %	0.4	0.4	0.3	0.3	0.3	0.2	0.2	0.2

For Breeding Beef Cattle:

	Dry pregnant cow, middle third of pregnancy	Dry pregnant cow, last third of pregnancy; lactating cow — late lactation	Lactating cow, first 3 months (average milking ability)	Lactating cow, first 3 months (high milker)	Lactating heifer, first 3 months	Bull (moderate work)
Dry matter lb per 100 lb of body weight	1.6	2.0	2.4	2.7	2.3	1.9
Crude protein %	7.0	8.0	10.0	12.0	12.0	7.0
Crude protein lb per 100 lb of body weight	0.10	0.16	0.24	0.32	0.28	0.13
TDN %	50.0	57.0	50.0	67.0	64.0	55.0
TDN lb per 100 lb of body weight	0.8	1.1	1.2	1.8	1.5	1.1
Energy Mcal per lb of feed	0.8	0.9	1.0	1.4	1.1	0.9
Calcium %	0.18	0.26	0.28	0.45	0.35	0.20
Phosphorus %	0.18	0.21	0.23	0.30	0.25	0.20

TDN = total daily nutrients.

* Feed required for moderately fast growth.

Note: Dairy animals require significantly higher levels of feed than beef cows. For example, protein levels need to increase between 20 and 40 percent over the levels shown above for lactating animals, depending on the cows' production. If you plan to milk animals commercially, you must study a good text on feeds and feeding (see appendix E, Resources). For homestead production, increase protein by 25 percent and double the calcium and phosphorus.

Chickens

Scientific name: *Gallus gallus*

Terms:

Broody = Hen that's ready to sit on a clutch
Capon = Castrated male
Chick = Newborn
Clutch = Group of eggs
Cockerel = Immature male
Hen = Mature female ("Down Under," a hen is a *chook*)
Molting = Regular shedding of feathers
Pullet = Immature female
Rooster = Mature male

Anatomy of a hen and cock, shown on *Gallus gallus,* the "original" chicken.

Reproduction:

Gestation = 21 days
Normal birth = up to 16 chicks successfully hatched from one hen, though typical clutches run 6 to 8 per hen

Tons of manure produced annually per 1,000 pounds of stock: 4.75
Kilograms of manure produced annually per 1,000 kilograms of stock: 9,500

Top sense: Sight

BREEDS

Major	Minor
Brahma (meat)	Barred Rock (dual)
Cornish (meat)	Brown Leghorn (layer)
White Leghorn (layer)	Jersey Giants (meat)
White Rock (meat)	Wyandotte (dual)

Fencing: If you want to fence chickens into a small yard, chicken wire is the way to go. Electroplastic Net (Premier Fencing, Washington, Iowa) also works.

Housing: Chickens are one class of animals that does need some type of housing, primarily for predator protection. There are lots of designs for small chicken houses available — check with your local extension agent to see if he or she has some free material. Once chickens become accustomed to roosting in a particular building or portable shed, they will return there each night. If you allow your birds to free-range during the day, just close the door right after the last birds go in to roost right before it gets dark out.

Other thoughts: Chickens deserve a place on every farm. They are economical and relatively easy to deal with. Good, fresh chicken meat is a delicacy that has strong market appeal, but even if you aren't selling chicken, keep a small flock of dual-purpose hens and a rooster (one rooster per twenty hens is a good ratio). You'll be kept in plenty of fresh eggs most of the year, and at least one or two hens will brood up a batch of chicks for you to butcher in fall.

Table A.2

Chicken Feed

For Layers:

Age in weeks	<2	2–4	4–6	6–8	8–10	10–12	12–14	14–22	22–24	24–26	26–30	30–40	40–50	50–60	60–70
Typical egg production (percent of hens laying each daily)	0	0	0	0	0	0	0	0	10	38	64	88	80	74	68
Feed consumption per hen in oz (lb) per week)	1.6 (0.1)	3.2 (0.2)	6.4 (0.4)	9.2 (0.6)	11.5 (0.7)	13.6 (0.9)	15.1 (0.9)	16.2 (1.0)	18.5 (1.2)	21.0 (1.3)	23.4 (1.5)	27.1 (1.7)	26.9 (1.7)	26.6 (1.7)	26.0 (1.6)
Crude protein %	18.0	18.0	18.0	15.0	15.0	15.0	15.0	12.0	14.5	14.5	14.5	14.5	14.5	14.5	14.5
Crude protein per hen per week	0.29 (0.02)	0.58 (0.04)	1.15 (0.07)	1.38 (0.09)	1.73 (0.11)	2.04 (0.13)	2.27 (0.14)	1.94 (0.12)	2.68 (0.17)	3.05 (0.19)	3.39 (0.21)	3.92 (0.25)	3.90 (0.24)	3.85 (0.24)	3.77 (0.24)
Energy Mcal per lb of feed	1.31	1.31	1.31	1.31	1.31	1.31	1.31	1.31	1.31	1.31	1.31	1.31	1.31	1.31	1.31
Calcium %	0.8	0.8	0.8	0.7	0.7	0.7	0.7	0.6	3.4	3.4	3.4	3.4	3.4	3.4	3.4
Phosphorus %	0.40	0.40	0.40	0.35	0.35	0.35	0.35	0.30	0.32	0.32	0.32	0.32	0.32	0.32	0.32

For Broilers:

Age in weeks	1	2	3	4	5	6	7	8	9
Feed consumption per cockerel in oz (lb) per week	4.2 (0.26)	9.2 (0.57)	13.7 (0.86)	18.8 (1.18)	26.1 (1.36)	34.5 (2.16)	38.5 (2.40)	42.6 (2.67)	46.5 (2.90)
Feed consumption per pullet on oz (lb) per week	3.9 (0.24)	8.4 (0.53)	12.5 (0.78)	17.6 (1.10)	22.7 (1.41)	28.2 (1.76)	32.0 (2.00)	34.1 (2.13)	35.6 (2.22)
Crude protein %	23	23	23	20	20	20	18	18	18
Crude protein per cockerel per week	0.97 (0.05)	2.12 (0.13)	3.15 (0.20)	3.76 (0.24)	5.22 (0.33)	6.90 (0.48)	6.93 (0.43)	7.67 (0.18)	8.37 (0.53)
Crude protein per pullet per week	0.90 (0.06)	1.93 (0.12)	2.88 (0.18)	3.52 (0.22)	4.54 (0.28)	5.64 (0.35)	5.76 (0.36)	6.12 (0.38)	6.41 (0.40)
Energy Mcal per lb of feed	1.45	1.45	1.45	1.45	1.45	1.45	1.45	1.45	1.45
Calcium %	1.0	1.0	1.0	0.9	0.9	0.9	0.8	0.8	0.8
Phosphorus %	0.45	0.45	0.45	0.40	0.40	0.40	0.35	0.35	0.35

Goats

Scientific name: *Capra hircus*

Terms:

Billy or buck = Mature male
Chevon = Goat meat
Doe or nanny = Mature female
Kid = Newborn
Wether = Castrated male

Reproduction:

Gestation = 150 days
Estrus cycle = 21 days
Normal birth = 1 to 3 kids per doe

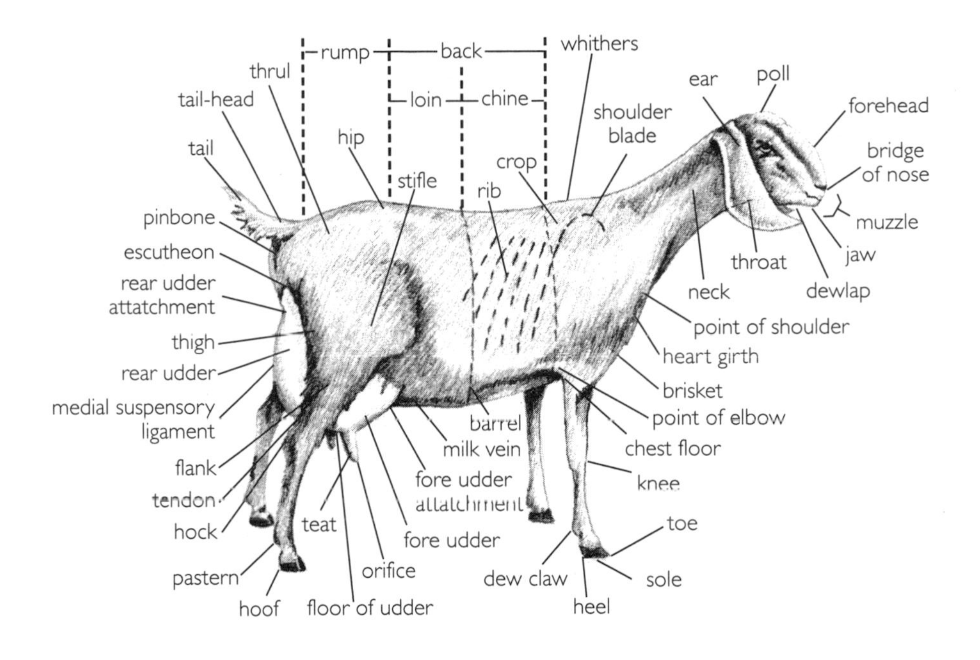

Anatomy of a goat.

Tons of manure produced annually per 1,000 pounds of stock: 9.00
Kilograms of manure produced annually per 1,000 kilograms of stock: 18,000

Top senses: Taste, smell

Fencing: Goats are about the hardest animals to fence. They can climb, they can jump, they can crawl; and if they can, they will! I've seen goats climb up on a parked vehicle to jump over a fence. The best approach is four to five strands of permanent electric fence, very well grounded, and with nothing (such as a parked vehicle) near the perimeter. If you only keep a few goats, Electroplastic Net (Premier Fencing, Washington, Iowa) will work well.

Housing: Goats are scrappy little things that only need a building if they'll be kidding during the winter. Otherwise, they'll find some little nook or cranny to wait out bad weather on their own.

Other thoughts: We've never kept goats commercially, though we've had a couple of pet goats over the years. They are fun, in part because they are so clever, but they can be a challenge. Our first pet goat, Giles, was an abandoned Nubian wether; I'm not sure if we adopted him or if he adopted us. Either way, Giles learned to open the back door to the house and, until we devised a better latch, would let himself in to have a party while we were at work. And goat parties are a messy situation.

Goats have real market potential if you live near a city with a large variety of ethnic populations, because in many cultures goat meat is a delicacy. We also know several small-scale, commercial goat-cheese manufacturers who do quite well producing and direct-marketing their product from a relatively small herd of dairy goats. (See the story of Lanie Fondiler on page 67.)

BREEDS

Major
Angora (fiber)
Boer (meat)
Cashmere (fiber)
Nubian (dairy)
Saanen (dairy)
Spanish (meat)
Toggenburg (dairy)

Minor
Alpine LaMancha (dairy)
Fainting goat (meat)
Nigerian Dwarf (dairy)
San Clemente (meat)

Note: The distinction between major and minor may be a fine one in goats. Since goats have never developed into a large-scale commercial product in the United States, their numbers are fairly low, currently estimated at 4.5 million animals total.

Table A.4

Goat Feed

For Growing Goats:

Live weight, in pounds	50	75	100	125
Dry matter lb	2.0	2.8	3.5	4.0
Crude protein %	10.0	9.0	8.5	8.0
Crude protein lb	0.2	0.25	0.28	0.32
TDN %	57	57	56	56
TDN lb	1.14	1.60	1.96	2.24
Energy Mcal	1.2	1.2	1.3	1.3
Calcium %	0.40	0.30	0.28	0.28
Phosphorus %	0.28	0.20	0.21	0.21

For Breeding Goats:

	First two-thirds of gestation	Last third of gestation	First 10 weeks of lactation	Last 14 weeks of lactation	Billies at moderate work
Dry matter lb per 100 lb of body weight	2.5	3.5	4.4	3.5	4.0
Crude protein %	9.0	10.0	11.7	10.0	9.0
Crude protein lb per 100 lb of body weight	0.23	0.35	0.51	0.35	0.36
TDN %	55	56	69	56	55
TDN lb per 100 lb of body weight	1.38	1.96	3.04	1.96	2.20
Energy Mcal per lb of feed	1.3	2.0	2.6	2.0	2.1
Calcium %	0.28	0.35	0.42	0.35	0.28
Phosphorus %	0.21	0.25	0.27	0.21	0.20

TDN = total daily nutrients.

Horses

Scientific name: *Equus caballus*

Terms:

Stallion = Mature male
Mare = Mature female
Filly = Immature female
Colt = Immature male
Foal = Newborn
Chunk = Cross between draft horse and saddle horse

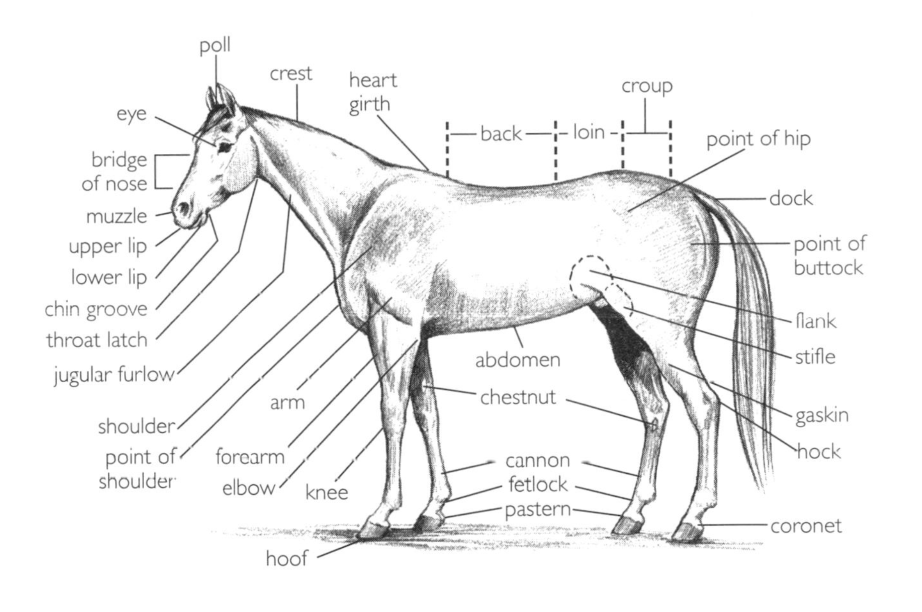

Anatomy of a horse.

Reproduction:

Gestation = 11 months
Normal birth = 1 foal
Foals are generally born between 3 A.M. and 6 A.M.

Tons of manure produced annually per 1,000 pounds of stock: 12
Kilograms of manure produced annually per 1,000 kilograms of stock: 24,000

Top senses: Smell, sight

Fencing: A horse that's been trained to electric fencing and knows where the fence is located can easliy be kept by a single strand of polywire. Avoid barbed wire with horses — we know of several people over the years who lost a horse when it became entangled in barbed wire and cut itself so badly it couldn't be saved. If you purchase a farm with existing barbed-wire fencing, run a single strand of polywire inside the existing fence to keep the horses away from it.

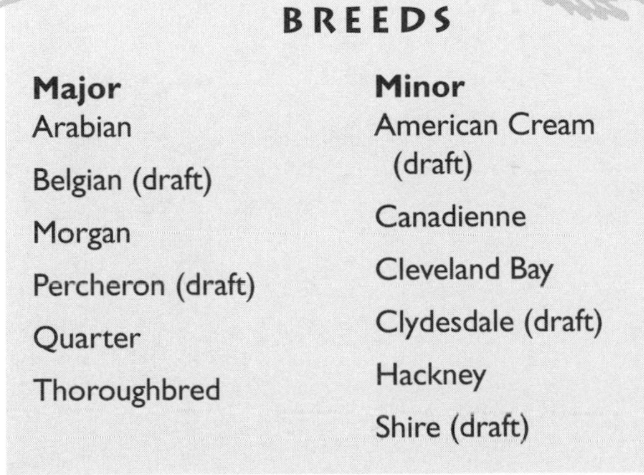

BREEDS

Major	Minor
Arabian	American Cream (draft)
Belgian (draft)	Canadienne
Morgan	Cleveland Bay
Percheron (draft)	Clydesdale (draft)
Quarter	Hackney
Thoroughbred	Shire (draft)

Housing: Horses are some of the hardiest creatures alive and require no housing at all, just a windbreak. Horses wintered outside put on a really thick coat, so owners who plan to show horses during the winter may need to keep them in a barn so they keep that slick, shiny coat that show people look for. Otherwise, don't baby these big, strong critters.

Other thoughts: Pet horses can be a large expense, so if you're planning on making a living at farming, either figure out a way to make money with equines or don't get one. If you really want horses and you want to make a living farming, then the good news is that there may be several ways to make money — using them for work, breeding and selling young animals, providing stud service, developing a lease agreement with townfolks who want their child to ride in shows or 4-H . . . use your imagination.

Table A.5

Horse Feed

For Growing Horses:

Age in months	3	6	12	18	24
Dry matter lb per 100 lb of weight	4.5	2.6	1.9	1.6	1.3
Crude protein %	20	15	12	11	10
Crude protein lb per 100 lb of weight	0.90	0.40	0.23	0.18	0.13
TDN %	62	62	62	62	62
TDN lb per 100 lb of weight	2.8	1.6	1.2	1.0	0.8
Energy Mcal per lb of feed	1.25	1.25	1.25	1.25	1.25
Calcum %	0.6	0.8	0.4	0.4	0.3
Phosphorus %	0.4	0.5	0.3	0.3	0.2

For Mature Horses:

	At rest	Light work	Medium work	Late pregnancy	Peak lactation
Dry matter lb per 100 lb of body weight	1.21	1.63	2.13	1.31	2.12
Crude protein %	10	10	10	11	13
Crude protein lb per 100 lb of body weight	0.12	0.16	0.21	0.14	0.28
TDN %	63	63	63	63	63
TDN lb per 100 lb of body weight	0.76	1.03	1.34	0.83	1.34
Energy Mcal per lb of feed	1.25	1.25	1.25	1.25	1.25
Calcium %	0.32	0.25	0.20	0.37	0.47
Phosphorus %	0.25	0.18	0.15	0.29	0.38

TDN = total daily nutrients.

Note: Ponies need about 125 percent more feed relative to their size, and draft horses only need about 90 percent relative to their size. In other words, ponies at rest need about 1.5 pounds of dry matter per 100 pound of body weight (1.21 x 1.25), and draft horses at rest need about 1.1 pounds per 100 pounds (1.21 x 0.9). All other figures can be adjusted the same way.

Pigs

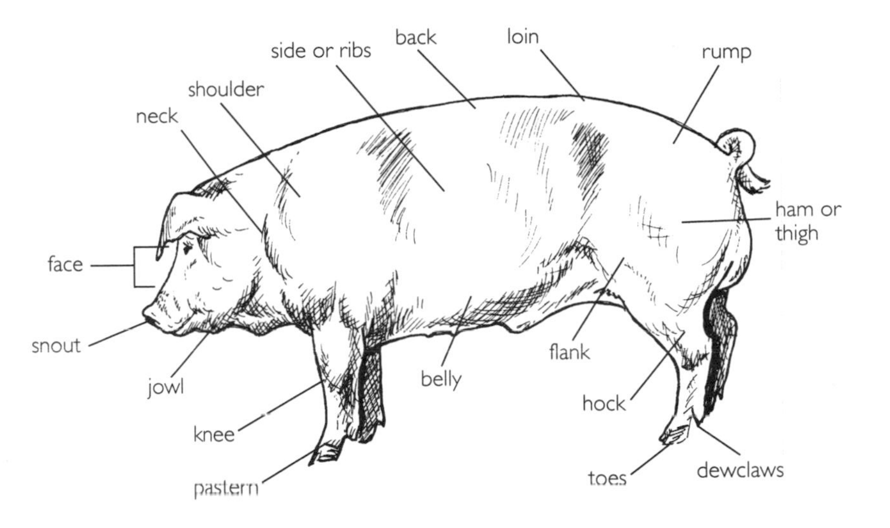

Anatomy of a pig.

Scientific name: *Sus scrofa*

Terms:

Barrow = Castrated male
Boar = Immature or mature male
Farrowing = Act of giving birth
Gilt = Immature female
Shoat or piglet = Newborn
Sow = Mature female

Reproduction:

Gestation = 114 days (3 months, 3 weeks, and 3 days)
Litter size = 8 to 13

Tons of manure produced annually per 1,000 pounds of stock: 18.25
Kilograms of manure produced annually per 1,000 kilograms of stock: 36,500

Top senses: Taste, smell

BREEDS

Major	Minor
Berkshire	Glouchester Old Spot
Chester White	Guinea Hog
Duroc	Large Black
Hampshire	Mulefoot
Landrace	Poland China
Yorkshire	Tamworth

Fencing: A single strand of electric wire about 1 foot (30 cm) off the ground will hold pigs trained to it — but they are notorious for rooting the dirt around the edge of the fence up and over the wire, thus shorting it out. To protect your pasture and your fence, use humane nose rings on hogs (see NASCO Catalog in appendix E, Resources).

Housing: Portable steel huts work very well and are inexpensive. If sows will be farrowing during the winter, they must do so in a heated structure.

Other thoughts: Pigs are an animal people either love or hate — there's no in-between crowd. We love pigs because they are intelligent and entertaining, but they can also be a challenge. They love to get into things (a result of their curious nature), so they can be quite mischievous. Since they are omnivores, they'll sometimes eat things you don't want eaten — like a live chicken.

Table A.5

Pig Feed

For Growing Pigs:

Live weight in pounds	1–10	10–20	20–40	40–70	70–120	120–240
Dry matter lb	0.6	1.1	2.2	3.3	4.4	6.6
Crude protein %	27.0	20.0	18.0	16.0	14.0	13.0
Crude protein lb	0.15	0.22	0.40	0.53	0.62	0.86
TDN %	85	85	79	79	74	74
TDN lb	0.47	0.94	1.74	2.61	3.26	4.88
Energy Mcal	0.93	1.75	3.37	5.06	6.74	10.11
Calcium %	0.90	0.80	0.65	0.60	0.55	0.50
Phosphorus %	0.70	0.60	0.55	0.50	0.45	0.40

For Breeding Swine:

	Bred gilts and sows; young and adult boars, summer	Bred gilts and sows; young and adult boars, winter	Lactating gilts and sows, summer	Lactating gilts and sows, winter
Dry matter lb per 100 lb of body weight	1.3	1.8	3.5	4.0
Crude protein %	12.0	12.0	13.0	13.0
Crude protein lb per 100 lb of body weight	0.15	0.21	0.47	0.51
TDN %	75	76	74	76
TDN lb per 100 lb of body weight	1.0	1.4	2.6	3.0
Energy Mcal per lb of feed	2.8	3.0	3.0	3.2
Calcium %	0.75	0.75	0.75	0.75
Phosphorus %	0.6	0.6	0.5	0.5

TDN = total daily nutrients.

Sheep

Scientific name: *Ovis aries*

Terms:

Ram or buck = Immature or mature male
Ewe = Immature or mature female
Wether = Castrated male
Lamb = Newborn

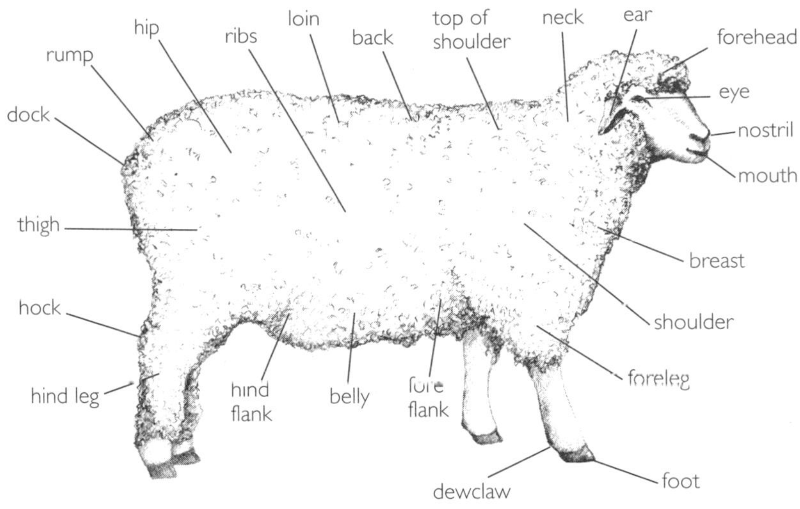

Reproduction:

Gestation = 148 days
Estrus cycle = 16 days
Normal birth = 1 to 3 lambs, though some breeds may have up to 6 lambs

Anatomy of a sheep.

Tons of manure produced annually per 1,000 pounds of stock: 9.75
Kilograms of manure produced annually per 1,000 kilograms of stock: 19,500

Top sense: Smell

Fencing: Sheep are the second most difficult species — right behind goats — to keep fenced. Two to three strands of electric fence will hold most sheep that are trained to it. Wool is a great electrical insulator, though, so the trick to training sheep to electric is to do it right after they have been sheared, or else wet their wool so that it will conduct the shock. If you only run a small flock of sheep, then Electroplastic Net (Premier Fencing, Washington, Iowa) works well.

BREEDS

Major	Minor
Corriedale (dual)	Cotswold (wool)
Dorset (meat)	Delaine-Merino (wool)
Hampshire (meat)	Gulf Coast Native (meat)
Polypay (dual)	Lincoln (dual)
Rambouillet (wool)	Oxford (meat)
Romney (wool)	St. Croix Hair (meat)
Suffolk (meat)	Tunis (meat)

Housing: Sheep have their housing on their backs. The only time you need a building is if you're lambing in winter, or shearing in winter.

Other thoughts: Sheep are high strung and more difficult to handle than many other types of animals. They frighten easily and move in a tight bunch, so working sheep requires a great deal of patience. But the reward can be worth the effort: Sheep are inexpensive to get involved with, they can be raised in small places and on pieces of land that aren't suitable to other types of livestock, and you can build up a flock from just a few purchased animals relatively quickly. A sheep that does bond to you will run up to visit as soon as it sees you coming and will follow you everywhere. If you just want one or two pet sheep to keep down the weeds in your yard, offer to buy some bottle lambs from a sheep farmer. Bottle lambs become completely bonded to the person who feeds them!

Table A.6

Sheep Feed

For Growing Sheep:

Live weight in pounds	50	75	100	125
Dry matter lb	2.2	3.5	4.0	4.6
Crude protein %	12.0	11.0	9.5	8.0
Crude protein lb	0.26	0.39	0.38	0.37
TDN %	55	58	62	62
TDN lb	1.21	2.03	2.48	2.85
Energy Mcal	1.14	1.18	1.27	1.27
Calcium %	0.23	0.21	0.19	0.18
Phosphorus %	0.21	0.18	0.18	0.16

For Breeding Sheep:

	First two-thirds of gestation	Last third of gestation	First 10 weeks of lactation	Last 14 weeks of lactation	Rams at moderate work
Dry matter lb per 100 lb of body weight	2.5	3.5	4.2	3.5	3.5
Crude protein %	8.0	8.2	8.4	8.2	7.6
Crude protein lb per 100 lb of body weight	0.20	0.29	0.35	0.29	0.27
TDN %	50	52	58	52	55
TDN lb per 100 lb of body weght	1.25	1.82	2.44	1.82	1.93
Energy Mcal per lb of feed	1.0	1.1	1.2	1.1	1.2
Calcium %	0.24	0.23	0.28	0.25	0.18
Phosphorus %	0.19	0.17	0.21	0.19	0.16

TDN = total daily nutrients.

Note: Don't feed sheep feed formulas or mineral mixtures that are not specifically recommended for them. The amounts of some trace minerals, such as copper, that are in feed for other classes of livestock may be toxic to sheep.

APPENDIX B

Composition of Common Feedstuffs

Feed, common name	Description	Typical % dry matter (DM)	Crude protein %, DM basis	Crude fiber %, DM basis	Calcium %, DM basis	Phos-phorus %, DM basis	Total digestible nutriens %, DM basis	Digestible energy, Mcal/lb
Forages								
Alfalfa	Fresh, vegetative	21	20.0	23	2.19	0.33	57–61	1.01–1.22
Alfalfa	Hay, early-bloom	90	18.0	23	1.41	0.22	55–60	1.00–1.31
Alfalfa	Hay, mature	91	13.0	38	1.13	0.18	50–55	0.90–1.10
Alfalfa	Silage	38	15.5	30	1.30	0.27	55–58	1.06–1.17
Bermuda grass	Fresh, vegetative	34	12.0	26	0.53	0.21	50–60	0.82–1.32
Bermuda grass	Hay	90	6.0	31	0.43	0.20	45–49	0.94–1.10
Bird's-foot trefoil	Fresh, vegetative	24	21.0	25	1.91	0.22	63–66	0.99–1.50
Bluegrass	Fresh, vegetative	31	17.4	25	0.50	0.44	56–72	0.92–1.40
Brome	Fresh, vegetative	34	18.0	24	0.50	0.30	68–80	0.90–1.26
Brome	Hay	89	10.0	37	0.30	0.35	54–55	0.99–1.29
Clover, red	Fresh, vegetative	20	19.4	23	2.26	0.38	57–69	0.92–1.39
Clover, red	Hay	89	16.0	29	1.53	0.25	49–60	0.91–1.37
Clover, ladino	Fresh, vegetative	19	27.2	14	1.93	0.35	58–68	1.13–1.57
Clover, crimson	Fresh, vegetative	87	18.4	30	1.40	0.20	49–57	0.92–1.39
Fescue	Fresh, vegetative	28	22.1	21	0.53	0.38	70–73	0.79–1.24
Fescue	Hay	92	9.5	37	0.30	0.26	48–62	0.82–1.24
Oat	Hay	92	4.4	40	0.24	0.06	40–47	0.81–1.22
Orchard grass	Fresh, vegetative	23	18.4	25	.58	0.54	55–72	0.93–1.34
Orchard grass	Hay	91	8.4	34	0.26	0.30	45–54	0.86–1.38
Redtop	Fresh, vegetative	29	11.6	27	0.46	0.29	60–65	0.84–1.24
Redtop	Hay	94	11.7	31	0.63	0.35	54–57	0.90–1.15
Reed canary	Fresh, vegetative	23	17.0	24	0.36	0.33	47–75	0.91–1.10
Ryegrass, annual	Fresh, vegetative	25	14.5	24	0.65	0.41	50–60	0.79–1.24
Ryegrass, annual	Hay	88	11.4	29	0.62	0.34	52–57	0.70–1.12
Ryegrass, perennial	Fresh, vegetative	27	10.4	23	0.55	0.27	60–68	0.80–1.35
Ryegrass, perennial	Hay	86	8.6	30	0.62	0.32	45–60	0.80–1.20
Sudan grass	Fresh, vegetative	18	16.8	23	0.43	0.41	63–70	0.83–1.40
Sudan grass	Hay	91	8.0	36	0.55	0.30	55–56	0.87–1.12

Feed, common name	Description	Typical % dry matter (DM)	Crude protein %, DM basis	Crude fiber %, DM basis	Calcium %, DM basis	Phos-phorus %, DM basis	Total digestible nutriens %, DM basis	Digestible energy, Mcal/lb
Forages *(cont.)*								
Timothy	Fresh, vegetative	26	18.0	32	0.39	0.32	61–72	0.76–1.34
Timothy	Hay	89	9.1	31	0.48	0.22	45–60	0.78–1.31
Vetch	Fresh, vegetative	22	20.8	28	1.36	0.34	55–57	1.02–1.23
Vetch	Hay	89	20.8	31	1.18	0.32	67–72	0.91–1.10
Wheatgrass, crested	Fresh, vegetative	28	21.5	22	0.46	0.34	70–75	0.95–1.26
Wheatgrass, crested	Hay	93	12.4	33	0.33	0.21	50–53	0.85–1.11
Other feeds								
Barley	Grain	88	13.5	6	0.05	0.38	80–84	1.34–1.75
Beet pulp	Dried with molasses	92	9.0	13	0.56	0.08	68–70	2.99–3.07
Brewer's grain	Dehydrated	92	30.0	14	0.33	0.55	64–68	1.14–1.60
Corn	Shell (grain)	86	9.0	2	0.03	0.27	78–79	3.45–3.48
Corn ears	Ground	87	9.0	9	0.07	0.28	74–83	1.36–1.70
Corn	Distiller's grains	94	23.0	12	0.11	0.43	70–86	1.25–1.75
Corn	Silage	33	8.1	24	0.24	0.22	66–71	1.32–1.42
Cotton	Seed hulls	91	4.1	48	0.15	0.09	33–42	0.65–0.97
Cotton	Seeds	92	24.0	21	0.16	0.75	90–96	1.05–1.57
Cotton	Seed meal	93	44.3	13	0.21	1.16	75–78	0.97–1.72
Oats	Grain	89	13.0	12	0.07	0.38	76–77	1.29–1.54
Rye	Grain	88	11.3	2	0.07	0.34	71–78	3.15–3.43
Soybean	Seeds	92	43.0	6	0.27	0.65	56–64	1.67–1.88
Soybean	Meal	89	50.0	7	0.33	0.71	82–86	1.22–1.71
Sunflower	Seeds, no hulls	93	47.0	11	0.53	0.50	61–68	2.67–3.01
Turnip	Roots, fresh	10	1.0	1	0.03	—	7–8	0.32–0.37
Wheat	Middlings	87	16.0	3	0.08	0.50	73–78	3.21–3.45

Note: The ranges for total digestible nutrients and digestible energy that are available in a feedstuff vary by species. As a rule of thumb, monogastric animals are at the low end of these ranges and ruminants are at the highest end of these ranges.

Figuring Feeds & Feeding

When mixing feeds, you can use this math trick (often called the square method) for finding the amount of each constituent needed to make a desired mixture:

Step 1. Draw a square.
Step 2. Insert the desired value in the center.
Step 3. Place the known values for each constituent in the upper and lower left-hand corners.
Step 4. Subtract the differences and place in the opposite right-hand corners.
Step 5. Total the figures in the right-hand corners and calculate the percentage from those figures.

As with the rules of algebra, these steps can be worked forward and backward.

Now for some real examples.

Example 1. You want to mix shell corn (crude protein = 9%) and soybeans (crude protein = 43%) to yield a mixture containing a crude protein of 18%. How much of each feed do you need? First draw the box, then place the known proteins in the left-hand corner, and 18 in the middle.

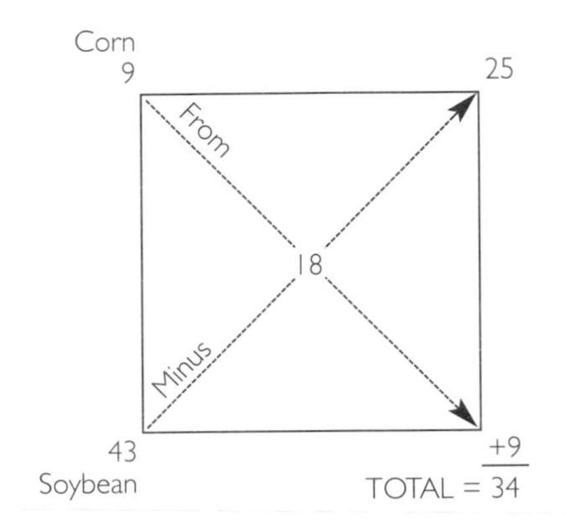

Now put the differences in the opposite corners. Add those numbers together (this becomes the divisor), then figure the percents of each.

Corn %:

$$\frac{25}{34} \times 100 = 73.5$$

Pounds of Corn per Ton of Mix:

$$.735 \times 2{,}000 = 1{,}470$$

Soybean %:

$$\frac{9}{34} \times 100 = 26.5$$

Pounds of Soybeans per Ton of Mix:

$$.265 \times 2{,}000 = 530$$

Example 2. You are feeding a grass-based hay with about 10 percent crude protein in it. You want to supplement with a grain mixture so your 50-pound lambs are receiving 12 percent crude protein. From the feed chart for sheep (Table A.6), you see that the lambs need about 2.2 pounds per day of dry matter. The hay is fed out free-choice, so let's make the assumption that the lambs can each get 85 percent of their dry matter from the hay, and you're shooting for about 15 percent from the supplement (0.33 pounds). In this exercise you work the square backward to discover the crude protein in the grain mixture. Again draw the box, but this time the desired protein (12) goes in the center, and the known protein of the hay (10) goes in the upper left-hand corner. Place the difference of these two numbers in the lower right-hand corner (2).

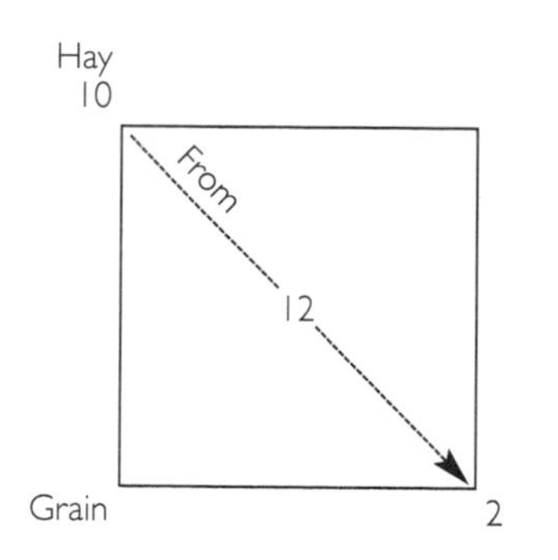

Now, you have to figure out what total would make 2 equal to 15 percent of the mixture. Here's the math for that:

$$0.15 = \frac{2}{\text{TOTAL}}$$

$$0.15 \times \text{TOTAL} = 2$$
$$\text{TOTAL} = 2 \div 0.15$$
$$\text{TOTAL} = 13$$

Now you know that the total equals 13, so 11 is the figure that goes in the upper right-hand corner. Add 11 and 12 to find that your grain mixture needs 23 percent crude protein.

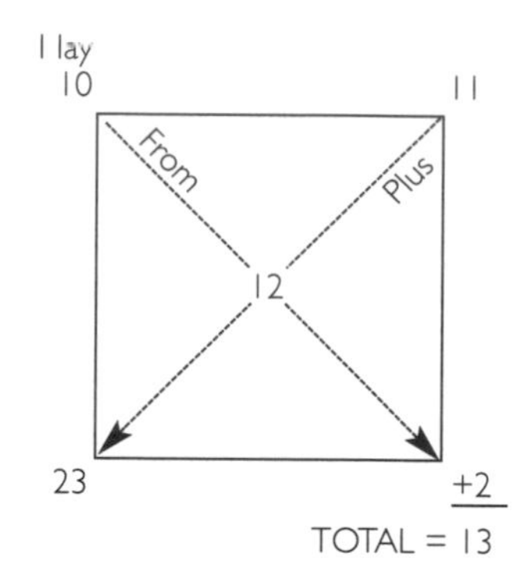

Example 3. For a final example, let's look at a mixture of more than two constituents that will yield the 23 percent crude protein grain mixture you need for your lambs in Example 2. Assume that you feed a mixture of 2 parts corn (9% crude protein) to 1 part oats (13% crude protein), and you will add soybean meal (50% crude protein) to make your mix.

First, figure out the crude protein of the corn-and-oat mixture using the following approach:

2 parts corn × 0.09 = 0.18

plus 1 part oat × 0.13 = $\underline{0.13}$
0.31

$\frac{31\%\ \text{crude protein}}{3\ \text{parts}}$ = 10.3% crude protein per part, rounded off to 10% crude protein

Now set up the box, as we did in Example 1.

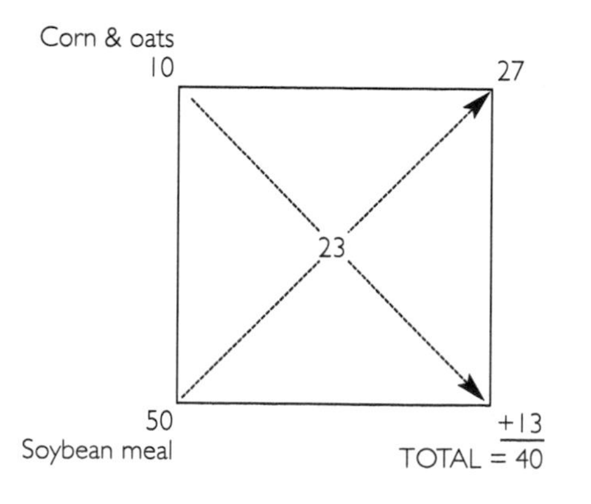

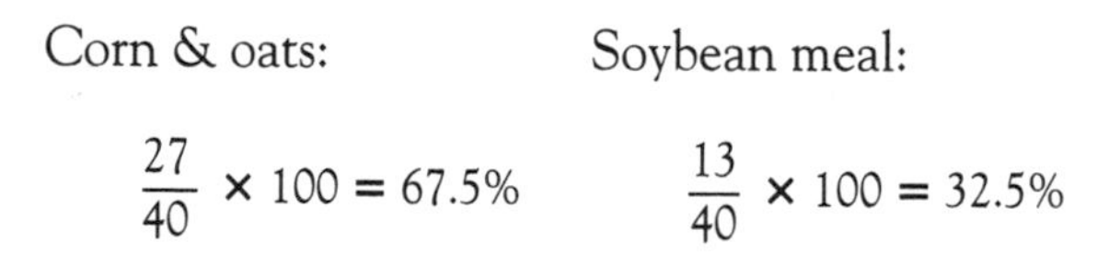

Corn & oats:

$$\frac{27}{40} \times 100 = 67.5\%$$

Soybean meal:

$$\frac{13}{40} \times 100 = 32.5\%$$

APPENDIX D

Grasses & Legumes

Crop	Seeding Rate Per Acre	Seeds Per Pound in Thousands	Germination Time in Days	Cool or Warm Season	Growth Type
Alfalfa	10–20	220	7	Cool	Perennial
Bermuda grass	6–8	1,800	21	Warm	Perennial
Bird's-foot trefoil	8–12	375	7	Cool	Perennial
Bluegrass, annual	15–25	1,200	21	Cool	Annual
Bluegrass, Kentucky	15–25	2,200	27	Cool	Perennial
Bluestem, big	15–20	150	28	Warm	Perennial
Bromegrass, smooth	15–20	137	14	Cool	Perennial
Clover, alsike	6–8	680	7	Cool	Perennial
Clover, crimson	15–25	150	7	Cool	Annual
Clover, ladino	5–7	860	10	Cool	Annual
Clover, red	8–12	260	7	Cool	Annual
Clover, white	5–7	700	10	Cool	Perennial
Cowpea	75–120	2–6	8	Warm	Annual
Dallisgrass	3–5	5,450	42	Warm	Perennial
Fescue, meadow	10–25	230	14	Cool	Perennial
Field pea, large-seeded	120–180	4	8	Cool	Annual
Grama grass, blue	8–12	900	28	Warm	Perennial
Grama grass, sideoats	15–20	200	28	Warm	Perennial
Lespedeza, common	25–30	343	14	Warm	Annual
Lovegrass, weeping	0.25	1,500	14	Warm	Perennial
Meadow foxtail	15–25	540	14	Cool	Perennial
Millet, pearl	16–20	85	7	Warm	Annual
Orchard grass	20–25	590	18	Cool	Perennial
Redtop	10–12	5,100	10	Cool	Perennial
Reed canary grass	8–12	550	21	Cool	Perennial
Ryegrass	25–30	330	14	Cool	Perennial
Sudan grass	20–35	55	10	Warm	Annual
Timothy	8–12	1,230	10	Cool	Perennial
Vetch, common	40–80	7	10	Cool	Annual
Vetch, hairy	20–40	21	14	Cool	Annual
Wheatgrass, crested	12–20	190	14	Cool	Perennial
Wheatgrass, western	12–20	110	35	Cool	Perennial

Note: The seeding rates are based on a straight seeding of individual species. When combining two or more species for a mix, seeding rate can drop by 50 to 75 percent.

Resources

Books

Beck-Chenowith, Herman. *Free-Range Poultry Production and Marketing*. Creola, OH: Back Forty Books, 1966.

Herman has done a lot of work on raising and marketing free-range birds and has put his experiences together to help other small-scale farmers with a less labor-intensive approach to poultry production. His detailed instructions include plans for skid houses.

Bingham, Sam, and Allan Savory. *Holistic Resource Management Workbook*. Covelo, CA: Island Press, 1990.

If you need the full holistic management planning model, or to continue your studies, read this workbook.

Damerow, Gail. *Fences for Pasture & Garden*. Pownal, VT: Storey Books, 1992.

In my book I didn't really have space to do justice to fence construction, so if you need some more information on the how-to aspects of building fences, Gail's book is very good. There are thorough discussions of design, materials, and lots of construction tricks that make for both a neater and a sturdier fence.

Haynes, N. Bruce. *Keeping Livestock Healthy*. Pownal, VT: Storey Books, 1994.

Anyone raising livestock should have a copy of this book.

Logsdon, Gene.

Read any books by Gene Logsdon (many of Gene's books are out-of-print, but they are worth looking for at the library or used bookstores). Gene is a superb advocate for small farmers, with lots of good how-to information in all his books. He is articulate, but best of all he's also funny!

Macleod, George. *A Veterinary Materia Medica and Clinical Repertory*. Essex, Engl.: The C. W. Daniel Company Ltd, 1992.

———. *The Treatment of Cattle by Homeopathy*. Essex, Engl.: The C. W. Daniel Company Ltd, 1992.

These two books are good resources if you want to learn more about using homeopathic preparations around the farm.

Mettler, John J. *Basic Butchering*. Pownal, VT: Storey Books, 1989.

A good primer for anyone who wants to do their own butchering.

Morrison, Frank B. *Feeds and Feeding*. Ithaca, NY: The Morrison Publishing Company, 1950.

All livestock farmers should have at least one good feed book. Morrison's is the classic on the topic. Despite its age, this is an excellent reference and is user friendly for farmers and ranchers.

If you can't find Morrison's, the following books are currently in print:

Jurgens, Marshall. *Animal Feeding and Nutrition*. Dubuque, IA: Kendall/Hunt Publishing, 1996.

Church, D. C., and W. G. Pond. *Basic Animal Nutrition and Feeding*. New York: John Wiley & Sons, 1995.

Rombauer, Irma S., and Marion Rombauer Becker. *Joy of Cooking*. New York: Macmillan, 1975.

If you plan on taking up butchering, the Joy of Cooking has the best instructions I know of for sharpening knives.

Salatin, Joel. *Pastured Poultry Profits*. White River Junction, VT: Chelsea Green, 1996.

———. *Salad Bar Beef*. White River Junction, VT: Chelsea Green, 1996.

———. *You Can Farm*. White River Junction, VT: Chelsea Green, 1998.

Any of Joel's books are well worth having. He's a phenomenally successful direct-marketer, and he outlines just how he does things in excellent detail in his books.

Savory, Allan. *Holistic Management*. Covelo, CA: Island Press, 1998.

Savory's textbook is the bible of holistic management.

Schwenke, Karl. *Successful Small-Scale Farming*. Pownal, VT: Storey Books, 1991.

A good general small-farm book, with information for the garden, on raising crops, maintaining a woodlot, and more.

Spaulding, C. E. *A Veterinary Guide for Animal Owners*. Emmaus, PA: Rodale, 1996.

Anyone raising livestock should have a copy of this.

Voisin, Andre. *Grass Productivity*. Covelo, CA: Island Press, 1988.

Voisin was the original "thinker" on grass farming.

Periodicals

Small Farm Today (bimonthly)
3903 W. Ridge Trail Road
Clark, MO 65243
1-800-633-2535
<http://www.datasys.net/edpak/small.html>
Good all-around small-farming magazine.

Small Farmer's Journal (quarterly)
P.O. Box 1627
Sisters, OR 97759
503-549-2064
Lots of horse-farming information, plus good general small farm information. Many reprints out of old out-of-print publications. Editor Lynn Miller writes a good commentary on American agriculture and society.

The Stockman Grass Farmer (monthly)
P.O. Box 2300
Ridgeland, MS 39158
1-800-748-9808 (ask for a free sample!)
If we could afford only one agricultural magazine, this would be the one we'd keep. "Al's Ob's" (editor Allan Nation's monthly column) is worth the price of admission.

Internet Sites and Organizations

The Internet is a great research tool, and there are thousands of sites of interest. Some of my favorites are:

Alternative, Complimentary, and Holistic Veterinary Medicine
<http://www.altvetmed.com>
Operated by two veterinarians, this site has good information on the different approaches to animal health, and maintains a reference list of veterinarians around the United States who incorporate one or more alternative approaches into their own practices.

American Livestock Breeds Conservancy
<http://www.albc-usa.org>
ALBC is an organization dedicated to helping maintain the domestic gene pool, by helping to keep minor breeds alive. If you are interested in raising minor breeds, they can help you identify other breeders or breed organizations for the breed you are interested in.

ALBC also publishes an excellent reference book, A Rare Breeds Album of American Livestock, by C. J. Christman, D.P. Sponenberg, and Don Bixby (ALBC, 1997). The album has good color photos.

Appropriate Technology Transfer for Rural America (ATTRA)
PO Box 3657
Fayetteville, AR 72702
1-800-346-9140
<http://www.attra.org>
ATTRA is a great resource. They provide technical assistance to farmers, market gardeners, and other agricultural professionals on any of the broad areas that fall under the classification of sustainable agriculture, alternative enterprises, or innovative marketing. They have a variety of preprepared publications, or if they don't have a publication that can answer your question, then the staff will research the question for you and get you an answer, usually within a week or two. All ATTRA materials and services are available free of charge.

Center for Holistic Management
1010 Tijeras NW
Albuquerque, NM 87102
505-842-5252
<http://www.holisticmanagement.org>
The site of the nonprofit Center for Holistic Management, which helps promote the adoption and use of holistic management techniques; numerous publications are available from this site.

Forage Information System at Oregon State University
<http://forages.orst.edu/>
Oregon State's done a good job with their site. Find out about any forage crop, including pictures of most plants, be it grass, legume, or forb. This is also the place to look for certified forage testing laboratories.

Heifer Project International (HPI)
PO Box 808
Little Rock, AR 72203
1-800-442-0474
<http://www.heifer.org>
Nonprofit; HPI helps feed hungry people around the world by providing breeding stock and sustainable agriculture training to those in need.

Livestock Breeds
Oklahoma State University
<http://www.ansi.okstate.edu/breeds>
Oklahoma State has put together a great site with information on livestock breeds from around the world. If you're searching for more information on a particular breed or just want to do some general breed-related research, stop here!

Organic Trade Association
PO Box 1078
Greenfield, MA 01302
413-774-7511
<http://www.ota.com>
The OTA is a membership organization for groups that are working in the organic industry, including organic certifying organizations. If you are interested in pursuing organic certification, the folks at OTA can help you identify a certification organization that's operating in your state.

Owenlea Holsteins
<http://www.bright.net/~fwo>
This site is run by F. W. Owen, whose biography reads, "I am a real dairy farmer, who actually milks my own cows (and have for the last 40 years)." Where Fred finds time to keep up this truly amazing site is beyond me, but kudos to him for his effort. He has extensive information on grass-dairying, and some eclectic stuff to go along with it (like the Internet Public Library). He has a search engine for searching the archives of GRAZE-L, an on-line dialogue of grass farmers from around the world.

Kevin Powell — Mulefoot Hogs
12942 338th Street
Strawberry Point, IA 52076
319-933-2252
E-mail: powellk@squared.com
Kevin is the farmer who has set a goal of helping save the Mulefoot hogs. If you know of some existing herds or are interested in raising Mulefoots, contact Kevin.

Sustainable Farming Connection
<http://metalab.unc.edu/farming-connection/>
This site is a labor of love for Craig Cramer. Craig used to be the editor of New Farm *magazine, until Rodale stopped publishing it. This site has great information on just about any sustainable agriculture topic you can imagine, including links to other sites and organizations.*

Sustainable Ranching Research and Education Project
University of California Cooperative Extension Service
<http://www.foothill.net/~ringram/>
Another good site, particularly for folks in a brittle climate.

United States Department of Agriculture
<http://www.usda.gov>
USDA's site can provide a wealth of information on agriculture in general. Statistics, programs, etc.

United States House of Representatives
<http://law.house.gov>
The House's site provides access to all federal laws and regulations as well as links to states that have their laws and regulations accessible via the Internet. You can also track what the "fellas" in Washington are doin' at this site!

Commercial Providers

Farmstead Health Supply
PO Box 985
Hillsborough, NC 27278
916-643-0300
<http://www.farmsteadhealth.com>
Linda Phillips offers herbal parasite controls and do-it-yourself fecal test kits.

McCarville Dairy
820 Center Street
Mineral Point, WI 53565
1-608-987-2416
Supplies nipples for barrel-feeding calves.

NASCO
901 Janesville Avenue
Fort Atkinson, WI 53538
800-558-959
<http://www.nascofa.com>
NASCO is the Sears Catalog of farming and ranching. If you don't have a well-stocked farm supply store near you, the NASCO catalog is a must. They have just about everything you could think of for a livestock operation.

Polywinder
174 Cane Creek Farm Road
Alexandria, AL 36250
205-820-3729
For famers who are running lots of temporary wire, David Wright's Polywinder can really speed up operations.

Fencing Suppliers

Call for catalogs! Most of these suppliers' catalogs have lots of good fencing advice, as well as the equipment they supply.

Kencove Farm Fence
111 Kendall Lane
Blairsville, PA 15717
1-800-KENCOVE
<http://www.kencove.com>

Premier Fencing
2031 300th Street — SG
Washington, IA 52353
1-800-282-6631

Southwest Power Fence
26321 Highway 281 N
San Antonio, TX 78260
1-800-221-0178

Twin Mountain Fence
PO Box 2240
San Angelo, TX 76902
1-800-527-0990

Calculations, Equations, & Equivalents

Equations Used in Text

Biological Planning

- Forage required (i.e., daily intake) = body weight x intake factor
- Forage production = hay equivalents/acre x acre
- Carrying capacity = $\frac{\text{(animal units × forage produced)}}{\text{(forage required × 1.25)}}$
- Total paddocks = (recovery period/grazing period) + 1
- Stock density = $\frac{\text{(available forage × utilization rate)}}{\text{(daily intake × grazing period)}}$
- Paddock size = animal units/stock density

Financial Planning

- Assets = liabilities + equity
- Opportunity cost = equity x 2 x T-bond rate
- Profit = Income – (variable costs + fixed costs)

Interest and Payments

- Payment =

$$\text{balance owed} \,/ \left(\frac{1-(1+\text{monthly interest})^{-\text{number of payments}}}{\text{monthly interest}} \right)$$

Ex. Payment on $50,000 borrowed for 30 years at

$$10\% = 50{,}000 \,/ \left(\frac{1-(1+\frac{.1}{12})^{-360}}{\frac{.1}{12}} \right) = \$438.79$$

Note: To use this calculation requires a computer or a calculator that is capable of raising a number to a negative power. Most "scientific" calculators will do the trick.

- To calculate the time it takes to double your money at different interest rates, divide 70 years by the rate.

$$Ex.\ \frac{70}{4\%} = 17.5 \text{ years} \quad \frac{70}{10\%} = 7 \text{ years}$$

Other Worthwhile Equations

*Area (A) and Perimeter (P)**

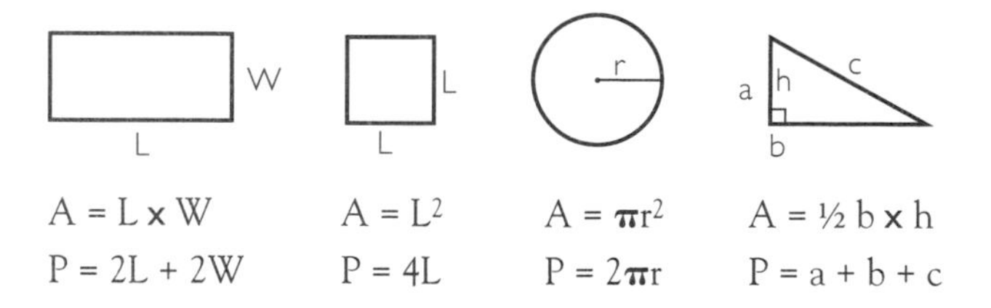

$A = L \times W$	$A = L^2$	$A = \pi r^2$	$A = \frac{1}{2} b \times h$
$P = 2L + 2W$	$P = 4L$	$P = 2\pi r$	$P = a + b + c$

*Note: L = length, W = width, r = radius, b = base, h = height, π = 3.1417

Equivalents

1 acre = 43,560 ft^2 = 0.4047 hectares (ha)
640 acres = 1 section = 1 sq mile
1 mile = 5,280 feet = 1,609 meters (m)
1 gallon = 8.34 pounds = 3.79 liters (L)
1 cu ft = 7.48 gallons = 28 liters
1" of grass in a well-sodded field = 300 pounds of available forage per acre (*or,* 1 cm = 132 kg/ha)
1 pound = 453.6 grams (g) *or* 0.4536 kg

Index

Note: *Numbers in* italics *indicate figures; numbers in* **boldface** *indicate tables.*

Other Storey Titles You Will Enjoy

Basic Butchering, by John J. Mettler Jr., D.V.M. Provides clear, concise, and step-by-step information for people who want to slaughter their own meat. 208 pages. Paperback. ISBN: 0-88266-391-7.

Fences for Pasture & Garden, by Gail Damerow. The complete guide to choosing, planning, and building today's best fences: wire, rail, electric, high-tension, temporary, woven, and snow. 160 pages. Paperback. ISBN 0-88266-753-X.

A Guide to Raising Beef Cattle, by Heather Smith Thomas. A bovine expert, Thomas explains facilities, breeding and genetics, calving, health care, and advice for marketing a cattle business. 352 pages. Paperback. ISBN 1-58017-037-4.

A Guide to Raising Chickens, by Gail Damerow. Expert advice on selecting breeds, caring for chicks, producing eggs, raising broilers, feeding, troubleshooting, and much more. 352 pages. Paperback. ISBN 0-88266-897-8.

A Guide to Raising Pigs, by Kelly Klober. Practical advice for buying, feeding, and caring for hogs, plus modern breeding and herd management. 320 pages. Paperback. ISBN 1-58017-011-0.

Keeping Livestock Healthy, by N. Bruce Haynes, D.V.M. A complete guide to preventing disease through good nutrition, proper housing, and appropriate care. 352 pages. Paperback. ISBN 0-88266-884-6.

Making Your Small Farm Profitable, by Ron Macher. Whether you're buying a new farm or jump-starting an old one, with Ron Macher's down-to-earth advice on planning, farming, and marketing you'll soon make your farm profitable—and find satisfaction doing it. 288 pages. Paperback. ISBN 1-58017-161-3.

Raising a Calf for Beef, by Phyllis Hobson. This no-nonsense how-to guide for beginners offers detailed information on choosing a calf, building and maintaining housing, nutrition, feeding, and daily care. Also includes instructions for slaughtering and butchering. 128 pages. Paperback. ISBN 0-88266-095-0.

Successful Small-Scale Farming, by Karl Schwenke. Contains everything small-farm owners need to know, from buying land to organic growing methods and selling cash crops. 144 pages. Paperback. ISBN -0-88266-642-8.